MANUFACTURING ENGINEERING:
Economics and Processes

manufacturing engineering: economics and processes

KENNETH C LUDEMA

Professor of Mechanical Engineering and Applied Mechanics
The University of Michigan

ROBERT M. CADDELL

Professor of Mechanical Engineering and Applied Mechanics
The University of Michigan

ANTHONY G. ATKINS

Professor of Mechanical Engineering
University of Reading, United Kingdom

Prentice-Hall, Inc., Englewood Cliffs, New Jersey 07632

Library of Congress Cataloging-in-Publication Data

LUDEMA, K. C
  Manufacturing engineering.

  Includes bibliographies and index.
  1. Production engineering.   I. Caddell, Robert M.
II. Atkins, Anthony G.,   .   III. Title.
TS176.L84–1987   670.42   86-21282
ISBN 0-13-555582-5

Table 3-1
Used with permission of The American Society of Mechanical Engineers from "Surface Texture"
(USASI-B46.1-1986)

Figure 4-3
From K. H. Moltrecht and R. M. Caddell, "How to Determine Production Tolerances–Part 1," *The Tool Engineer* (Oct. 1957), pp. 81–85. Reprinted courtesy of The Society of Manufacturing Engineers.

Continued, page 405

Printed in the United States of America

10 9 8 7 6 5 4 3 2 1

Cover design: George Cornell
Manufacturing buyer: Rhett Conklin

ISBN 0-13-555582-5 025

PRENTICE-HALL INTERNATIONAL (UK) LIMITED, *London*
PRENTICE-HALL OF AUSTRALIA PTY. LIMITED, *Sydney*
PRENTICE-HALL CANADA INC., *Toronto*
PRENTICE-HALL HISPANOAMERICANA, S.A., *Mexico*
PRENTICE-HALL OF INDIA PRIVATE LIMITED, *New Delhi*
PRENTICE-HALL OF JAPAN, INC., *Tokyo*
PRENTICE-HALL OF SOUTHEAST ASIA PTE. LTD., *Singapore*
EDITORA PRENTICE-HALL DO BRASIL, LTDA., *Rio de Janeiro*

contents _____

preface _____

Because of the diversity of disciplines involved in what is called manufacturing, it is not possible to give a simple definition of a "manufacturing engineer." Yet, in our opinion, a manufacturing engineer is more concerned with the design of manufacturing systems than with their operation, per se. Depending upon the academic program in which courses related to manufacturing are offered, certain topics are presented in some depth, while others are ignored completely. Although subjects such as business practices, classical economics, and the like all address content that falls under the broad umbrella of manufacturing, our concern in this book addresses, primarily, subjects that are of an engineering nature. Historically, such coverage has usually involved the materials and processes used to manufacture products; in most instances a detailed description of hardware and a qualitative approach to problem solving were considered. Although useful, this approach does not address the fundamental aspects of the individual processing methods, nor does it pose the kind of problem solving that typifies most other undergraduate engineering subjects. Probably because of these reasons, manufacturing courses are often held in low esteem by those whose specialties are narrower in scope but more analytical in nature. This is truly ironic, since to analyze any of the major processes in depth demands the use of a number of those specialties. Such intensive study of any individual process must be left to those specialized texts that provide an in-depth coverage of a given process such as casting, machining, and the like. Our intent is to address the major principles of processing methods plus certain other topics of concern that engineers face in manufacturing industries, and to include *numerical* problem solving wherever possible. Any essential descriptive coverage is kept to a necessary minimum. We note that this field is currently laced with acronyms and catchy phrases, but acronyms come and go, while fundamental concepts remain.

This book is divided into three major sections, each of which carries a brief fore-word. Part A includes some philosophical aspects of manufacturing concerns and stresses the importance of economic considerations. The latter does not require any formal background in business procedures or classical economics per se. A concise coverage of surface finish and tolerances, which find importance in all processes, is also included here. Part B starts with a review of mechanical properties and engineering materials and then uses this background as it applies to the four major and traditional processing methods. It concludes with a chapter on what has come to be called nontraditional or special processing techniques. Part C includes a number of types of case studies in which the principal goal is to study the economics of producing parts by different manufacturing methods. It concludes with some ideas related to integrated design and manufacturing.

Since courses of this type are usually offered in the latter part of typical engineering curricula, we assume that users of this text will have earlier covered the topics of statics, strength of materials, mathematics, and a traditional course in materials science. If at all possible, we suggest that self-paced computer instruction packages be available to students for review of topics concerning the above. TV tapes and, possibly, plant trips can be used to demonstrate many of the hardware aspects of processes and systems, thereby negating the need for extensive descriptive coverage during classroom lectures.

With regard to units, a few comments are pertinent. Although most students are familiar with the SI System, much of industry in the United States has not followed this usage. Even the traditional metric system has met with much resistance. It is likely that a number of graduates who enter a manufacturing industry will encounter situations where the units involved may be SI, metric, or English. For that reason we have interspersed different units throughout this text with the hope that the reader will feel comfortable regardless of which system is posed. Following this preface is a table of useful conversion factors that relate to the subject matter in this text.

To give equal and uniform coverage to all chapters in a typical one-semester course is unlikely. Depending upon which engineering department offers such a course, instructors may decide to extend the material in certain chapters while reducing or even skipping the content of others. Naturally, we think it best to address all topics, but time constraints may make this too difficult. In our experience mechanical engineers seem to think that the business and human factors of manufacturing are not their concern, whereas industrial engineers tend to regard processes and equipment as things that can be handled via catalogs and handbooks. Our hope is that this text will show each group that each attitude fulfills only a part of the complete picture.

1. Industrial engineering students usually have a strong background in statistics and economics but have a lesser exposure to mechanics and materials. Chapters 1 through 5 and selected parts of Chapters 13 and 14 may find greatest interest, and can be extended as an instructor sees fit. If less emphasis is to be placed upon processing methods the most logical groupings are Chapters 6, 8, and 9, *or* 7, 10, and 11.

2. Metallurgical or material engineers, who have a deeper background in materials, could consider a brief review of Chapters 1 to 5 and then proceed with detailed

coverage of Chapters 6 to 12; in fact, an extension of Chapters 7, 10 and 11 might be considered. Selected topics from Chapters 13 and 14 may then be covered.

3. Mechanical engineers, who have a deeper background in mechanics, should have at least a reasonable exposure to Chapters 1 and 2, and a definite coverage of Chapters 3, 4, and 5. Then Chapters 6 through 12 should be addressed in full, followed by selected portions from Chapters 13 and 14.

A number of individuals have provided information and suggestions to us as this book was being compiled; some was quite recent, while some was accrued over years of contact. Professors W. G. Ovens and W. R. DeVries reviewed the original manuscript and made a number of suggestions and constructive criticisms, almost all of which we have incorporated into the final version. We are grateful for their help. Our close association with Professor W. F. Hosford shows clearly in Chapter 8. The late Professor L. V. Colwell was our friend and colleague, and a teacher to two of us. A bit of what he taught us makes up much of Chapters 4 and 9, and we are grateful to have known him for so many years. A group of people supplied much information that has been included in Chapters 13 and 14, and we thank them collectively for their help. They include Professors J. M. Alexander, J. A. G. Kals, W. A. Knight, K. Lange, and G. W. Rowe, and Drs. A. Beevers, B. Lengyel, and C. Ruiz.

Kenneth C Ludema
Robert M. Caddell
Anthony G. Atkins

SOME BASIC UNITS AND THEIR ABBREVIATIONS FOR THE SI SYSTEM

Unit	Standard	Abbreviation
length	meter	m
mass	kilogram	kg
time	second	s
*force	newton	$N = kg\ m/s^2$
*stress	newton/meter^2	N/m^2
*stress	pascal = 1 N/m^2	Pa
*energy	joule	$J = Nm$

* These are derived from basic units.

MULTIPLICATION FACTORS USED IN THE SI SYSTEM

Factor	Prefix	Symbol	Factor	Prefix	Symbol
10^{18}	exa	E	10^{-3}	milli	m
10^{15}	peta	P	10^{-6}	micro	μ
10^{12}	tera	T	10^{-9}	nano	n
10^{9}	giga	G	10^{-12}	pico	p
10^{6}	mega	M	10^{-15}	femto	f
10^{3}	kilo	k	10^{-18}	atto	a

USEFUL CONVERSION FACTORS—ENGLISH TO SI UNITS

To convert from	to	Multiply by
inch (in.)	meter (m)	2.54×10^{-2}
feet (ft)	meter	3.048×10^{-1}
inch^2	meter^2	6.452×10^{-4}
feet^2	meter^2	9.29×10^{-2}
inch^3	meter^3	1.639×10^{-5}
feet^3	meter^3	2.832×10^{-2}
pound-force (lbf)	newton (N)	4.448
pounds/inch^2 (psi)	pascal (Pa = N/m^2)	6.895×10^{3}
kilopounds/inch^2 (ksi)	pascal	6.895×10^{6}
horsepower (hp)	foot-pounds/minute	33×10^{3}
horsepower	watts (W)	7.457×10^{2}
foot-pound (ft-lbf)	joules (J)	1.356

SOME OTHER USEFUL RELATIONS

1 micron = 10^{-6} meter
1 angstrom = 10^{-10} meter
10 angstroms = 1 nm

MANUFACTURING ENGINEERING:
Economics and Processes

FOREWORD TO PART A

The manufacture of consumer products is a dynamic profession. Winning or losing may involve "shaving pennies" from the cost of a product, and this requires much ingenuity. A slight redesign of the part, the choice of a better material, the use of a processing method that produces less scrap, and a more efficient arrangement of machinery are some of the considerations that can lead to lower costs of production. Aspects of these topics are addressed in Chapter 1.

Decisions in designing both products and manufacturing systems are really interdependent, and they have one common, vital relationship. This has to do with economic advantage. Chapter 2 covers some of the important factors that engineers should consider, and presents the importance of the time value of money. Engineers need not know business or accounting procedures to gain an appreciation of this topic.

Chapters 3 through 5 cover topics that are vital in manufacturing but often catch engineers by surprise. In Chapter 3 ideas on product quality attributes are discussed; although of great importance, the defining and measuring of "quality" are often considered to be the responsibility of "someone else." Certain aspects of tolerances are presented in Chapter 4. Again this is sometimes viewed as a problem that is handled by others, but both quality and assembly must be considered in the design of a manufacturing system. In Chapter 5, the use of automation and computers is introduced. This section is intended to illustrate, in a somewhat philosophical way, topics of importance to engineers who are involved with the production of consumer products—topics that are often viewed as being the sole responsibility of others. We disagree with such an attitude, since all designed products must be processed from selected materials, and the manufacturing processing of said materials is the gate through which all designs must pass before products reach the consumer. The production of quality items at a competitive price is of prime importance to manufacturing engineers.

_1_
engineering in the manufacturing industry _____

1.1 INTRODUCTION

Engineering is the art of distilling relevant scientific knowledge for the purpose of bringing useful items, materials, or services into being. The engineer seeks to quantify all of the variables that are germane to a project, makes informed estimates where hard data are not available, and sets down a logical progression of thought toward solving problems, in terms and language that others can readily understand and use as a basis for action. Obviously, engineers are important people.

Engineers in manufacturing industries have an added responsibility, and that is to include economics in their calculations. Many engineers are spared of economic concerns, such as those in government, in academia, and in the junior positions in design and analysis. Certainly everyone must be concerned about available funds for projects (or pay), but relatively few engineers need to give as much thought to company profits as do engineers who design manufacturing systems.

Manufacturing is, after all, done to make a profit. In fact, in planning the manufacture of products, there should be a prospect of higher profit than could be realized from investing the capital needed for the manufacturing facilities into stocks, for example. Only upper management has the authority to decide what level of profits are worth an investment, but they depend on others to provide the information for decision making. Manufacturing engineers are the major source of information on manufacturing costs.

A manufacturing enterprise can be very complex, involving many people of various skills. It may be surprising to know that many companies employ no professional engineers at all. The expertise of skilled tradesmen and hard-working managers is often sufficient to keep a company profitable, particularly if the products have a long _life cycle_.

However, some products require continual updating, which may involve mathematical skills (airplane design), materials expertise (jet engines and oil drill bits), chemical skills (plastics and oil refining), or computer skills (adaptive control and information systems) in their design or manufacturing. These are the skills in which engineers are proficient to varying degrees.

In the same way, manufacturing systems must be designed, and this often requires engineers. The design of a manufacturing system would seem to be merely the selection and arrangement of commercially available machines, conveyors, robots, and computers. Certainly a part of the work of manufacturing engineers is just that. However, success in business is not ensured by using ideas and items available to everyone. Innovation provides the vital margin, and engineers are the authors of most technical innovation.

Traditionally engineers concentrated on the hardware of products and processes; some have been directly involved in the manufacturing of products. A new and vital area for engineers is the design or specifying of *manufacturing information systems*. Manufacturing of products is done by the coordination of the efforts of a great number of people and machines. However, the system is prone to error, some of which are human in origin and others not. Each error is a threat to profitability. The minimizing of errors has usually been done by organizing people so that everything that is necessary to know is known by the proper people. It may be seen that there is a great amount of information concerning the operation, beginning with the design of the product and ending with the rate of sales of products, and it is widely distributed. Computers provide the capability of automating the transfer of such information, and engineers design these automated systems.

The sections following describe the manufacturing industry and the engineering challenges in it. There follows a description of the major new approach to the design of manufacturing systems, namely, to include the transfer of design information along with the design of the hardware of the system. Finally, the role of the computer in designing the systems and in the operation thereof will be discussed.

1.2 THE MANUFACTURING INDUSTRY

Manufacturing constitutes between 30 percent and 70 percent of the gross national product (GNP) in the United States, depending on what is defined as manufacturing. We can readily identify *manufactured* products. Both a nail and an automobile are manufactured. But the act of manufacturing requires defining. For our purposes it is the conversion of raw materials into *hardware,* both completed products and components thereof. This definition excludes the conversion of crude oil into gasoline, because gasoline is not hardware. It also excludes building houses, because in the building process there is little or no conversion of raw materials. The building of automobiles is not only a manufacturing operation, but a significant assembly activity as well. In fact, most large automotive companies are a large mixture of manufacturing and assembly operations. They may cast engines in iron, purchase sheet metal which they form into shapes for body panels, and purchase tires which they mount on wheels with no conversion at all.

Frequently manufacturing is an action defined as producing *discrete* parts, which is different from continuous processing as in making gasoline. But wire making is also a continuous process, in that the product is often miles long, hardly a discrete part in the usual sense of the term. Of course, we could argue that the making of gasoline and wire are continuous processes only while a batch of raw materials lasts, whereupon the process variables should be changed because two batches of raw material are seldom exactly alike.

Most manufacturing organizations begin with an idea for a product to sell for a profit. Someone finds the money to begin, and begins to make and market a product. Some manufacturers remain small, but others continually offer new products and become very large and diverse. Both types succeed and both types fail. Success depends on such mundane matters as having the right products and satisfying consumers. At times it would appear that the large corporation with its professional polish and elegantly appointed offices might be insulated from concern for consumers, but they cannot be.

New products must be suggested and reviewed regularly and continuously for a company to survive. The sequence of events and activities by which products come into being and enter the marketplace varies considerably according to the product and individual company practice. The concept or idea for a new or updated item (or product, or machine) may originate from any of many places, such as from market analysis groups, salespersons, individual entrepreneurs outside of industry, or from an engineer within a plant. Promising ideas are assessed by small groups of experienced people, often in a new product development department or equivalent. Most ideas are rejected for a variety of reasons. A few ideas, the "sure winners", are sent on to design groups and perhaps produced. Some of those sell well, and some do not.

The handling of the new design beyond the initial exploration stage varies by industry and company, but we may differentiate between two major classes of products, namely *capital* products and *consumer* products. Capital products include large airplanes, railroad cars, machine tools, cash registers, and other items and machines used in conducting commerce or business. The major concern in the making of capital items is that the item should perform to some well-defined specification. The buyer usually wants some guarantee that the item will not only perform a given task but will do so, over a specified period of time, below some specified cost; this includes the costs to buy, operate, maintain, and dispose of at the end of use.

Consumer products are treated very differently. Examples are lawn mowers, floor covering, automobiles, toasters, shoes, and small personal computers. With these items the major concern is first cost. Performance and life-cycle cost, though they should not be misrepresented, are not as clearly communicated from the manufacturer to the end user as in the case of capital items.

Capital products and consumer products are rarely made in the same factory. If one company makes both types of products, they are usually made in different divisions at least. The reason is that each type of product requires very different organization. The basis for the success of capital products is consistent quality and responsive service. This requires an emphasis of design and development. Capital products do not change model or style very much. They are usually made from a relatively small range of materials, by a staff of skilled craftsmen, and often with very specialized machinery.

Consumer products require good design and development as well, but competitive advantage is gained by efficient material use, material handling, and manufacturing processes. Consumer products are usually easier to make than are capital products, and thus more manufacturers will compete in the consumer market. The manufacturing system must then be both physically and economically flexible, and the people designing the systems must understand the entire system.

An outline of the steps required to design a manufacturing system for consumer products is shown in Figure 1–1 in terms of the evolution from the idea for a product to the preparation of a factory to make the product. The steps will vary according to industry and product, but the urgency to plan carefully is the same for all. An ill-conceived product line is a merciless sinkhole for money. Every step in the planning process must assure the feasibility of making a large profit (or minimizing losses in some cases). Figure 1–1 shows that the decision sequence involves the joint effort of people of many disciplines. The time dimension cannot be shown in the figure, because each product and each organization places different emphases on different steps. Indeed, the time dimension often depends on relationships between people and depends on the effectiveness in transferring information among them. The personal aspect is often minimized by engineers, but it is important. In a successful industry one finds a great deal of harmony and personal understanding among and between individuals in the sequence shown. Acrimony between groups and a confused organization structure tend to lengthen and garble the lines of communication, thus slowing the process and allowing errors to survive in the system. Losing sports teams often have the same problem.

Beside the purely human aspect of information transfer, the technical content of design information must also be efficiently transferred. Since a design document (for example, blueprint and specifications) contains information provided by people of very diverse technical skills and backgrounds, all participants in the design must understand each other adequately. This often requires a major shift in thinking from college, where engineering students learn how to specialize, and ignore broader areas of study.

It should be noted that Figure 1–1 shows close communication between engineers and business specialists. The reason is that many decisions in manufacturing are based on economic and business considerations. But many business decisions depend on technical input as well. Engineers usually avoid courses in economics and business, and business students rarely take courses in engineering. Thus, at the start of their careers there is little common means of communication between them. Yet they must communicate. Long-term prosperity in the manufacturing industry requires good communication, and when an industry is prosperous, its valuable employees also prosper.

Figure 1–1 can be viewed as an information flow network. Engineers do not usually think of their handiwork as an element in a flow of information. The engineer calculates and designs for the purpose of solving problems, and submits his conclusions in a report to his supervisor. Traditionally, designers acted as if their responsibility were ended when the design was first committed to paper, that is, the blueprint. The next group to deal with the design was usually the manufacturing group, whose responsibility was to make whatever was designed so that a third group could market it. There was little interaction between designers and manufacturers. The separation was *organic*, or planned organiza-

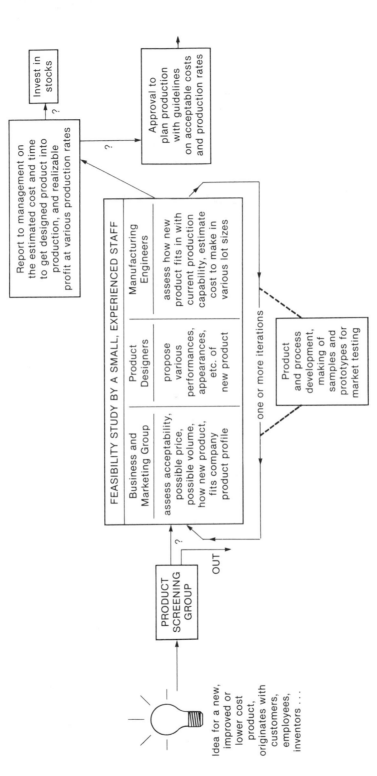

Figure 1-1 Chart of the steps in bringing product concepts into reality via engineering, business and marketing specialists, and labor.

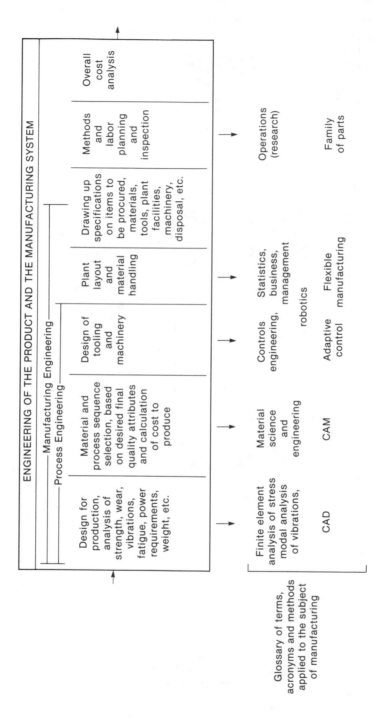

Figure 1-1 (con't.)

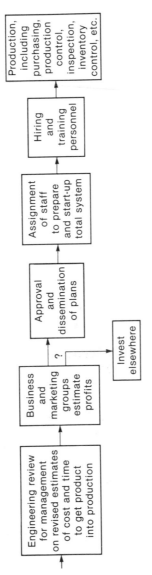

Figure 1–1 (con't.)

tionally by placing each group under a different vice president of the company. The result was a system that could not respond well to changes in the markets. Such companies fade from view, particularly during economic recessions. The modern view is that organizational walls should not be built, or conversely, all specialists who can contribute to the design of a product and manufacturing system are given free and orderly access to each other and to all relevant information. This means that each specialist must learn at least the basic terms and concepts of all other specialists, and each specialist must communicate his own conclusions and thinking in terms that others are likely to understand. Thus it is very important to regard ones work as an element in a flow of information. The flow of design information is as important to the survival of the company as is the flow of materials through the manufacturing processes. For that purpose, information, like materials, must be more completely *characterized* and its flow automated. This topic will be discussed at greater length in a later chapter. First, the technical role of engineers will be discussed.

1.3 ENGINEERING IN THE MANUFACTURING INDUSTRY

The manufacturing industry is very large, but a relatively small fraction of people in these industries are engineers. The reason is that many products do not need engineers to design or to make them. Clever and hard-working people can and do accomplish many things without engineers, and will continue to do so. Doubtless many products are made by companies that employ engineers but do not really need or use the capability of engineers. Doubtless too, there are others that need engineers but have not yet recognized the need.

The need for engineers is probably related to the complexity of the manufacturing operation as much as to the complexity of the product. In fact, we may differentiate between the product and the manufacturing operation as to the engineering need. Complicated products, such as an x–y plotter, require engineers for the design but skilled machinists to make them. The design of machinery to make shoes requires engineering, but the design of the shoe itself does not. The curious practice in many industries is that the design group is often called the Engineering Department. Most other engineers are in the Manufacturing Department or Plant Construction Department.

To place the work of engineers in the context of other people it is useful to distinguish between *operating* a manufacturing facility on the one hand, and *designing the product and designing the manufacturing system* on the other. Figure 1–2 illustrates the difference in expertise needed within the manufacturing system. The diagram shows the management staff to one side, but only to keep the illustration simple. It shows that the major engineering challenges in manufacturing are in the following areas:

1. Product design and engineering
2. Material and process engineering
3. Industrial engineering

The prominence of each of these groups depends strongly on the products produced. In fact, whether engineering will be prominent at all in a particular company may also

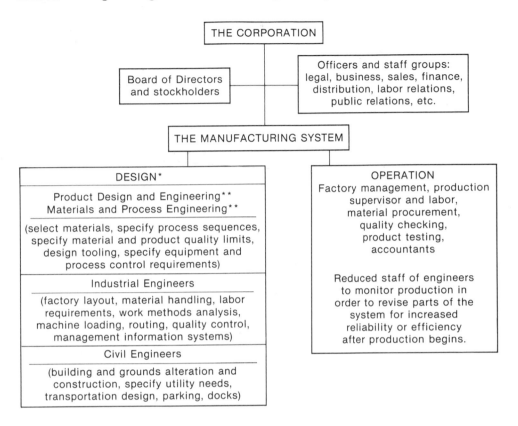

THE CORPORATION

Board of Directors
and stockholders

Officers and staff groups:
legal, business, sales, finance,
distribution, labor relations,
public relations, etc.

THE MANUFACTURING SYSTEM

DESIGN*

Product Design and Engineering**
Materials and Process Engineering**

(select materials, specify process sequences,
specify material and product quality limits,
design tooling, specify equipment and
process control requirements)

Industrial Engineers

(factory layout, material handling, labor
requirements, work methods analysis,
machine loading, routing, quality control,
management information systems)

Civil Engineers

(building and grounds alteration and
construction, specify utility needs,
transportation design, parking, docks)

OPERATION
Factory management, production
supervisor and labor,
material procurement,
quality checking,
product testing,
accountants

Reduced staff of engineers
to monitor production in
order to revise parts of the
system for increased
reliability or efficiency
after production begins.

*Any or all of these functions could be contracted to outside
  companies, particularly Civil Engineering.
**The technical background of these groups depends on the
  product. Some products may require electrical or mechanical or computer skills above others.

Figure 1–2   Diagram of specialties required in manufacturing.

depend on business cycles. It is common to find that a company with a good product in
a stable market decreases its engineering activity after some time. This has the benefit to
a company of decreasing both the payroll and other expenses associated with product
development. This may be followed by other steps to increase profit, such as to use raw
materials of lower cost and to shortcut some of the steps in production. This tactic gains
considerable credit for upper management, but only for a short time. Eventually product
quality will decline, or not keep pace with the competition, and sales will decline or
warranty costs increase very quickly. Management will call upon engineers for new ideas,
but the few remaining engineers will probably be overwhelmed with emergency problems. Very likely the most competent person(s) to solve a particular problem will have
been urged into early retirement, or some modern tools for problem solving will have not
been purchased.

If the company survives such folly, a later folly that sometimes occurs is to give excessive authority to engineers, to design and produce a high-quality product—which will cost more to make than can be recovered from sales.

1.4 COMPUTERS IN MANUFACTURING

Programmable computers and electronic processors are becoming widely used in the manufacturing industry. Computers stand in a long line of developments that have changed industry. Historians of technology cite previous developments to include:

1. The use of animal power to replace human power.
2. The use of water wheels and steam engines to allow larger-scale and higher-speed operation than is possible with animal power.
3. The use of machinery to generate electricity, with which forces can be multiplied, energy can be widely distributed, and high temperature can be generated.
4. The discovery of oil and the development of the chemical industry.
5. Labor unions, which raised the standard of living of workers.
6. Automation, which allowed mass production of consumer goods.
7. Computers, which, when applied to process and inventory control, allow manufacturers to make short production runs economically in order to make several options available in many consumer products.

So far, the complete potential of computers has not yet been realized. Thus there is a great amount of exploration and research under way in this field, which will doubtless spawn several new engineering activities relevant to the manufacturing industry.

The presence of computers as *aids* in design and manufacturing has popularized the acronyms CAD (computer-aided design), CAM (computer-aided manufacturing) and CAE (computer-aided engineering). The simplest meaning of the wide use of acronyms is that the computer has become more useful than previous aids to engineers. CAD usually refers to automated drafting and a set of software packages for force and stress analysis, dynamic modeling, modal (vibration) analysis, and other aids to designers. CAD does not usually include that vital and creative step in the design activity which converts a concept for a product into the representation of feasible parts and shapes on paper or other storage media.

The most visible aid rendered by computers to the manufacturing enterprise is in controlling the motions of various parts of process machinery. Since the electronic means for effecting mechanical control is by way of (a number of) pulses sent to stepping motors (or equivalent), control by microprocessors has become known as *numerical control* or NC. The NC machines are particularly effective for medium and small batch sizes of products where labor costs in *job shop* scale of manufacturing are too high and high capital cost of *transfer machine* for *mass production* is unwarranted. Older NC machines are programmed to make tools or parts move along some predetermined simple path, without measuring the quality of the work being done. Modern NC machines are made to optimize their motions along multiple complex paths, and a few are equipped with sensors for controlling the quality of the product being produced.

A technology somewhat separate but related to NC is robot technology. The construction of the robot is an extension of machine tool technology to a class of machinery that makes higher-speed movements in more degrees of freedom. Robots find particular application at present in material handling ($\approx$30 percent), spray painting (3 percent), spot welding (30 percent), overlay welding (11 percent), and other large-range but low-precision motions. Research in this field will probably yield robots that *learn* new movements from manual patterning and robots that are controlled from computers which will be programmed to recognize geometric images via television.

Production management in real time has gone beyond human capability and requires computers. In manufacturing of automobiles, for example, there may be 30 options available to customers, including color, type of radio, engine size, etc. There is little point in assembling the car unless all of the parts are available at the proper time (except minor items such as wheel covers or bumper guards). After production start-up the salespeople project, in general terms, how many four-door bodies, large engines, FM radios and reclining seats will be needed, and these are ordered into production. The production of components in-house and the supply of components from outside are usually done in batches so that buffer storage regions are planned to hold a limited number of parts. One attempts to store mostly the low-cost, small-volume, hard-to-add-on-later, discontinuous, type of parts. When all parts are in hand, the signal is sent out to start the major selected component down the line. This may be the floor of the auto body, for example. All parts are timed to meet the major component at appropriate points along the assembly line, so that the correct parts are added to the proper car.

This complicated operation is coordinated by computers. When properly done, the car involving the largest investment in parts is completed first so that the cost may be recovered early. In other words, a properly controlled system is optimized relative to some quantity, often highest profit. There are many factors to consider, such as the changing costs of one material compared with another, or the cost of paint versus the cost of upholstering, etc., but all of these details are recorded somewhere and should be used in the decision making process. Primarily, this activity engages the industrial engineer, but with considerable initial input from manufacturing engineers.

It is apparent from this illustration that the economic aspects of production are of great concern to the manufacturing engineer, but there is yet one necessary interaction. That is the management of design information in the factory. This is the next area for aid by the computer.

1.5 TRANSFER OF DESIGN INFORMATION

Information on the design of products is difficult to control. The problem is not secrecy primarily, but rather it is a matter of insuring that everyone working on a new product or process is using exactly the same information. Very often, after an initial design is completed and circulated in a plant, a great number of changes must be made. Perhaps a different material or bearing should be used somewhere in the design because material costs have changed, or because the company happens to have a large supply of leftover material from a failed product line, or perhaps because an old machine must be replaced in order to fabricate parts with the material specified in the new design. All of this information should be available to the designer, but frequently it is not. Success or failure of a manufacturing plant often depends upon good management of such information. Formerly this information was managed by the organization of people into appropriate groups, with hierarchies of coordinators, and liaison personnel. Recently the technical content and volume of such information has greatly increased. This is due mostly to increased competition but partly also to high energy costs, to sudden variations of materials prices and availability, and to the emergence of many new processes. One emerging reality is that human organizations are not dynamic enough to channel or transfer the appropriate information to appropriate individuals with adequate lead time. Computers have been very helpful in this area. One example may be cited in which design information is transferred with minimum error through a plant. That example is the engine plant of General Electric at Evendale, Ohio. The jet engine is a product that is steadily upgraded part by part; thus there is a steady flow of new design details to coordinate with the entire system. Formerly a new design for a part was committed to paper for communication throughout a plant. The new design was reviewed by material buyers, process engineers, plant engineers, and by the designers of adjacent parts. Frequently someone suggests a way to revise a new design for increased versatility or convenience, so changes are made. Unfortunately, not all who have seen the early versions of a new design are immediately aware of the recent changes. The consequence is that materials may be purchased or tooling may be designed on the basis of unfinished designs. Good organization minimizes this risk but does not eliminate it. General Electric designers store their designs electronically. All who have need to know about the new design are notified, and they must respond with a note of the action they will take. When a design is revised, those that have requested and seen the earlier design are interactively notified of changes. All such transactions are recorded so that the responsibility may be fixed for any inaction or action.

In the GE plant, when a design is reasonably fixed, tooling is made, equipment is prepared, and raw material is purchased. Production is begun by checking the quality of incoming material. If the material has the correct alloy content and hardness, it receives a laser-etched number. The part is transferred to a machine where its number is read by automatic reader. If it is at the correct, previously prepared station, approval to proceed is given on a TV screen on the machine. On the same screen there is a *print* of the part, with a list of all of the tools and gages to be used at that station. If the processing machine is automated, it will make the proper motions to produce the shape shown on the screen. After the part is finished, it is moved to an automated inspection station, where its shape is compared with the original design; if all is well, the part is cleared for scheduling into subassemblies.

The GE technology is not directly applicable to all products or manufacturers, but the concept of design information control has considerable potential. An important beginning point in establishing such a system in a factory is to link all of the computers at design stations, processing stations, and inspection stations together. Through such links information can be gathered that would simply overburden human information channels. Consider two examples of information control that are immediately possible in computerized plants. The first is to keep a running record of the accuracy of parts from each machine, and this serves two purposes: to monitor machine condition and to continually assess the cost of holding manufacturing tolerances. The second example is to instrument machines to detect material properties or process behavior that would produce out-of-specification parts. When such conditions are found the process can be stopped before more cost is added in later processing steps.

The full potential of the computer information systems now awaits several major developments. To date the computer has been used to automate the simpler tasks of humans, in a way that is suited to human senses and limitations. New systems must be designed so that computer based systems complement humans. In process control, for example, there are three such needs, namely,

1. To determine what other forms of data are obtainable from a process, in addition to what is now obtained.
2. To develop new methods of direct sensing of product quality, rather than simply to adapt existing methods.
3. To develop fundamental models of processes.

Designers are said to make between 80 and 90 percent of the first choices of materials and processes for a product. Within this figure, there is considerable latitude on the definition of a designer, but optimal choices on processes are not made until very late in the progression toward the start up of manufacturing. The exact value of having early and accurate information for making the final decisions on processes and materials is not known. Managers feel very vulnerable in this area. Most would welcome progress in gathering accurate information, but few are able to express the form of aid required. The core of the problem is that competing processes are not well enough modeled or described so that a short analysis can be made on their relative costs and capabilities. It would be very useful if some of such information were available on a computer, but linked with a data base under the jurisdiction of specialists in various areas for information beyond the understanding of the designer. In other words, the designer should have access to a computerized consultant (either human or an expert system) in materials and processes in such a way that he is compelled to ask well-focused and vital questions of the proper specialists. This aim serves the dual purpose of formalizing material and process data, which has not been done to a significant extent, and it will relieve designers of excessive dependence on familiar materials and processes.

The latter point is not widely recognized by designers as a need. Perhaps the selection of materials and processes is not considered a part of the design process by very many, but this topic has its obverse side. Process engineers and production engineers too often discover that a design is finalized or frozen before consideration for the best materials and processes is included.

1.6 ECONOMIC AND MANAGEMENT CONCERNS IN MANUFACTURING

The engineering of a manufacturing system is a very broad field and it includes some consideration of every aspect of a manufacturing enterprise, including the financial and marketing activities. The engineering covers every detail of products, from the conceptual design to the shipment of a product to the customers. It involves intermediate redesign of a product for economic manufacture, material monitoring and handling, manufacturing process control, and some aspects of plant engineering and waste management.

Of major importance in manufacturing is consideration of the effective use of capital assets such as buildings and machinery. If a product will expend the useful life of a machine over the expected production run of the product, an optimum has probably been achieved in the economic use of capital. More often, a simple optimum is not possible. For example, a machine will usually outlast a production run of one product, be altered and adjusted for a different product, etc. If a very expensive and versatile machine is purchased such that it can be adapted to many products, the engineer faces the possibility that the machine will never be properly amortized if a product line changes drastically. Specifically, a company may purchase ten large presses expecting to produce stamped metal parts for ten years. However, after five years it may become necessary to produce the parts in plastics, which will require retiring the presses early and purchasing plastic molding machines. Or it may occur that after the purchase of ten presses of a particular capacity, it may be found that for competitive reasons, larger and more accurate parts are required, rendering the purchased machines obsolete earlier than expected.

The same applies to the use of buildings. At one time it may seem best to store materials at the ground floor and assemble products at the second floor level, but the product line may change in such a way that it would be most economic to intersperse storage and production on the same floor. A third example, one in material handling, involves trucks and railroads. A plant may be constructed with little storage area for raw materials or finished products, expecting materials to be transported by highway truck. However, freight rates may change over a short period of time in such a way that rail transportation would be recommended. This may require construction of a rail bed and more storage space than is currently available.

Engineers must make decisions that involve materials, plant facilities, and in-house capability of people as well. Take as an example the manufacture of coffee makers. Several of its parts could be made of plastics; others must be made of metal or glass. Some important points to consider in making this product are the shipping weight, the method of packaging in order to protect different types of materials, differing warranty considerations, and different production methods. Parts may be made in-house or purchased outside. The decision on which parts to produce in-house depends upon what machinery is either available or may be purchased with the expectation of future products requiring the same machines. For the water heater tank it may be cheaper to use the services of a supplier who has both the experience and the machinery for deep drawing of metal. If your company is not skilled in deep drawing, it may require six months to get up to production. On the other hand, when you rely on a supplier, you must be assured that the supplier's labor problems will not halt production of your parts. By this example it is seen that manufacturing engineering includes consideration of labor relations and perhaps an analysis of the economic climate of the city or of the entire nation.

The economic analysis and decisions of engineers are very little different from that of the individual consumer, except in the scale of concern. In deciding whether to replace a troublesome but operable TV set the consumer is mindful of the possibility that new and revolutionary developments in TV may be available in one or two years; they may be worth waiting for. Likewise, before purchasing an automobile it is useful to consider whether the maker of a particular brand will maintain convenient dealerships in the future, or if there may be a problem in parts supply.

In manufacturing it is paramount that all decisions be tempered by considerations of economics. Everyone must participate in lowering costs, at every stage of manufacturing. For example, when the Ford Motor Company of Europe was planning the plant in Valencia, Spain for building the Fiesta, they considered it necessary to reduce the cost of "redesigning components, sorting out the plant and tooling, small improvements in the product, changes in material . . . ," from the usual "18% of the launch cost . . . (to no) more than 7 or 8%." To this end the management stated that "(We) must have . . . total discipline, . . . can't have any nonsense . . . (about lowering) the cost of changes, in putting things right _after_ production has begun . . ." (John McDougall, chairman, Ford-Europe).* On the latter point management usually expects about 1000 changes of one kind or another during the first year of production of automobiles.

Most designs are not immediately convertible to profitable products. Even where a product is useful, reliable, and attractive, a low-cost substitute may capture the market. There are many reasons why a product may be too expensive. Some of the reasons are out of the range of technical solution. Some of these include an excessive overhead rate because of large staff groups or other high expenses to maintain a company in existence. More direct causes of high prices may be that employees are paid too much or are underutilized. Or the design may be too difficult to make economically. The specified materials may be too expensive to purchase and/or to process, process control may be inherently poor, or the production yield may be low if the part is made as designed. Or, again, the product may deteriorate in service more quickly than expected.

Redesigning a product for efficient and economic manufacture requires a direct and delicate interaction between two major groups in any company, the conceptual designers and the manufacturing personnel. Each group performs a vital function in the company, and until a product is on the market each can argue for the importance of their own position with some impunity. To lower the cost often requires a slight change in appearance, shape, function, smoothness of operation, or some other feature of the product.

---

* E. Seidler, _Let's Call It Fiesta_ (Lausanne, Switzerland: Haessner Publishing Co., 1976).

Designers resist such compromises and urge manufacturing groups to *dig* the economies out of their own domain.

In deciding between the concerns of designers and manufacturers, management often makes decisions that involve the lower risk among several alternatives. For example, if a proposed product is elegant but overpriced, it may be projected that production costs can be reduced after production begins. If the company is in good financial condition, it may be able to temporarily *carry* the difference between the first manufacturing cost and the return from the price at which the product must be sold. If production were held up until the manufacturing group is satisfied that the product can be made to meet market price, it may be of such low quality that it quickly establishes a poor reputation on the market. It requires a well-informed and resolute management to arrive at the proper compromise in product design.

To a great extent, the strategy used by a manufacturing enterprise may depend upon one's position in an industry. For example, in the early 1920s, General Motors was recovering from a total financial failure. Alfred P. Sloan, then president of the company, adopted a strategy for recovery using the philosophy, "In order to gain market shares against a competitor, it is not necessary to have greater than competitive quality."* This philosophy is now widespread and was restated, critically, in 1981 in the words, "if a product works well for an extended period it may be overdesigned in some of its aspects. If such is the case, consider downgrading the materials or processing where you can."†

The compromise between designers and manufacturers is often carried out in a spirit of caution, faithfully reflecting organizational divisions. It has been found that the most profitable products are likely to come from a group consisting of a mature conceptual designer, a mature manufacturer, a person with experience as a shop foreman, and one or two younger people with analytical skills. Product designs developed by such groups are often ready for production with little alteration. Such products go into production months and perhaps years earlier than those from divided activities simply because there are fewer errors to purge from the system and because the latest information on most aspects of product production was used in the design.

PROBLEMS

1–1. What engineering skills are required to bring the following products from the idea stage to the market place?

    (a) Wooden lead pencils.
    (b) Pads of paper.
    (c) Cans of beverage.
    (d) Electrical power.
    (e) Bicycles.

* Alfred P. Sloan. *My Years with General Motors* (Anchor Books, Doubleday, 1963).
† H. E. Chandler, *Metal Progress* (Feb. 1981), p. 9.

1–2. How much more would *you* pay for the following improvements in consumer products? Justify your answer in terms of both subjective and economic terms.

    (a) A rustproof automobile.

    (b) A quiet vacuum cleaner.

    (c) A TV set that is only 30 mm thick.

    (d) A personal computer that operates at twice the speed as present models costing $1000.

__ 2
economic principles applied to manufacturing ___

2.1 INTRODUCTION

The simplest expression for the economics of manufacturing is (on a unit basis):

$$\text{profits} = \text{selling price} - \text{marketing cost} - \text{cost to manufacture} - \text{overhead cost}$$

Many details are left out of the equation, but in practice, there are many costs assessed against the selling price. Only the *cost to manufacture* will be of major concern in this book. The other variables in the equation are the concern of management, and for their purposes some very intricate accounting schemes are used. Management is continuously adjusting accounting and information systems, in order to be completely informed on the status of every activity that contributes to the profit and loss of the company.

The cost to manufacture is determined most confidently after a product has been in production for some time. But a thriving business must continually develop new products, using new materials and new processes. One of the main responsibilities of manufacturing engineers is to *predict* the cost to make such products, and this is done when a manufacturing system for the products is being designed. The goal is not always to find the absolute minimum cost at the moment, but rather the lowest range of cost for several alternatives which depend on ever-changing markets. Some of these alternatives involve considerations of the size of production lots, production rates, choices of materials and processes, and various amounts of plant automation. Predicting product cost requires a detailed study of both the hardware and the software (i.e., the information) of a proposed system and also requires contact with many specialists in business and engineering. It is from the others that manufacturing engineers obtain the interest rates, depreciation rates, overhead rates, market projections, and new developments in material properties that affect the cost to manufacture.

2.2 THE DIRECT COST OF MANUFACTURING

The factors in the direct cost to manufacture products include the following:

a. The cost to buy, inspect, store, move, and inventory raw material or purchased components.
b. The cost of production machinery and material-handling equipment.
c. The cost to prepare, install, and maintain the entire facility.
d. The cost to operate and service equipment.
e. The cost to assemble, package, store, and inventory finished products.
f. The cost of utilities, transportation, disposal, and protection.

The sum of all these costs divided by the number of units the system will make should give the cost to make each unit of product. But such an overall approach does not provide information with which costs can be *shaved* from products. A more thorough approach is to determine the *value added* to the product by each operation. If the cost of a particular operation exceeds the value added by that process, then the operation should be studied carefully. Some engineers, usually the design-materials-process engineers, will study the materials and processes themselves. Industrial engineers will probably determine whether the labor content or material-handling aspects of the operation can be revised, and civil engineers will consider the cost of space and facilities in the equation.

The cost of a manufacturing operation is not a fixed or intrinsic amount. That is, the cost to bend a sheet of metal into an electronic chassis, 100 mm × 300 mm, is not simply $3.50, for example. If process costs are given in handbooks, or as examples in textbooks, they will usually be given in ranges of cost, because the actual cost depends on many local conditions. The cost of equipment, personnel, and facilities, plus the time value of investment, are added to products with the passage of time, whether the facilities operate or not. To avoid the buildup of cost it may be necessary to cease manufacturing and dispose of the facilities. To reverse the buildup of cost it is necessary to sell products. Management regards the liquidating of the business as a real possibility at all times. They have little loyalty to a product line or to the plant or equipment. Rather, management is required, by the owners, to provide a return on the money invested in the company, by legal means, of course, and with consideration of the social consequences of their decisions. Thus career opportunities for engineers and all employees are inextricably tied to the success of the products produced. This is more often discovered in times of economic recession than during full employment and prosperity.

2.3 THE TIME VALUE
OF MONEY

The specific time value of money is the interest rate. Borrowing money entails paying back the principal plus interest; investing or lending money entails receiving it back again after some time with interest added. Discussions of interest are complicated by the several methods of calculating interest, and by the terminology used by business specialists.

Interest may be calculated either by the simple method or the compound method. With simple interest one assesses a fixed amount of interest over a period, for example, 5 percent per year. Thus a present loan of 100 dollars is paid back in one year plus the 5 dollars interest; or it can be paid back in two years plus 10 dollars interest, and so on. Compound interest is a series progression in interest accumulation; it may be compounded quarterly, monthly, or any other time period. If, for example, 100 dollars is borrowed at 5 percent interest compounded quarterly, the interest builds up as follows: 5/4 percent or $1.25 interest is owed after three months, making the amount to be repaid at that time $101.25; over the next quarter 5/4 percent is charged on the $101.25, making the amount owed after six months $102.52; over the next quarter 5/4 percent is charged on the $102.52, making an amount owed after nine months of $103.80, and finally 5/4 percent is charged on the new amount, so that $105.09 is owed at the end of the year. This is equivalent to a simple interest of ≈ 5.1 percent. If the interest is compounded monthly, the final amount is $105.12, or an interest equivalent to ≈ 5.12 percent.

Since most interest-bearing transactions cover several interest periods, it is convenient to use equations instead of the method used above. There are two common problems in calculating the relationship between a present and a future sum as altered or influenced by interest. One is for calculating the single payments to make at the end of one interest period, and the other is for calculating payments each interest period so that repayment of a loan occurs over a given number of interest periods. The converse of borrowing money is to invest it in order to receive a return. An investor may wish to know how much to invest now to receive a desired amount at the end of one interest period, etc. The equations for calculating these amounts are:

To calculate what sum of money S will be available after a periodic rate of interest i accumulates for several interest periods n on a single present investment, P use:

$$S = P(1 + i)^n \qquad (2\text{-}1)$$

(referred to as the single-payment compound amount)

The reciprocal calculates the amount P that must be invested now in order to receive a desired sum S at the end of n periods:

$$P = S\{1/(1 + i)^n\} \qquad (2\text{-}2)$$

(referred to as the single-payment present worth amount)

To calculate what sum S will be available at the *end* of n years when an amount R is invested at the beginning of each interest period, use:

$$S = R \frac{\{(1 + i)^n - 1\}}{i} \tag{2-3}$$

(referred to as the uniform annual series compound amount)

The reciprocal calculates the amount R that must be set aside at the beginning of every year to accumulate a desired sum S by the end of n years:

$$R = \frac{Si}{\{(1 + i)^n - 1\}} \tag{2-4}$$

(referred to as the sinking fund deposit amount)

To calculate the uniform end-of-the-year payment R that can be realized for n years from an investment of P, use:

$$R = P \frac{\{i(1 + i)^n\}}{\{(1 + i)^n - 1\}} \tag{2-5}$$

(referred to as the capital recovery amount)

The reciprocal gives the amount P that should be invested now so that n uniform payments of R can be received each period to dissipate the investment:

$$P = R \frac{\{(1 + i)^n - 1\}}{\{i(1 + i)^n\}} \tag{2-6}$$

(referred to as the uniform annual series present worth amount)

Equation (2–5) for capital recovery amount is particularly useful for calculating how much return a machine should generate in order to "pay for itself." Usually the value and the production capability of the machine are known. The process engineer should estimate, where possible, how much value is added to a part being processed by the machine in question; if that value is greater than the cost of the operation, then the operation will contribute to the profitability of the company.

Occasionally a question arises as to whether investment in a machine is a good business decision. In such a case the expected investment in the machine may be inserted into the equation as the value P. The likely return from the machine in production is R, and some projection can be made on the number of interest periods n that the machine will be used. Either Eq. (2–5) or (2–6) can then be used to calculate a value of i, which is the return on the investment. If this value of i is greater than the interest rate offered in the money or stock markets, it would be wiser to invest in and use the machine rather than invest in stocks.

2.4 THE INFLUENCE OF PRODUCTION LOT SIZE ON COST

The cost of manufacturing includes the cost of the facilities prorated or spread over a production lot. For example, if a factory and equipment cost 10^7 and 10^7 parts will be made in these facilities, leaving worn-out facilities that have sufficient value to cover only the salvage cost, the amortized cost of facilities is $1 per part. Such simple economics are rare outside of the sidewalk lemonade industry. In most industry the 10^7 has time value. That is, the 10^7 could earn 10^6 per year, if the interest rate is 10 percent, without the bother of setting up a manufacturing facility. If one chooses to set up a manufacturing plant, some of the 10^7 must be allocated for plant construction up to three years before production begins. In that three years, whatever money is expended is not available to yield a return; perhaps it was borrowed and interest must be paid for its use. This time value of invested money must be added to the price of the product. For example, assume a linear investment rate in plant construction and equipment purchasing over three years, at an annual interest rate of 12 percent. The accumulated cost of the money at the time production begins may be calculated in several ways. If the money is paid at a linear rate at the time of monthly billing by contractors and suppliers, the time span will be 36 months with a monthly interest rate of 1 percent per month. The added cost of money then may be seen by the use of Eq. (2–3), where $n = 36$, $i = 0.01$ and $R = 10^7/36$. $S = 1.197×10^7.

Further, the invested 1.197×10^7 is unavailable at the beginning of the year for profitable investment elsewhere, so the production run must absorb the interest again. If we assume a linear monthly recovery of the invested money, by Eq. (2–5) we can calculate the monthly payback required from sales. For this calculation, $n = 12$, and $i = 0.01$. Then $R = 0.106×10^7. This amount must be recovered from $10^7/12 \approx 825,000$ parts per month, which is $1.289/part. Now assume that the cost of material per part is 55¢ and the cost of labor is expected to be $2.00. The cost to manufacture, then, is $3.839/part. It is simple to project what the cost/part would be if it should occur that sales are greater or less than projected.

The selling price will be much higher than the manufacturing cost, depending on the method of storing, transporting, and distributing the product. The markup from manufacturing cost to selling price is in the vicinity of a factor of 3 for one brand of home-appliances and a factor of 20 for some automotive replacement parts. One cost to be covered by the revenue from selling a product is the cost of doing business, and the cost of planning new products and production facilities. Ideally, all of the costs connected with bringing product A to market should be covered by the price of product A. In industry, the time of engineers, lawyers, business specialists, and upper management is so diffusely spread over all of the products that precise accounting is not possible. Furthermore, there are municipal taxes, insurance, lawsuits, and board meetings to cover. Often it is most reasonable that all costs other than materials, labor, and facilities directly chargeable to a product are added up and allocated to the products sold. This cost is called ''indirect cost,'' or ''burden,'' or ''overhead.'' (Teachers are in this category.)

The indirect cost may be assigned in a simple way, as a figure to be added to the direct cost of manufacturing a part. With complicated items or in the event of inefficient management, the indirect cost may be applied as a multiple factor, as much as 3 or 5. The $3.839 direct cost of producing a product will have its price increased by the overhead cost. Assuming the overhead rate to be twice the direct cost of manufacture (that is, the overhead multiple factor is 3), the total cost associated with manufacturing the item is $11.517.

A profit margin must now be added. Recall that there is little point in setting up or continuing to run a manufacturing operation unless the financial return exceeds the potential return from investment in stocks or bonds. Recall also that the cost to manufacture includes the current cost of money. The profit margin is the *extra* return; however, profit margins cannot be set at will. The produce price must be set so that the product will sell in the marketplace at approximately the expected rate of production.

Price setting is a delicate art. Let us assume that an appropriate wholesale price in the case of our example would be $12.645, which is a profit of $1.128. This amount would appear to be an 8.9 percent return. Actually, the return can be applied only to the investment, which constituted $1.289 within the total manufacturing cost. The profit, $1.128, then becomes 87.5 percent of the investment share per part, and is worthy of some attention. In reality, a profit of 15 to 20 percent is to be expected.

But the selling price cannot be set by the seller. The marketplace sets the final price. Sometime the profit margin becomes unattractively small. At such times ways are explored to reduce the total cost of the product, direct cost first (because direct cost is leveraged by indirect costs) and then the indirect cost (requires sacrifices by managers). The limit on reducing direct cost is the point at which the buyer detects reduced quality or reliability. The limit on reducing indirect cost (engineering and management time) is the point at which smooth functioning of the plant is in danger or when future product plans receive inadequate attention. These are difficult trends to detect and very much depend on the projections one makes for future markets, future material, and labor cost and even future tax and inflation trends.

In fact, it sometimes happens that a price will be set below total cost to manufacture plus overhead. This may occur where a manufacturer wishes to *penetrate* a market. This may require three years to do. One make of small cars was estimated to lose between $700 and $800 per car in 1981, all in anticipation of a favorable sales position in the future. It seems foolish to deliberately manufacture cars at a loss, but if there are no products to present to the market, there will be no penetration of a market. The auto maker may not have expected to lose as much as $700 per car, but his annual loss may have been greater with an idle plant than with an operating plant. For the same reason, old established manufacturers experiencing severe competition may sell goods at a loss in order to maintain market share, or in order to test the strength of the competition (that is, to try to drive them out of the market).

2.5 THE INFLUENCE OF PRODUCTION
RATE AND PRODUCTION METHODS

Marketing people have the responsibility to estimate the expected sales rate of products, but the estimates cannot be assured. Thus the manufacturing system designer must provide alternatives in processes and manufacturing methods to provide a range of profitable production rates. The most flexible economic system is one which has a minimum of capital investment and a minimum of indirect cost. This means a heavy, or almost exclusive, use of either direct labor or of outside suppliers. The main problem with outside suppliers is that some of the control of production schedules is lost. A secondary problem is that proprietary information is not guarded as carefully by outside suppliers as by the owner of the information.

The problem with nearly exclusive use of direct labor is that little labor-saving or automated machinery would be used, which can readily be shown to result in high manufacturing cost. An example of the costs for various degrees or levels of automation is given below, for making parts with four types of lathes. The machines, their costs, and the times to make the same part on each machine are given in Table 2–1.

The accounting can become very complicated if all possible factors are included, so the following assumptions are made in order to get to the point:

1. One normal 8-hour shift consists of 1500 production hours per year.
2. Each machine will be worn out after making a million parts. Thus the engine lathe could produce 9000 parts per year on one shift. It should therefore last 111 years, or 37 years on three shifts. The fully automatic lathe should last about 22 years on one shift, or about seven years on three shifts.
3. The annual interest rate is 10 percent.
4. Direct labor cost is $15/hour and applies only while a part is in production. (Overhead applied to direct labor equals direct labor cost and is included in Table 3–2 as labor cost.)
5. Material and cutting tool cost is 50¢ per part.
6. Setup cost, i.e., installation, initial tooling, and start-up cost of the machines, is 5 percent of the cost of the machine.
7. Since different machine lives introduce a factor that is very difficult to amortize, for a first approximation the actual rate of wearing out of machinery can be ignored in calculating the cost of money invested in machinery and an arbitrary linear 10-year depreciation life can be applied.

With the above figures we can calculate the cost/piece to make the parts in various annual lot sizes. With Eq. (2–5) we calculate an annual amount R that must be returned from the sale of the annual production. The investment in the machine is P; also, $i = 0.1$ and $n =$

Table 2–1

Machine type	Cost to purchase*	Labor time per piece	Production rate/hour
a. Engine lathe (manual)	$ 15,000	10 min	6 pieces
b. Semi-automatic lathe	20,000	5 min	12 pieces
c. Automatic lathe	60,000	1 min	30 pieces
		(the operator serves two machines)	
d. Automatic lathe with robot to load and unload	120,000	0	30 pieces

* 1985 prices.

Table 2–2

Machine type	Interest (R)	Setup cost	Labor cost per piece (inc. ov'hd)	Material and tool cost/pc	Cost per part at various annual prod. lot sizes		
					100	10^4	10^{6*}
a.	$ 2,441.18	$ 750	$5.00	50¢	$ 37.41	$5.82*	$5.50
b.	3,254.91	1000	2.50	50¢	45.55	3.43	3.00
c.	9,764.72	3000	.50	50¢	128.65	2.28	1.01
d.	19,529.45	6000	0	50¢	255.79	3.05	.53

*Beyond the capacity of the machines.

10. The interest cost and the setup cost must both be recovered; both are shown in Table 2–2.

Note that with large annual production lots and low labor cost, the cost of material becomes a large fraction of the cost to make the part. Material cost will be discussed further.

We can also calculate the cost per piece for full annual production of each machine for one shift and for three shifts by using the figures in Table 2–2 (see Table 2–3). With alternative "a" the cost per piece is nearly the same over a wide range of production rates; any advantage could be offset by a second and third shift "premium" for labor. With alternative "d" there is a clear advantage in using the machinery many hours per day.

The cost picture often becomes confused by the rate of depreciation applied to equipment. In engineering terms, the rate of depreciation is the rate at which the machine life is dissipated, but in accounting terms it is the rate at which the Internal Revenue Service allows one to declare a reduction in value of the machine. The IRS rate is usually faster than the actual rate for general machinery. Fast write-off does serve three purposes, but it adds one problem. The first advantage is that a fast depreciation reduces property values for local taxes. Second, depreciation amounts are subtracted from profits when income is calculated; thus high rates of depreciation reduce income taxes. Third, management is less reluctant (different from *more likely*) to dispose of a well-depreciated machine than others in order to purchase an upgraded variety. The main disadvantage of an artificially high depreciation rate is that the final cost to manufacture is set arbitrarily

Table 2–3

Machine type	1 shift		3 shifts		Difference in cost
	production	cost/piece	production	cost/piece	
a.	9000	$5.85	27,000	$5.62	− 4%
b.	18,000	3.24	54,000	3.08	− 5%
c.	45,000	1.28	135,000	1.09	− 7%
d.	45,000	1.07	135,000	.69	−36%

high (depreciation is included in overhead rates). In the above examples, recall that the engine lathe would have lasted much longer (say ten times longer) than the assumed depreciation life even when operated three shifts per day. The cost per piece for a full production rate would have been less than $5.62 if a lower overhead rate were used.

With reference to the machinery in the above example, the best choice for very large annual production lots is clearly the automated system. However, relatively few items are made in optimum annual lot sizes. First, the optimum annual lot size may exceed the needs of the market, for example, size 16 shoes. Second, the American consumer, at least, wants a very wide range of options to choose from. Thus for many products a production lot may be an uneconomic 10,000 per year or even 1000 per year. There are then two alternatives available for keeping costs low.

The first alternative is to buy components from an outside supplier. That supplier may be making similar components for others, and together all of the components can be made at low cost. This is a common practice. Consider the many manufacturers of lawn mowers; the great majority buy engines from only two sources. The same applies to refrigeration compressors, electric motors, integrated circuits, light bulbs, computer printers, tires, and components in most other consumer products.

The second alternative is to contract work from the outside and use your idle machines. This means that your company will be making products that do not carry your brand name, but recall that the purpose of the company is to make a profit, not to hold to high principles of exclusiveness from supplier status.

2.6 INFLUENCE OF MATERIAL AND PROCESS SEQUENCE ON COST

The choice of a material for a product will usually influence the needs in manufacturing facilities. Take, as an example, a portable tape recorder, which may be marketed with one of three types of covering:

1. With a sewn leather case, which, in this example, would be purchased from outside. This decision would be based on the reliability of outside suppliers and on a policy that your company could not become competitive in the area of leathercraft. Assume that the suppliers offer two prices, depending on the number purchased.

2. With a removable case of molded polypropylene, molded in-house and available in two colors. Plastic molding is a very competitive business, but you expect to take advantage of the availability of a fully amortized molding machine and low-cost dies.

3. With a bonded coating of polyurethane, available in several colors. The urethane would be applied by a new process which has been patented by your company. It makes a more durable and attractive product than the other options, and no competitor is likely to offer a similar product in the near future. The cost for the urethane option involves an investment in machinery and therefore is expected to be higher

than for the polypropylene option if only tape recorders are coated, but lower if other products beside the tape recorders will be coated.

Assume that the cost per covering for each of the options will be a function of annual production, as shown in Fig. 2–1. Note that two prices are given for the urethane option.

Note that purchasing items from a supplier insulates one from the effect of sales volume within certain ranges, and also that the price for the urethane option depends on the sales of other products, an intangible quantity for the manufacturing engineer to work with in predicting production costs. Information of the above type must now be considered in the light of the probable price that the market will bring for each option.

Curves of the type in Fig. 2–1 assume that the employees and machinery are devoted to a single product for the year. Frequently it is good strategy to make products at a high rate for a short time, stop production for a time, and wait until the inventory decreases. These decisions are usually not made by the engineer. It is sufficient to provide management with the type of information shown in Fig. 2–1.

An important type of information that the design-materials-processing engineers must provide for management is the influence of changes in material and energy prices on processing costs. It happens frequently that some prices change drastically and quickly due to political upheaval in foreign lands. For example, the price of foreign crude oil increased by a factor of 3 in one day in 1973. This increase in price was *passed on* to the final user progressively as the suppliers sorted out what fraction of their crude oil had to be purchased from expensive sources versus the fraction that would be supplied from their own wells, the prices of which were controlled by the government. Crude oil prices directly affect the cost of plastics such as polyurethane, polyethylene, and polypropylene. Many companies that made plastic toys, milk cartons, sheet, and fiber could no longer compete with similar products made of wood, cotton, or coated paper and were forced out of business within a few months. The same increase in oil prices raised the cost of industrial fuel, and suddenly made energy-intensive processes expensive. Examples of such processes include heat treatment of steels, reduction of bauxite, and electroplating processes.

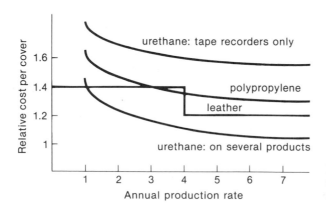

Figure 2–1 The relative costs of three options in covering for portable tape recorders.

The market in metals varied considerably as well in the 1970s. The price of cobalt increased from $4/lb to $14/lb because of civil war in Zaire. This suddenly increased the cost of the cemented carbide cutting tools, which made it *economic to slow the rate of production* of some products (see section 13–9). Chromium was another element that became expensive when unrest occurred in (then) Rhodesia, which increased the cost of stainless steel and some wear-resisting irons, both of which contain large fractions of chromium. At a time of such price changes, engineers must find alternative materials and processes, but always with the prospect of closing down an existing process for which the machinery and tooling are not amortized as planned. When stainless steel became expensive, there were several alternatives available for some applications. One was to use plain carbon steel and coat it or paint it; this required new operations. For other applications, glass may be substituted, but this requires not only new processes but also a redesign of some products.

One example in energy conservation is choosing between lasers and oxyacetylene flame for surface hardening. Lasers are expensive, and only about six percent efficient in terms of delivery of energy from the power line to the target surface, whereas the gas flame is a low-cost installation and 30 percent efficient. But upon review of the surfaces being processed, it was found that a few passes of the laser beam in selected areas was sufficient for the end use, and the general hardening with the flame was not necessary. Lasers therefore became preferred for many applications on the basis of energy costs.

Process cost estimating is thus seen to be a rather complicated activity, but it is also challenging. An engineer involved in cost calculations becomes aware of a broad range of concerns in the company, in the manufacturing industry at large, and in world affairs.

2.7 COST ACCOUNTING

Cost accounting is a means of arriving at the cost, and therefore value, of performing manufacturing operation. It is not simply the tabulation of the costs as an end in itself. Rather, cost accounting involves several closely connected activities which, together, are properly referred to as a *management information system*. Engineers have a prime responsibility for identifying the data that are both necessary and possible to acquire from the manufacturing system, and the means of acquiring those data. The data to be obtained are those necessary for checking on a regular basis that the actual costs of manufacturing follow the base-line prediction developed when the manufacturing system was being planned. In some cases it is sufficient to check once a week or once a day, but with the advent of computers it is possible to check some operations every minute or even continuously.

If there are variances from the norm, either real profits decrease (or even vanish), or profits increase. Variances arise for the several reasons given in Secs. 2.5 and 2.6. Negative variances may be due to unexpected incidences of breakdowns, work stoppages, and adverse weather conditions. Positive variances may occur when new processes are *learned* faster than expected.

Management information systems vary considerably in complexity according to the size of a company, the variations of the products, and the personal preferences of upper management. Smaller enterprises often succeed with a very informal system, and some larger ones have failed though they have very elaborate systems. The goal of all is the same, however, and that is to determine as precisely as possible how profitable each and every employee and machine is. Some can be carried at a loss for a time if it fits in with a long-range plan. It would be expected that profitable elements would *cover* for the losing elements. But management must know precisely how much one element covers for another, and what the trend of such costs is.

The basic ingredients of a comprehensive cost accounting scheme are:

1. Identifying the physical areas of factories called *cost centers* for which costs can be reasonably well isolated and in which responsibility for cost incurrence can be assigned to responsible individuals.
2. Comparison of the actual costs in a center with the forecast results for a planned period.
3. The measured variances and the reasons for them.
4. The action needed to control and reduce costs to manufacture products.
5. Profit or loss, and cash availability, forecast at regular intervals.
6. Regular comparison of sales prices with unit or product costs in order to determine the competitive position of the company.
7. Planning for replacement of existing facilities, or provision of new facilities.
8. Planning for allocation of raw material and material sources.

Engineers are usually expected to work in harmony with the cost accountants, but not to be deeply involved with the methods of cost accounting. This section was provided for perspective on the continuum of expertise required to successfully operate a manufacturing enterprise.

PROBLEMS

2–1. Assume that the market value of automobiles driven 15,000 miles per year declines each year to 72 percent of its value at the beginning of that year, and that the operating costs remain substantially constant for the life of the car, but the annual maintenance costs to keep it in top condition increase linearly to equal the market value of the car at the end of the fifth year. Further assume that you can invest money or borrow money at the same rate, say 10 percent for simplicity. What is the most economic strategy for buying a car for personal use (at 15,000 miles per year)?

2–2. Financial institutions often advertise that they offer higher interest rates than others do. A common method is to offer monthly or even weekly compounding of interest, but in fine print an interest rate is given to three decimal places to make sure that there is no misunderstanding. Provide a table of interest rates that yield the same annual return for:
(a) Simple interest (assume 5 percent).

(b) Compounded quarterly.

(c) Compounded monthly.

(d) Compounded weekly.

(e) Compounded daily.

2–3. Assume that the material cost of hamburgers is 24 cents each, and all other costs per hamburger at a sales rate of 600 per hour is 36 cents. Normally the hamburger sells for 69 cents each, but to boost lagging sales the management advertises hamburgers at 39 cents each.

(a) At what rate must hamburgers be sold to avoid losing money?

(b) At what rate must hamburgers be sold to avoid a decline in hourly profits?

_3_
product
quality
attributes _____

3.1 INTRODUCTION

The quality of a product need not be high, only adequate, in order to sell well. Consumers usually abhor this attitude in principle, but they endorse it in their buying practices.

Manufacturers, like cooks, know approximately how well their product is made but do not always know how to assess the tastes and desires of the consumer. Some do so by watching the inventories in the warehouses rise or fall. Others measure _their_ fraction of the market for similar products, and strive to hold or enlarge it. Still others monitor quality by determining customer loyalty, that is, the fraction of repeat sales to customers, if they can be identified. The auto industry, for example, uses numbers of repeat sales in the range of 60 to 70 percent as a guide. If fewer than 60 percent of previous customers buy the same brand, or another in the same family, of the new model cars, then special attention must be paid to new products. There may be a problem of styling, price, or quality. If customer loyalty exceeds 70 percent, perhaps too much is being paid for manufacturing the product. In other words, the product may have excessive quality.

Product quality is a broad subject, encompassing such attributes as the appearance of a product, its reliability, the noise it makes, how it _feels_, whether or not it builds up a charge of static electricity, or whether or not it fails gracefully. To get a perspective on some of the measures of product quality, it is instructive to sample the various magazines available on newsstands that cater to the fisherman, the sports enthusiast, the auto racer, the electronics and computer club, the flyer, the muscle builder, the debutante, and many other consumers. Most of such magazines contain, as a regular feature, an analysis of new products, assessed and criticized in intricate detail. Clearly, there is some bias in some evaluations, but there is also considerable truth. Ultimately there is limited benefit in attempting to separate bias from truth, because consumers will buy what they want to buy. Thus if the consumer thinks that a noisy vacuum cleaner is better than a quiet one, then by all means supply noisy ones. If customers *need* FM tape player sets of 200 watts power and with assured frequency response far beyond the capability of the human ear to receive, supply it. If car buffs insist that rack-and-pinion steering aids cornering ability, by all means build on that attitude with products and advertising. If consumers insists on buying a 10-megabyte computer for letter writing (word processing), then satisfy them quickly before they raise their *needs* to a 15-megabyte computer.

3.2 THE CONTROL OF QUALITY IN MANUFACTURING

Quality is usually designed into a product; it is strongly influenced by the choice of manufacturing process, and it is *controlled* by properly organizing the people within a manufacturing facility. If there is a deficiency in any of the three areas, costs can be high. One company found in 1983 that 25 percent of their manufacturing assets were tied up in reacting to quality problems, and about 20 percent for the inventory. It required company-wide action for a year to control the problem.*

The design-materials-process engineers write specification sheets for product parts in such terms as "part dimension," "flatness," "hardness," "surface roughness," "reflectivity," "density," and the like. These specifications are taken from some knowledge of the intended use or function of the part. The engineers also write guidelines on how to measure the *objective* quality attributes, and specify the tools and instruments to use. Some attributes must be evaluated *subjectively*, such as uniformity, texture, color, and the like. These are sometimes defined by the marketing group, and their measurement may involve psychologists and others who are not engineers.

During the design of a manufacturing system, industrial engineers usually determine the location in the system at which quality can and should be checked. They determine how many people will be required, what human factors are involved, the training the inspectors should receive, and the methods that the inspectors will use to report data on quality to management. When production begins, the factory managers effect the plan and supervise the inspectors.

Quality control managers, or people of equivalent title, analyze the data from inspectors. They frequently use statistical methods to determine a number of very useful quantities, such as:

1. Whether the system is drifting toward more or fewer rejects.
2. The cause and source of the most rejects, for example, worn machinery, poor raw materials, operator carelessness, and so on.

* Young, *Wall Street Journal* (1981).

3. Whether the tolerances on the dimensions specified by the design engineers can be economically achieved.

4. Whether inspection is too rigorous or too lax.

5. Whether the current inspection stations are optimally located in the system.

In parallel with a continual activity to check the quality of manufactured parts, there is usually a careful study of the reasons for products to be returned for refund or repair. If product durability is a problem, or if consumers frequently complain that a product does not function as expected, a new study is done by designers. Perhaps some part will be redesigned or perhaps a manufacturing process must be revised to improve the product.

3.3 QUALITY STANDARDS AND COST

Some products can no doubt be made at a profit whatever the quality. But for most products there must be an appropriate balance between quality and cost to manufacture. Thus it is difficult to set down fixed standards of quality for all parts for all time.

The balance of cost and quality varies with the type of market and the economic times. For example, the price of a Rolls-Royce sedan is ten or more times that of a Chevrolet sedan, whereas the quality ratio is far less than 10:1. But the Rolls-Royce continues to sell enough to remain in production. The reason is that the two brands are not in the same market; that is the same people are not *competing* for both products.

Difficult though it may be to obtain, some absolute measure of quality must be used by inspectors. Where an instrument is used to measure a quality attribute (e.g., with a scale to measure weight), an inspector has only to adjust the settings and assure the proper operation of the instrument. However, much inspection is subjective in nature, such as rubbing the hand over a surface to detect *bumpiness*, or other quality. With experience, human inspectors become very adept at finding even miniscule defects on very large or intricate parts. Examples are pinpoint-size defects on large sheets of glass, or disorder in the array of wires and terminals on computer boards.

In the subjective area, as inspectors become better trained to spot defects, they can and do begin to reject parts that few consumers would reject or complain about. Consumers vary considerably in their opinion of the quality of products, depending on many personal and societal factors. It is evident that, to prevent a gross mismatch of perceptions of product quality, there must ultimately be a *meeting of the minds* of the inspectors and consumers in some way. The ideal is to establish direct communication between them, which does occur between manufacturers selling finished or semi-finished components to one another. But communication with the final arbiter, the consumer, is not practicable. Indirect communication does occur, but through the very imperfect information network consisting of market analysts, factory management, and quality control supervisors. If this network is efficient, then good and flexible quality standards will be developed and used. Otherwise, profits will eventually and surely decline, either because parts are made too expensively, or because consumers will detect undesirable quality and not buy.

3.4 THE DESIGN-MATERIALS-PROCESS ENGINEER
AND QUALITY CONTROL

Consumer attitudes are frequently dismissed by engineers, who prefer, and were trained, to be methodical and logical, at least in their professional lives. But the reality of the consumer product industry is that a number of subjective factors often govern the quality of products.

In defense of engineers it must be stated that the primary responsibility for quality control has not been theirs in many industries. Inspection is most thoroughly done near the end of the manufacturing sequences, just before shipping. It has therefore become the jurisdiction of the manufacturing managers, and engineers are thereby somewhat removed from such concerns. But again, design-materials-process engineers have not been well equipped to impinge upon the quality problem. In fact, they do not adequately understand quality themselves. Consider the usual method of specifying a quality attribute, for example, surface finish, on a new engineering drawing. This specification is frequently copied from the drawings of previous successful parts, not because it is known to be appropriate, but because this practice usually works well, probably for two reasons. The first is that surface finish is almost always overspecified in order to be safe. Second, a familiar specification will *signal* the processing people to use established processes. The importance of this practice cannot be overemphasized. Most everyone in industry knows of numerous examples of good parts being produced by an old machine or by a prototype machine under the watchful eye of process engineers, but when a new machine is installed numerous problems appear. One type of problem is that the new machine may make parts to specifications, but the parts do not function satisfactorily. The converse is found as frequently: The parts fail inspection, but testing shows that they function well. The genius of technicians and engineers eventually makes the process function satisfactorily, but not without some price in mental turmoil, cost overrun, and delay in start-up of the system.

Many problems could be averted if the processes were well understood, and if the product attributes resulting from the process were measured in terms that have some connection with the function of the part. Take the example of grinding of a cylindrical shaft. The finished product must meet specifications of diameter, hardness, straightness, and surface finish. Most grinding machines produce out-of-round and slightly bowed parts. After grinding, only the shaft diameter and surface roughness are measured, and the part is either rejected or accepted. Out-of-roundness is not measured explicitly, and yet this attribute is important in some applications. Surface roughness in terms of average asperity height is measured, but skew and kurtosis (discussed in Sec. 3.7) are usually not, though they are thought to be important attributes of lubricated surfaces. The list of such examples is endless, and most designers plead for relief from such concerns. These details are thought to be the domain of materials people or the manufacturing department. However, here is the problem in quality control: It is a complicated field, and requires good communication between specialists.

It will be increasingly necessary for engineers to formalize their understanding of product quality attributes because of the trends toward complete automation. There are two aspects of automation of consequence to quality. One is the automation of inspection, to escape some of the limitations of human inspectors. The other is the in-process inspection of parts for real-time control of automated production machinery. The first is obvious, but the second requires some explanation.

In-process inspection is some combination of taking measurements from a part while a manufacturing operation is taking place on that part, or taking measurements from a part that is between operations. One purpose of such inspection is to insure that the failure of one operation does not damage a part or the processing machinery. But there are two other important purposes. One is to find defects in parts as early as possible, to avoid investing further processing cost in the part. The other is to determine very quickly when a process needs adjustment to maintain quality, before several unacceptable parts are made. This is the way soup is made, and this is the procedure of processing parts in a job shop. Human operators of machinery realize when a process variable has changed during the process, by such senses as smell, sight and hearing. These senses have meaning to experienced machine operators. However, in high-production-rate factories, more of the processes are semiautomatic. The machines are attended by less skilled people and often by people with minimal experience with any particular machine. Thus the detailed conditions of the materials in-process, and important nuances in the processes are not fully perceived until formal inspection occurs by designated inspectors. Inspection stations are often widely distributed, and the result is that, depending on the inspection procedure, several bad parts could be made before the defect is caught and the process altered. Or worse, damage to the machine or its fixtures often results from unattended broken tools, or parts that have slipped from a fixture. Process sensing is often prescribed for such problems, but in this case the sensing is done to avert serious consequences to the machinery and is not aimed primarily at the problem of product quality.

3.5 RANK ORDERING OF PRODUCT QUALITY ATTRIBUTES

In the area of quality control and subjective inspection, the design-materials-process engineer will have increasing responsibility to provide much more formalized information than has been done in the past. Two major areas are:

1. The instrumentation and automation of inspection. This consists of two parts, namely,
    a. To find ways to use available sensors of light, temperature, pressure, and the like to detect product quality attributes now perceived by human senses.
    b. To identify areas in which new sensing techniques should be developed.
2. To use the output from the sensors to control processes in real time.

One of the first steps in formalizing the subject of product quality attributes is to translate common expressions for quality into technical terms. Product quality is often expressed in quaint ways. For example, when an auto enthusiast claims that car A *corners better* than car B does, this apparently has reference to some quality in a car by which a sharp curve in a road can be satisfactorily negotiated. The *ordinary* car may do so successfully, but not satisfactorily. The difference apparently is a hitherto unquantified quality of a car providing appropriate *feedback* whereby the driver purports to perceive when the limit of traction or friction between the tires and road is being approached.

The formal approach to quality control consists in developing a rank ordering of *product quality attributes*. That is, the designer should first determine which are the primary attributes of a part, or, in other words, those attributes that are vital for functioning or consumer acceptance of the product. Then, since primary attributes are often difficult or expensive to measure, one or more secondary attributes for each primary attribute should be given, with perhaps tertiary attributes as well. With each listed attribute there should be one or two methods given by which it may be measured, and there should be a comment on how practical it would be to measure that attribute. Also, with each listed secondary and tertiary attribute there should be some comment on the extent to which it connects or fails to connect with the higher-order attribute. Finally there should be an analysis of each process to determine the urgency of measuring primary attributes. Recall that there are two concerns in the engineering of process control. The one is the control of product quality, and the other is prevention of damage to the machine and fixtures. It would be useful to list all of the ways that a process can go wrong and to devise protection according to the probability and seriousness of such damage.

Practical examples of rank ordering of quality attributes are now given. To begin, the primary attributes of a product such as silk cloth may be appearance, or *feel*. But it may not be possible to automate the measurement of these attributes. In such cases relevant secondary attributes must be found that correlate satisfactorily with primary attributes. For example, part of the desirable appearance of a silk surface may be its shininess. This then could be a useful secondary attribute, and could be measured by reflectivity. However, reflectivity deals only with the intensity of reflected light as compared with the intensity of incident light. The intensity of light is measured with a photocell, but the photocell and associated electronics does not *see* the surface in the same way the eye does. The difference may be due to surface chemistry or composition which alters the polarity of light in a way not detected by photometers but in some instances is perceived by the human eye as a general surface attribute. Thus the secondary attribute, reflectivity, may not be a complete measure of the primary attribute, appearance, but it may detect a majority of the variations in the primary attribute. In the case of appearance it would seem that physicists and psychologists could already have defined all of the necessary variables so that an equation could be available to relate the relevant attributes, but that is not yet in hand.

Another example using reflectivity is found in the coffee industry. The ultimate test of coffee is the drinking thereof, apparently. Whereas advertisements refer to the *taste test,* that quality known as odor may also be important. But these are not the attributes used in the first quality check of coffee beans. Rather, *every* bean from the roaster is slid down a chute, past an incandescent light bulb and a photocell. If the bean is too light or too dark, it is rejected. It is difficult in this case to determine whether the measured attribute has rational meaning relative to the primary attribute. The measured attribute may not be a secondary attribute as much as it is a *substitute attribute*, defining some quality pleasing to wholesale buyers, but not sought in the *taste test*.

Another example is in metal cutting. One problem of great importance is the progression of tool wear. It is sometimes found that after severe *crater wear* the tool tip breaks off and the part becomes damaged. If a tool could be taken out of service before it fails, it would reduce scrap rates and allow economic regrinding of the tool. However, there are many changes in the process that occur long before the tool is near the point of fracture. In particular, the surface finish changes considerably, and the residual stress state or the depth of work-hardening changes. In some instances a dull tool exerts a larger force against the part than does a sharp one, which bends the part away from the tool during the cut. This would affect the dimension of the finished part. The primary attributes listed so far for this problem are surface finish, residual stress state and dimension, but none of these are measured in process. Work on this problem has focused on the measurement of forces on the tool during cutting, and the analysis of noises generated during wear, to see if trends can be detected that correlate with tool wear. Some results are available from the noise analysis, but they are obscured by noises from many sources other than those that can be directly connected with cutting.

The primary attributes connected with tool wear are not measured because of the difficulties in doing so, particularly in process. Cutting force and noise are easy to measure, but these are not obviously connected with any of the attributes of a product, or of a process. In this case the ready availability of some sensing technologies obscures the reason for process sensing in the first place. Something of value will inevitably be found from research on force measurement and noise analysis, but a more direct approach to in-process sensing and design of real-time control begins with an analysis of relevant quality attributes.

A clearer example of rank ordering of attributes may be found in the assembly of electronic components. Soldering is a common method of connecting a wire to a connector tab, and this must be done with 99.999 percent reliability, because there are hundreds of such soldered joints in one board or subassembly. A failed solder joint ultimately is manifested as a high electrical resistance, the solder may have surrounded the wire but did not wet and bond metallurgically. In a poor joint the wire and the solder are separated by less than a micron of oxide and soldering flux, and the electrical resistance of such a joint usually varies with humidity and temperature and with the way the soldering was attempted. Therefore, measuring the electrical resistance of the joint may not reveal its true state. Consequently, it is common to measure a secondary attribute of the joint, namely, some evidence of wetting. The evidence of a liquid solder wetting a vertical wire is that the solder will *climb* the wire about 3 mm or so; lack of wetting is manifested by the solder's being depressed around the wire. If complete wetting is the only condition that produces the climb, and if there are no known cases of complete wetting without solder-to-wire bonding, then the secondary attribute is sufficient for quality measure. The climb of the solder can be determined by visual inspection, or it can be automated by analyzing the image from a TV camera with a shape analysis program in a computer.

The important point in the above examples is that the measurement of a secondary attribute in the place of a primary attribute involves some risk. Unfortunately, the attraction of presently available instruments obscures the study of *needed* instruments with the result that most processes are not controlled as directly as they could be.

3.6 METHODS OF MEASURING PRODUCT QUALITY ATTRIBUTES

Most of the primary quality attributes, such as size, roundness, straightness, cylindricity, taper, reflectivity, and electrical resistance, can be measured with simple instruments which require little technical background. Some attributes, such as residual stress and submicroscopic cracks, are more difficult to measure and require specialized equipment. Some of the secondary attributes which are measured for process control involve the measurement of forces, vibration modes, and machine speed, but again require fairly simple equipment. Most of these measuring devices are available with the proper electronics and data display stations. However, surface texture is a much more complex attribute, both to measure and to understand. Its measurement will now be described.

3.7 SURFACE ROUGHNESS

Line drawings of objects often leave the impression that surfaces are perfectly smooth, but this is never achieved in practice. Except for cleaved materials, few surfaces are atomically smooth. Most surfaces are composed of arrays of small bumps, known as *asperities* or *protuberances,* which were formed in the manufacturing process. Furthermore, many processes *damage* the substrate beneath a surface to a small depth, which may affect the performance of the part. For example, severe grinding can reduce the fatigue life of a steel part by a factor of 10 or more.

Some processes such as polishing produce very smooth surfaces, but the asperities are still over 30 atoms high. Such surfaces are formed on hard memory disks and optical surfaces, but they are far too expensive for most other products. Economics, then, will limit the relative roughness or smoothness of manufactured parts, and economics will demand orderly ways to specify and verify surface smoothness. Table 3–1 shows the relative costs of surface smoothnesses as achieved by various common metal cutting and finishing processes. (See page 46 for Table 3–1.)

One surprising fact is that designers rarely know how smooth a surface should be, or how to make a surface of a particular quality. Useful surfaces have been manufactured for many years, but success usually continues only as long as every detail in making the surface is done as it has always been done. A major problem often occurs when it is necessary to change suppliers of tooling or materials during a production run. Though the product is made exactly the same *way* as in the past, and though the product passes inspection, some will suddenly fail in the hands of the consumer. An opposite type of problem can also occur, when a process is changed so that it produces a measurable change in surface quality without affecting the failure rate of the product. This produces the suspicion that close control of product quality may not have been necessary all along, and that too much money was spent in making the product.

Surfaces are too complicated to describe with a single word or number. To illustrate, take the example of the highways on which we travel. We can describe at least five scales of geometry of roads, most of which are regular and apparently periodic, some of which are irregular. The largest scale relates to following the grand features of the earth surface, such as large hills and mountains. On a smaller scale we see roadways that follow the smaller contours of rolling countryside. A third and smaller scale of geometry has a wavelength of the order of the length of a car, which at 60 mph can produce uncomfortable oscillations of about two or three cycles per second. A fourth smaller scale, of the order of 50 to 100 mm, has little affect on ride comfort, but produces a loud rumble from vehicle tires. The fifth and smallest scale, of the order of order of 1 to 10 mm, is important for wet skid resistance of tires on the road. We see that, in describing road surfaces, the size scale of interest depends on the function or use that one has in mind for the road.

The topography of manufactured surfaces may be visualized as being similar to the contour of the earth surface. Surfaces contain irregular features called *flaws*. The regular features, minus the flaws, are referred to as the *surface texture*. For some purposes it is useful to define two scales of texture, a fine scale known as the *surface roughness* and a coarser scale known as the *waviness*.

Some aspects of *surface roughness* can be assessed either visually or by running a fingernail over a surface. Such subjective methods have limits but are still widely used. A less subjective method uses comparators which are four or five samples of surfaces made to different roughnesses. A visual comparison is made between a production part and the comparator or standard surfaces. The inspector declares that the production surface looks more like standard "c" than "d" for example, and thus may reject or accept the part.

The most common objective method of measuring surface texture uses a stylus tracer. It consists of a diamond or other hard projection much like a needle or stylus of a record player. The stylus moves along a solid surface, rising over the peaks and descending into valleys as it moves. The rise and fall of the stylus, that is, its vertical motion, is detected by an electronic system and amplified for various purposes. The stylus is moved horizontally along the surface to be traced by one of two different systems. The first and simplest is the *sled* arrangement, which involves two spheres of about 6 mm radius that ride on (slide over) the same surface as that being measured, to the sides of the tracer. The sled also rises and falls while sliding over the asperities, following the waviness mostly. The radius of the stylus tip is much smaller than that of the spheres. The actual effect of the vertical motion of the sled on measured roughness is rather small, but the sled is nevertheless not recommended for careful work. Its major appeal, however, is that a tracer with a sled can be set on any surface by hand and data can be obtained in less than a minute.

The second and superior (at least, more expensive) mechanical system guides the stylus on a remote sliding surface of high precision. To use the latter system it is necessary to bring a specimen surface to the stylus and align the specimen to be fairly parallel to the surface that guides the stylus. A sketch of a precision stylus tracer unit is shown in Fig. 3–1.

The vertical motion of the stylus is detected by a sensor connected to an amplifier.

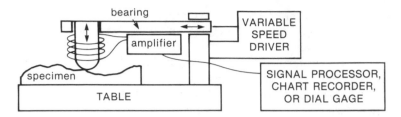

Figure 3–1   Sketch of a stylus tracer system for measuring surface roughness.

There are four end uses for the output from the amplifier. The simplest and most common older method is to connect the amplifier output to an averaging circuit and a dc voltage meter. Vertical oscillations of the stylus deflect the needle of the meter. The dial on the meter is calibrated in terms of an average asperity height for direct and convenient readout in the production environment.

A second objective is to record an amplified profile on a chart paper for visual characterization. If profiles are obtained along closely spaced parallel lines of tracer travel on the surface, one can construct a three-dimensional surface map. To map accurately it is necessary to move the stylus at the same speed during each traverse, to trace each pass in the same direction, and to know the location of the start and stop of each trace very accurately.

The usual output from a tracer system to a chart recording is shown in Fig. 3–2a. Such a recording is deceptive, in that the asperities appear as jagged peaks. This is the result of the common practice of amplifying the vertical motion from the stylus to the chart about 100 to 1000 times more than that for the horizontal displacement of the tracer carrier. When the two dimensions are properly represented, as shown in Fig. 3–2(b), the profile seems very smooth, but it is realistic. The slopes of most profiles are in the range of 3 deg to 10 deg. Actually, most surfaces have the approximate profile of the surface of a lake swept by a gentle breeze. On such a surface at least three scales of roughness are visible.

A third objective is to process the signals from the amplifier to produce numbers that characterize the roughness of a surface. The analog signals can be processed directly, or they can be digitized before processing, either with or without compensation for the nonlinearity of the electronics. This nonlinearity results in signal amplification and phase shift that varies with the frequency of the processed signal. Each manufacturer of surface tracer equipment has its own way of compensating for these effects.

The signal to be processed may be represented as a seemingly random wavy line,

a. Trace of rough surface due to high vertical
    amplification of the signal from the tracer.

b. Realistic trace of a rough surface

Figure 3–2   Strip chart recording from a stylus tracer system.

as shown in Fig. 3–2. The instrument "selects" a sample length, nominally about 0.030 in. or 0.75 mm.* This length should include a minimum of 30 crossings of the *mean*. The electronics are then designed to pass all wavelengths less than the sample length and *cuts off* the longer wavelengths to avoid extraneous data. Next the electronics selects a *mean line* such that equal areas are enclosed by the wavy line above and below the mean line. All vertical measurements are taken as absolute values from the mean line.

Signal processing is done in a number of ways, and more than 20 results or *surface profile parameters* have been proposed to describe a surface. These include parameters for asperity height, spacing, and slope, expressed as extreme values, average of selected values, and average of all values within a sample length. There are also some *statistical parameters* which provide a distribution of values and a measure of the randomness of such distribution. The most common parameters will now be described.

The ordinary parameters in use characterize asperity height, and they are listed with the most useful parameter first. The parameters are defined in terms of quantities shown in Fig. 3–3: y is the absolute value of a vertical dimension from the mean line, y_i indicates that each of N values of y is taken in turn, and the values p and v are the values of y of selected high peaks and deep valleys. The parameters are

$$R_q = \sqrt{\frac{1}{N} \sum (y_i)^2}$$ and is the *root mean square (RMS) average roughness height*

$$R_a = \frac{1}{N} \sum |y_i|,$$ and is the *arithmetic average roughness height*

$$R_z = \frac{(5 \text{ highest peaks } + \text{ 5 deepest valleys in one cutoff length})}{5}$$

and is the *ten point roughness height*

* Amstutz and Hu, "Surface Texture: The Parameters," a publication of Warner & Swasey, Dayton, Ohio, # MI-TP-003-0785.

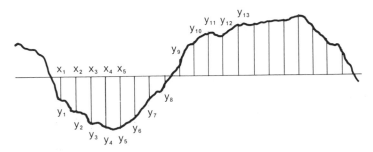

Figure 3–3 Topographic waveform and coordinates for analysis.

$R_t = P_{MAX} + V_{MAX}$ over 5 cutoff lengths, and is the *maximum height, over five cutoffs*

$R_p = P_{MAX},$ or $R_v = V_{MAX},$ and are the *maximum peak height or maximum valley depth in one cutoff*

The statistical parameters tabulate the y values at regular sample spacing shown as x values in Fig. 3–3(b). The amplitude density function (ADF) is often plotted to the right of the surface profile, as shown in Fig. 3–4. Two useful parameters relate to the shapes of the ADF curves. Figure 3–4(a) shows a Gaussian ADF: Fig. 3–4(b) shows an ADF *skewed* upward, and Fig. 3–4(c) shows an ADF that is skewed downward. The corresponding waveform illustrates the reason for the skew.

Mathematically *skew* is defined as:

$$S_s = \frac{1}{(R_q)^3} \times \frac{1}{N} \sum_{i=1}^{N} (y_i)^3$$

The ADF in Fig. 3–4(a) has a $S_s = 0$; in Fig. 3–4(b) $S_s < 0$, and in Fig. 3–4(c) $S_s > 0$. An $S_s < 0$ is said to be desirable as a bearing surface, whereas an $S_s > 0$ is preferred for electrical contacts.

A further characterization defines the peakedness or sharpness of the Gaussian curve in Fig. 3–4(a), and that is *kurtosis*. Mathematically it is

$$S_k = \frac{1}{(R_q)^4} \times \frac{1}{N} \sum_{i=1}^{N} (y_i)^4$$

A flattened curve has $S_K < 3$, a sharp one has $S_K > 3$, and a Gaussian curve has $S_K = 3$. A waveform with $S_s > 0$ automatically has $S_K < 3$ and vice versa.

The only wavelength parameter in use is the *correlation length*. It is a measure of repetitiveness of waveforms. A perfectly random waveform displays no repetition, whereas

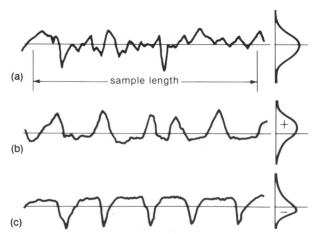

(a)

sample length

(b)

(c)

Figure 3–4 Three types of topographics and then amplitude density function, a. Gaussian, b. + skewed, and c. − skewed.

other surfaces may display a semblance of repetition every millimeter (for example) on average.

There is no advice available on which parameters one should use. The reason is that there is almost no connection between any of the parameters and the likelihood of a surface to function properly. Each corporation has different practices, probably based on the personal preference of a group leader. However, recall that surface roughness measurement is done mostly to control processes. Each process is done in a unique way in each plant, and this is the best reason to accept individual ways of characterizing surfaces. But virtually all of these methods can be improved considerably. For example, in principle, a single parameter cannot describe more than one surface feature or cannot adequately describe the shape of an ADF. At least two unique and independent parameters should be used, and perhaps as many as five.

In all cases, however, certain practical matters should be noted. In using the stylus tracer a single trace should not be taken because surface roughness is very directional, and it varies by as much as 20 percent across a surface. A minimum of five traces should be done, at least 2 mm apart and in various directions.

When working with soft materials the tracer can leave considerable damage behind. The stylus has a load applied to it, usually about 0.5 gram or 5 mN, which produces a contact stress between the stylus and sample surface sufficient to produce plastic flow of most materials. With the following equation one can calculate the contact radius a between a sphere and a flat plate by

$$ a = \left\{ 0.75 W r \left(\frac{1 - \partial_1^2}{E_1} + \frac{1 - \partial_2^2}{E_2} \right) \right\}^{1/3} $$

where W = the load applied to the sphere.
  ∂ = the poisson ratio of the materials 1 and 2.
  r = the radius of the sphere.
  E = Young's modulus of the materials 1 and 2.

Most styli are diamond, and if we take the E of diamond to be four times that of the average metals, assume $r = 10$ μm, and take poisson ratio of all materials to be 0.3, then $a \approx 0.00695/E$, where E is that of the surface to be traced. Now, the mean contact pressure between a sphere and a flat plate is $p_m = W/\pi a^2$, which finally is $p_m \approx 6.6 E^{0.67}$. For steel the average contact stress is about 0.02 times E, which is about twice the tensile strength of the hardest steel. Clearly the stylus will scratch a groove in virtually all metals. These scratches are hardly visible on very hard surfaces but constitute defects on softer surfaces.

Another practical problem results from the mass of the stylus and mounting lever. At high rates of horizontal traversal the stylus jumps over some peaks and gives erroneous readings. Most commercial stylus tracer systems provide a range of traversal speeds of the order from about 0.1 to 1.5 mm/s.

There are several other commercial devices for measuring surface roughness. These methods include contact resistance, capacitive methods, and light scattering. Each is useful in some specialized area, but in every case the results are compared with the results of the stylus tracer. The stylus method is the standard of the world, and it was invented at the University of Michigan.

PROBLEMS

3–1. Express the surface roughness of your home county.

3–2. Explain the particular surface attributes that provide the basis for the resistance, capacitive, and light-scattering methods of surface characterization.

3–3. Give the primary, secondary, and tertiary product quality attributes for;
    (a) A steel shaft being cut in a lathe.
    (b) A silicon wafer being etched.
    (c) A computer hard disk surface being coated with media.
    (d) A plate of glass being ground.
    (e) A watermelon that you are considering for purchase.
    (f) A pad of paper that you are considering for purchase.

TABLE 3–1 AVERAGE RANGE OF SURFACE ROUGHNESS, R_A, ACHIEVED BY VARIOUS PRODUCTION PROCESSES, AND THE RELATIVE COST TO ACHIEVE THE CENTRAL VALUE OF ROUGHNESS GIVEN. THE FULL RANGE OF ROUGHNESS SEEN IN PROCESSED PARTS MAY BE THREE TIMES AS WIDE AS THOSE GIVEN. (TABLE 14–3 ON PAGE 386 GIVES VALUES FOR JOB SHOP OPERATIONS.)

Process	Average Range of $R_a(um)$	Relative Cost
(Reference)	50	1
Sand casting, hot rolling, flame cutting	25 to 12	$\approx$ 1.5 to 2
Forging, sawing, planing and shaping	12 to 3	$\approx$ 1.8 to 2.5
Permanent mold casting, extruding investment cast, die casting, cold rolling, drawing, drilling, electrical discharge milling, chemical milling, broaching milling, reaming, boring, turning	5 to 2	$\approx$ 2 to 3
Barrell finishing, roller burnishing, electrolytic grinding, honing grinding	7 to 2	$\approx$ 3 to 5
Polishing, lapping	3 to 0.5	$\approx$ 4.5 to 7.5

4
tolerances _____

4.1 INTRODUCTION

Regardless of the method by which parts are processed, dimensions of concern *cannot* be produced to exact sizes. This is so regardless of the number of parts to be made. The use of *tolerances* is both a recognition and acceptance of that fact. Consider the following questions. If exact sizes could be produced, would that be necessary? From a *functional* viewpoint the answer is a qualified no, since most mechanisms, assemblies, and the like will perform satisfactorily if the desired *basic* sizes of individual components are bounded by permissible variations. Tolerances define such variations. It is admitted that in most cases, the smaller the tolerance, the better will be the quality of the product; however, a balance between quality and cost of the product must be considered. As shown in Fig. 4–1, costs escalate rapidly as tolerances are decreased. This leads to the next question. Is it economical to strive for *exact* sizes? Except in rare cases, the answer is definitely no.

One such rare case is the production of gage blocks, which industry uses as reference measuring standards. Essentially, each gage block consists of two flat and parallel surfaces, where the distance between these surfaces is produced to a certain basic size that constitutes the reference dimension; tolerances are maintained to a few *millionths* of an inch. No simpler geometry can be envisioned yet to maintain such exactness; the manufacturing methods necessitated plus the special measuring devices needed to verify this exactness lead to a cost of thousands of dollars for a set of such books. Because of their intended use, such close tolerances are justified, but this is an *exception* rather than a general rule. For most products, such exactness is not only unnecessary but also uneconomical in terms of putting competitive products in the marketplace. A sound rule to follow is that tolerances should always be as large as possible, consistent with func-

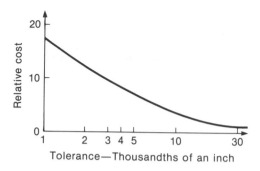

Figure 4–1 Schematic of the trend of relative cost versus specified tolerance.

tional demands. An old story, attributed to Sir Isaac Newton, is appropriate. When asked how he had his many experimental devices produced, his purported reply was that he made a freehand sketch of all dimensions to the closest inch and then turned it over to his assistant. Using a pencil and straight edge, the assistant refined the sketch and added tolerances of one-quarter inch to all dimensions. This was given to a junior draftsman who made an elegant drawing to full scale. To avoid future criticism, the draftsman assigned his own tolerances. He considered the smallest number he could think of and divided it by two. The veracity of this story is open to question, but its message is important. Designers, wanting assurance of proper function, believe that tight tolerances will achieve this aim, but this can lead to such high costs that the end product is not economically competitive.

There are three discrete attitudes involving tolerance specifications. The first concerns the *designer,* whose goal is to insure proper function. Next are those who must see that the item is *manufactured;* their goal is to produce the item as economically as possible. Finally come those responsible for *assembling* individual components into a unit; their main concern is to complete the assembly without problems. It is most sensible and economically efficient if these different viewpoints are considered *concurrently* during the initial design stage; otherwise, problems may often arise and redesigning for efficient manufacturing and assembly will be required.

4.2 THE DESIGN ASPECT

In certain situations, such as press fits or the allowable clearance between a shaft and journal bearing, proper tolerances can be calculated, but these are exceptions rather than a general rule. For the overwhelming majority of cases, there seems to be no *scientific* approach used to determine appropriate tolerances. The authors of this book have for many years sought an answer to the question of *how* tolerances are determined, and the conclusion drawn is that *past experience* provides the basis for such a decision.

Considerations of size, weight, strength, and the like are often studied initially. As an example, suppose that a gearbox must fit into a certain volumetric space and is to transmit certain loads or torques over a range of speeds. Maximum levels of noise may be of concern, in addition to strength and allowable deflections. Using equations related to gear design, stress analysis, beam theory, and so on, an initial design that satisfies functional demands can be accomplished in a *reasonably* direct manner. Then tolerance specifications must be considered, and several questions arise here.

1. To satisfy noise levels, what surface finish and, in essence, what tolerances are required for the gears?
2. Where are the best locations for any bearings to satisfy deflection specifications? By what extent can these locations vary?
3. If shafts are supposed to be parallel, to what extent can their centerline distances vary and still provide proper meshing of gears?

The answers to such questions are not simple, so it is not surprising that a designer faces a degree of uncertainty in arriving at tolerance assignments. In numerous cases, some of the components to be used in the gearbox may be purchased from other companies who specialize in the manufacture of such items; in the above situation, bearings are an obvious example. Proper sized bearings might be selected from a vendor's catalog where the bearing tolerances are specified. A similar comment could apply to the gears. Thus, the designer must often consider the basic dimensions and manufactured tolerances of such purchased items in the overall design.

One final observation pertains to those situations where a finished product could be made by practically *hand-fitting* components as assembly proceeds. For example, a craftsman could make his own bolt with his own version of a thread, and then make an individual nut to fit that bolt, but think of the problems that would arise if replacement parts are needed. To avoid such a problem, national and international standards have been developed for thread sizes as well as many other engineering components. Even then, control of tolerances is demanded if any bolt of a certain thread size is to mate directly with any nut of the same size. This leads to the concept of *interchangeable assembly,* which is the prerequisite for mass production. Consider as an example the cost of a Rolls-Royce, made in rather limited quantities, compared with a Chevrolet which is mass-produced. The Rolls involves more costly parts in the engine, transmission, body, and trim, and much tighter control of mating parts is maintained (almost like hand fitting). Because of these higher quality standards, it is no wonder that the Rolls costs an order of magnitude more than the Chevrolet.

4.3 THE MANUFACTURING ASPECT

Once a product is designed, those who specialize in manufacturing give consideration to selecting processing methods that might be used to make the product. Economic considerations are *always* involved at this point. Most often the initial design includes a specification of the materials to be used as well as the basic dimensions and tolerances. Not all materials can be processed with equal ease, and very close tolerances require high costs, as shown in Fig. 4–1. Because of these facts, the manufacturing personnel will often raise questions with design people in order to be convinced that the original specifications are really essential. It is possible that the designer was unaware of the potential problems (and excessive manufacturing costs is one of the crucial concerns) as seen by the manufacturing engineer. Often a modification of the original specifications is made that does not impair functional demands yet leads to lower production costs. Such give and take, involving both viewpoints, leads to an acceptable compromise.

4.4 THE ASSEMBLY ASPECT

After components are manufactured they are usually combined with other items to form an assembly.* Perhaps the major concern of those involved at this stage is that randomly selected parts should assemble on the first attempt. That is what *interchangeable assembly* is all about, and a careful analysis of the dimensions and tolerances of individual items is required if assembly concerns are to be satisfied.

Three distinct approaches can be considered. The first is termed *complete* or *100 percent* interchangeability, where assembly will result regardless of the combination of components chosen at *random* for each assembly. *Statistical* interchangeability is the second approach. There it is *probable* that assembly will result, but 100 percent assurance is not guaranteed. Finally, with *selective assembly,* the critical dimensions of components are first measured; then the components are segregated into groups or batches of particular size variations. From these batches, pairs of components are selected so that desired tolerance limits are satisfied. The initial preinspection or measurement plus the need to categorize components into pertinent batches make this approach more costly than statistical interchangeability. Selective assembly is used only where close mating tolerances are deemed essential. With appropriate examples, each approach is now discussed. In *all* of the examples cited, the basic sizes and tolerances carry units of inches. This negates the need for constantly repeating units in this chapter and has no effect on the concepts. If SI units are preferred, all basic sizes, tolerances, and answers can be multiplied by 25.4 to convert to millimeters.

4.5 100 PERCENT INTERCHANGEABILITY

Consider Fig. 4–2, where three blocks, *A*, *B*, and *C* are to be assembled in a channel of dimension *D*. Except for the tolerances to be assigned to *D*, all other basic sizes and tolerances are shown. It is emphasized here that for this entire chapter we shall always use *equal, bilateral* tolerances;† the reason will become evident as various examples are

---

* Throughout this chapter the words *component, item,* or *part* indicate individual items and are used interchangeably. Similarly, the words *assembly* or *unit* denote a combination of single items.

† Note, for example, that $1.000 \, {}^{+0.004}_{-0.000}$ shows a unilateral (one-sided) tolerance, whereas $1.000 \, {}^{+0.003}_{-0.001}$ indicates unequal, bilateral tolerances. In contrast, $1.000 \, {}^{+0.002}_{-0.002}$ gives equal, bilateral tolerances. For each case, the *basic or nominal* size is 1.000.

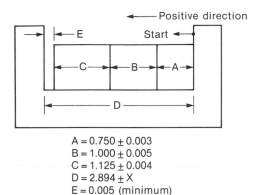

A = 0.750 ± 0.003
B = 1.000 ± 0.005
C = 1.125 ± 0.004
D = 2.894 ± X
E = 0.005 (minimum)

Figure 4-2 Schematic assembly of three blocks in a channel; all basic dimensions and tolerances are in inches.

presented. For a detailed coverage of other ways to assign tolerances as well as many aspects of dimensioning, the reader can consult the text by Spotts.*

Suppose it is essential that the minimum gap size E must *never* be less than 0.005 and this is the only restriction involved. It is then necessary to determine the tolerances that must be assigned to D. Because of the simplicity of this problem it is possible to state the answer directly; it is 0.002. However, with more complicated situations, a systematic approach provides greater efficiency, so that approach is introduced here. We employ what is called the *path equation*. Usually, two decisions are required. The first is to determine what is the most extreme combination of limiting conditions that could be encountered. Then one must decide how the various surfaces involved must contact each other once the limiting conditions are determined. For the assembly in Fig. 4–2, the extreme conditions occur when A, B, and C have their *maximum* possible sizes and are assembled in a channel where D is the *smallest* possible size. Tolerance X can then be found. The path equation can be started at *any* contact point between the various items or at a free surface corresponding to any of the dimensions involved. Positive values in the path equation are chosen arbitrarily in either of two directions. As shown in Fig. 4–2, the equation has been chosen to start at the interface between A and the right side of the channel, and moving left is considered positive. Using the limiting conditions, we find that the path equation becomes

$$(0.750 + 0.003) + (1.00 + 0.005) + $$
$$(1.125 + 0.004) + (0.005) - (2.894 - X) = 0 \qquad (4\text{-}1)$$

so $X = 0.002$. This means that $D = 2.894 \pm 0.002$ *since bilateral tolerances* were used. Although not a specified requirement, the maximum value of E occurs when an assembly consists of the smallest sizes for A, B, and C and the largest size for D. Start the path equation at the left-hand side of the channel and use movement to the right as positive; then

$$2.896 - 0.747 - 0.995 - 1.121 - E = 0 \qquad (4\text{-}2)$$

---

* M. F. Spotts, *Dimensioning and Tolerancing for Quantity Production* (Englewood Cliffs, New Jersey: Prentice-Hall, Inc., 1983).

thus $E = 0.033$. If any assembly consisted of a combination of extreme dimensions, E would vary from a minimum of five to a maximum of thirty-three thousandths of an inch. The *average* value for E, from the average sizes of the four components, is 0.019; thus

$$E = 0.019 \pm 0.014$$

Now consider a case where it is not a simple matter of arriving at an obvious answer. The components are shown in Fig. 4–3(a); both are flat plates, one containing two pins and the other two holes. Assembly involves fitting the pins in the top plate into the holes of the lower one. The problem is to assign tolerance X to the centerline dimensions so that assembly will always occur. The extreme combination of dimensions results when the pin dimensions are maximum, the holes minimum, one centerline dimension is maximum and the other minimum; Fig. 4–3(b) illustrates this extreme case. The path equation is started arbitrarily at the centerline of the left pin, and moving left is considered negative. Then

$$\begin{aligned} &-(0.500 + 0.001)/2 + (0.505 - 0.002)/2 + \\ (2 - X) + &(0.380 - 0.002)/2 - (0.375 + 0.001)/2 - (2 + X) = 0 \end{aligned} \quad (4\text{-}3)$$

Note here that the full tolerance X enters into the equation, whereas only one-half of the tolerances on pins and holes pertains. This should be kept in mind when this problem is reanalyzed in the next section.

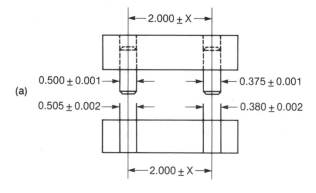

(a)

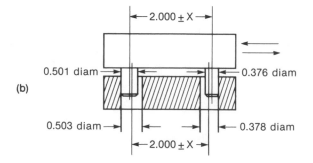

(b)

Figure 4–3 (a) Sketch of two components prior to assembly and (b) an assembly using one combination of extreme conditions.

Equation (4–3) reduces to $X = 0.001$, so the centerline dimensions are specified as 2.000 ± 0.001. In problems such as this, the path equation is most useful.

4.6 STATISTICAL INTERCHANGEABILITY

With regard to the two examples just discussed, it is appropriate to ask how probable it would be that one is likely to encounter a group of dimensions having extreme values occurring in any one assembly. The answer is not very likely, even though such a possibility *does* exist. In fact, as the number of items in a given assembly increases, the probability of such an occurrence can be shown to approach zero. For these reasons, the concepts of probability theory and the use of statistical methods for assigning tolerances should be considered since this leads to less stringent tolerances and lowered manufacturing costs. Rather than assuring positive assembly as discussed in Sec. 4–5, we are taking a chance that some assemblies will not be possible, but the *odds* are greatly in our favor that this is not a likely result.

In this book, our major intent is for the reader to gain an appreciation of why this approach is useful. We assume *no* prior background in statistical theory, nor do we expect that complete expertise will follow the minimal coverage to be presented. In addition, we point out that statistical concepts beyond those used here are available and used. But they are far beyond what is necessary for this basic introduction.

Although the type of problems of concern here will not require detailed calculations of certain basic statistical parameters, such parameters must be introduced if a physical understanding is to be meaningful. First is the term *average*. This is defined as the summation of the total number of individual measurements, regardless of what is being measured, divided by the total number of individual measurements. It is an *indication* of *central tendency*, or of the measurement that occurs most often. In equation form it is

$$\overline{X} = \sum X/N \tag{4-4}$$

where $\overline{X}$ is the average, $\sum X$ is the sum of all individual measurements, and N is the number of measurements.

The second parameter is called the *standard deviation*. This is related to the spread or dispersion of individual measurements both greater and less than the average. It carries the symbol σ and is defined as

$$\sigma = \sqrt{\sum (X - \overline{X})^2 /N} \tag{4-5}*$$

In words, the difference between each individual measurement and the average is squared; these are all summed, divided by the number of individual measurements, and

---

* In later chapters, σ is used to denote stress, but since each chapter is self-contained, confusion should not result.

then taken to the one-half power (square root). We note that the term σ^2 is called the *variance*.

Suppose a part is assigned a certain basic size and *bilateral* tolerance. In producing many such parts, those responsible for manufacturing will *aim* at the basic size so that parts both larger and smaller than basic will still fall within the acceptable tolerance range. It is likely, therefore, that more items will have a final size closer to basic than would those near the extremes of the tolerance range. If a plot were made of the size versus the number of times it occurred (that is, a *frequency distribution*), the results might look like Fig. 4–4, where the points are then connected and described by some continuous curve as shown. The most widely used curve is called a *Gaussian* or *normal* curve and is the only one we shall use in this regard. The distribution of occurrences described by such a curve is called a *normal distribution*. Distributions other than normal can occur, but to become involved with them is beyond our aims. It is also added that a normal curve is described by a particular equation that provides a plot from minus to plus infinity. The equation itself is of little interest here; rather, it is the *area* beneath the curve that is of importance. The total area under the curve and bounded by the abscissa is considered to have unit area, while the area bounded by any two *discrete values* of the measured quantity is a fraction of one (that is, the total area) and indicates the probability or likelihood that an individual value will fall between the two discrete values chosen. Consequently, the shape of the curve (see Fig. 4–4) shows that there is a greater probability that values closer to the average will result as compared with values further from the average. Since the standard deviation is a function of the individual dimensions (i.e., X), we can superimpose units of σ as shown in Fig. 4–4. With a normal distribution, the area under the curve between limits of $\pm\sigma$ from the average encompasses about 68 percent of the *total* area from minus to plus infinity while limits of $\pm3\sigma$ include over 99 percent of the total area. To be more precise, the probability is 99.73 percent, which means that if 10,000 measurements are made only 27 out of that entire number would be expected to fall outside the 3σ limits. So, although this could happen, the probability is extremely small. For our purposes, all bilateral tolerances will be related directly to $\pm3\sigma$ limits; this is the reason that bilateral tolerances were chosen earlier.

Before we use the above ideas for specific problems, there are certain assumptions to be added.

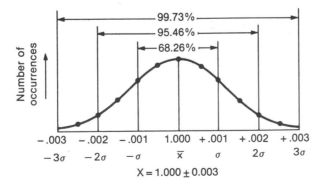

Figure 4–4 Illustration of a normal distribution showing the spread of tolerances and the corresponding multiples of the standard deviation σ.

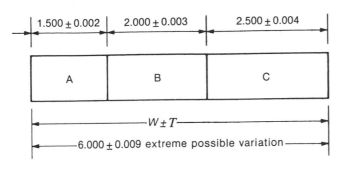

Figure 4–5 Stack of three blocks showing the extreme dimensions of the overall stack.

1. For *each dimension* involved, the variation of sizes displays a normal distribution.
2. The occurrence of the dimension of any individual component, due to the manner of manufacture, is *completely independent* of the individual dimension of any other component.
3. The selection of individual components to be assembled together is completely *random*.
4. The basic size of each component is equal to the average of all such parts, and the *total* tolerance spread is equal to 6σ for each individual part and its resulting distribution. Thus if a bilateral tolerance is given as ± 0.003, then 6σ equals 0.006.

As a first example, consider Fig. 4–5, where parts *A, B,* and *C,* whose basic sizes and tolerances are shown, are to be stacked together in an assembly whose basic size is shown as *W* with a tolerance of $\pm T$. The extreme variations that could *ever* occur fall between 6.009 and 5.991 and for that possibility, $W = 6.000$ and $T = \pm 0.009$. How likely is it that these extreme combinations might occur? This is where the idea of probability enters, and Fig. 4–6 will assist here.

Since each component displays a normal distribution and assembly is random in nature, there is no reason to expect that if a component of *A* happens to have a dimension less than average, it will be assembled with parts *B,* and *C,* having dimensions also less

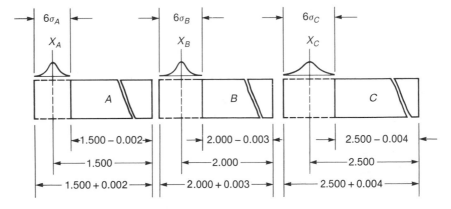

Figure 4–6 The three individual blocks from Fig. 4–5 showing the 6σ spread for each block.

than average. A mix of A, B, and C, where some dimensions are above and some below average, is more likely. Because the greatest probability of the occurrence of dimensions for each individual part is close to their average values, the most probable average value of W is the sum of those individual averages.

Here, A, B, and C are the independent variables, while W is the dependent variable and the probable variation of W can be found by using the following,*

$$\sigma_u^2 = \sigma_A^2 + \sigma_B^2 + \sigma_C^2 \qquad (4\text{-}6)$$

where σ_u is the standard deviation of the assembly or unit and the other three relate to the components. Recalling that the total tolerance spread equates to 6σ, we revise Eq. (4–6) to

$$(T_u/6)^2 = (T_A/6)^2 + (T_B/6)^2 + (T_C/6)^2 \qquad (4\text{-}7)$$

or simply

$$T_u^2 = T_A^2 + T_B^2 + \mathrm{T}_C^2 \qquad (4\text{-}8)$$

where T is the tolerance of each component and the overall unit, respectively.

This form will find major use and implies that the distribution of assembly measurements (i.e., W) will also display a normal curve. Returning to Figs. 4–5 or 4–6, and using Eq. (4–8), we have

$$T_u^2 = (0.002)^2 + (0.003)^2 + (0.004)^2 \qquad (4\text{-}9)$$

so $T_u = 0.0054$. Thus from a statistical viewpoint W would vary as 6.000 ± 0.0054 rather than 6.000 ± 0.009 in., as found earlier.

Now let us reassess the situation connected with Fig. 4–2. There, the average size of the three parts when they are stacked together is 2.875, and since the average channel size is 2.894, the average value for E is the difference, which is 0.019. With the minimum value for E of 0.005, the 3σ spread is ± 0.014, which is the dependent variable; therefore,

$$(0.014)^2 = (0.003)^2 + (0.004)^2 + (0.005)^2 + X^2 \qquad (4\text{-}10)$$

so $X = 0.012$ and D becomes 2.894 ± 0.012. Recall that X was found to be 0.002 for 100 percent interchangeability from Eq. (4–1), so assigning the larger tolerance has obvious economical advantages. Note that *if* the three largest possible blocks were combined with the smallest channel, interference would result; however, the probability of this happening is remote.

Next consider Figs. 4–3(a) and 4–3(b), where from Eq. (4–3), X was found to be 0.001. There are six independent variables, the tolerances, but note that only *one-half* of the values assigned to pins and holes applies in Eq. (4–3), whereas the full value of X pertains. When all dimensions are basic, the average clearance between either pin-hole combination is 0.005; that is the value of the possible horizontal translation. Since the minimum clearance must be zero, the 3σ value here is taken as 0.005. Then, with Eq. (4–8) there results

* This is given without proof. For a derivation, see M. F. Spotts, *Dimensioning and Tolerancing.*

$$(0.005)^2 = \left(\frac{0.001}{2}\right)^2 + \left(\frac{0.002}{2}\right)^2 + \left(\frac{0.001}{2}\right)^2 + \left(\frac{0.002}{2}\right)^2 + X^2 + X^2 \quad (4\text{-}11)$$

from which $X = 0.0033$ as compared with 0.0010 from Eq. (4–3). These introductory examples illustrate the benefits derived, in terms of larger tolerances, if one is willing to take the small risks involved. It is again stressed that if normal distributions do not exist, then σ is not equal to $T/6$ and different, but similar, factors must be used. Yet, even when the individual components do not display normal distributions, if enough components are involved in an assembly, the assembly dimension itself will show a distribution that approaches a normal one. Finally, we note that industrial concerns often use a multiplying factor as a ''factor of safety'' on tolerances to reconcile theory and practice, so complete reliance is not placed upon the results predicted by statistical procedures.

4.7 SELECTIVE ASSEMBLY

Occasionally, functional requirements are so demanding in regard to tolerance specifications that the method of selective assembly is used. As an introduction, assume that a shaft is to mate with a hole, that the minimum and maximum clearances have been specified by the designer, and that this restricted range of clearances *must* be maintained. To avoid possibly excessive manufacturing costs, the tolerances assigned to each component are of such a magnitude that random assembly would lead to many instances where the allowable clearance would not be attained. The following procedure is then necessary.

1. Establish the minimum and maximum mating sizes.
2. Compare the positions of each component relative to the basic size of each part.
3. Determine the interval size to be used for categorizing each part as to selective measurement. This is found by taking one-half of the difference between the maximum and minimum allowable variations.
4. Inspect *every* part and place each part in a particular group denoting maximum size variation of that group.
5. Select a pair of components from their respective groups such that the allowable variations in any individual assembly will not be exceeded.

Suppose the *clearance* between any shaft and hole must be within 0.001 to 0.003 in. In Fig. 4–7, the dimensions of the hole and shaft are given as 2.002 ± 0.002 and 1.998 ± 0.002, respectively, and the extreme clearances must be between 0.001 and 0.003 for satisfactory function. The interval size, as noted in step 3 above, is 0.001; this is shown on Fig. 4–7. Note that with random assembly many pairs of individual components, when assembled, would not only violate the stated requirements but could lead to interference (that is, no clearance). However, after the hole and shaft diameters are *inspected*, each is assigned to a particular group; these are indicated as *A* through *D*, as shown in Fig. 4–7. Then the assembly of *any* pair of components from similarly labelled groups will positively satisfy the acceptable clearance range. For example, two parts from group *A* would

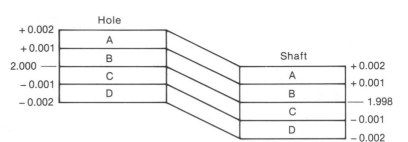

Figure 4–7 A selective assembly system for combining shafts and holes that insures a desired limit of clearance between any appropriate pair of components.

provide a maximum clearance of 2.002 − 1.999 or 0.003, whereas the minimum clearance is 2.001 − 2.000 or 0.001. Similar calculations pertain to the other three groups.

We note that the allowable tolerances on both parts could be increased in the above example and a greater number of groups, still having an interval of 0.001, would be used. It should also be realized that the added costs associated with the inspection of *all* individual parts, plus the need to segregate and store parts in specific groups, introduce expenses greater than those connected with statistical interchangeability. Therefore, a careful cost study should be made before it is decided to use selective assembly.

4.8 TRANSFER OF REFERENCE SURFACES

When a part is designed and dimensioned, the process by which the part is to be produced should be considered. For this discussion, assume that the various surfaces involved are to be *machined** to final sizes. The workpiece must be placed on particular surfaces of the machine tool and then clamped in place before operations can be performed. From these locating surfaces, the various tools are referenced; then the necessary cutting operations follow. If the reference surfaces selected for part location on the machine tool differ from those specified by the designer, then a *transfer* of references must take place. Inevitably, at least some of the tolerances available for the machining operations will be *less* than those specified by the designer. A cardinal rule to follow if such a transfer is necessary is that the sum of new tolerances that affect a dimension must not *exceed* the tolerance assigned to the original dimension. A few examples that illustrate this concept are now presented.

For the piece shown in Fig. 4–8(a), the original design specifications are shown. In locating the part on the machine, suppose surface *A* is first machined and surfaces *B* and *C* are to be finished by using *A* as the reference surface; thus dimensions *AB* and *AC* must be considered. To maintain the original tolerance of 0.002 in. between *B* and *C,* the *sum* of the tolerances on dimensions *AB* and *AC* cannot exceed that value when the transfer occurs. One solution is shown in Fig. 4–8(b); this will satisfy the original specifications, but notice the reduction of initial tolerances that results.

* Discussed in Chapter 9.

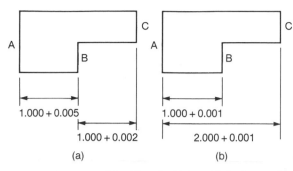

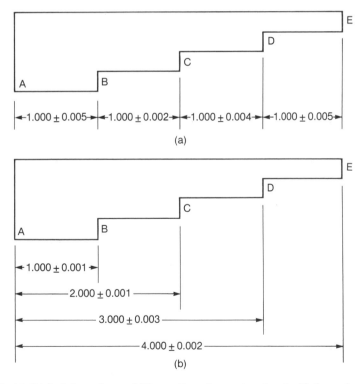

Figure 4-8 Transfer of references from initial dimensioning (a) where surface A is chosen as the reference surface for all dimensions (b).

Next consider Fig. 4–9(a), which shows the design specifications. If A is chosen as the reference surface, all new specifications are transferred with respect to A. The original tolerance on BC is the *smallest* and therefore is the one to consider *initially*. The transferred tolerances on new dimensions AB and AC cannot exceed the initial tolerance of ±0.002 on BC; the value of ±0.001 on each of the new dimensions satisfies that constraint as shown on Fig. 4–9(b). Now consider the next smallest initial tolerance of ±0.004 on CD. This dictates the sum of the new tolerances on AC and AD, and since AC carries a value of ±0.001 from above, the new maximum tolerance on AD is ±0.003. Finally the maximum tolerance on AE is given as ±0.002, since in combination with the

Figure 4–9 (a) Original dimensions and (b) new dimensions as transferred with A as reference surface.

value used on *AD,* it will not exceed the original tolerance of ±0.005 associated with *DE.*
Note that the basic sizes of the transferred dimensions are in agreement with the design
specifications on Fig. 4–9(a).

In many cases, a choice of reference surfaces is available, and each should be
studied to provide an optimal situation. Fig. 4–10(a) shows initial dimensions and tolerances of a plate that is to contain three holes to be machined. Using hole centerlines as
references for other dimensions is not the best practice, since the holes themselves do not
exist originally. Suppose surface *B* is finished first and is then to serve as the reference for
all other dimensions. Figure 4–10(b) shows one acceptable set of dimensions and tolerances. Since the smallest initial tolerance, ±0.002, was placed on the 0.500 dimension,
this must be shared by new dimensions *C* and *D;* each carries a value of ±0.001 as
shown. Now the maximum tolerance that could be applied to *E* is ±0.003 to meet the
original value of ±0.004 on the 1.000 dimension, but this would then require a tolerance

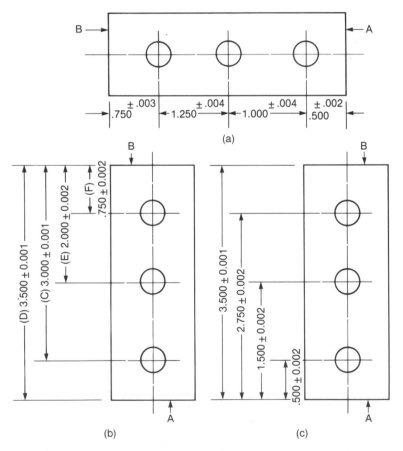

Figure 4–10 An illustration of transfer of references showing two possible results where (b) *B* is used as
the reference surface and (c) where *A* is used as the reference; initial dimensioning is shown in (a).

of ± 0.001 on F. It would be more sensible in general to share the available possibilities and use values of ± 0.002 on both F and E, as shown in Fig. 4–10(b).

What if A were used as the basic reference surface? One combination of transferred dimensions and tolerances that satisfied the initial specifications is shown in Fig. 4–10(c). Several points are noted. If A is used as the reference, the *sum* of all individual tolerances is greater than that which results when B serves as the reference, that is ± 0.007 versus ± 0.006. Finally, if a particular individual machining operation requires a larger tolerance than others (that is, it is more difficult to produce the desired dimension within restricted limits), it is always possible to revise the tolerances on the transferred dimensions to handle such a situation as long as the *sum* of any two new tolerances does not lead to possible dimensional variations that would not satisfy initial specifications.

As a final comment, designers must avoid the use of redundant dimensions other-wise, problems are apt to occur. Fig. 4–11 illustrates such a situation. In Fig. 4–11(a) only

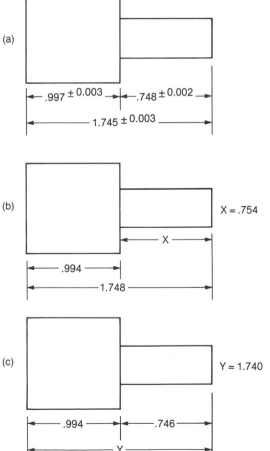

Figure 4–11   An illustration of redundant dimensioning (a), with two possible consequences shown in (b) and (c).

two basic sizes and tolerances should be specified, or the possible results in (b) or (c) could occur. The manufacturing people could certainly argue that they have correctly followed the original requirements in terms of two of the three dimensions, and, in fact, it is not always possible to satisfy all three dimensional specifications simultaneously.

PROBLEMS

4–1. Three blocks are to be assembled as shown in Fig. P4–1.
  (a) Determine the average size and the bilateral tolerance of dimension A that results, based upon the combinations of extreme possible conditions.
  (b) Using a statistical approach, determine the bilateral tolerances expected for dimension A.

4–2. Refer to Fig. P4–1. If blocks 1 and 3 are dimensioned as in Problem 4–1 and dimension A varies from 1.608 to 1.592, what tolerance must be placed on block 2 to satisfy any combination of worst conditions? Repeat this problem if the bilateral tolerance on block 2 is determined by using a statistical approach.

4–3. Refer to Fig. 4–2 in the text. Suppose dimensions A, B, and C remain as shown, while D is given as 2.899 ± 0.005.
  (a) Determine the extreme variations of E that could ever result.
  (b) Repeat part (a) from a statistical viewpoint.

4–4. Figure P4–4 shows a pair of typical components that are to be assembled by having the pins fit into the mating holes. The tolerances to be assigned are to be in the ratio of $x = y/4$, while the centerline tolerances are shown as ± 0.003.
  (a) For 100 percent interchangeability, determine x and y.
  (b) Repeat for statistical interchangeability, using all of the usual assumptions discussed earlier in the text.

4–5. Figure P4–5 shows three parts. For proper assembly, the pins must fit into the mating holes, and the bottom surfaces of parts A and B *must* remain in full contact with the reference surface. For 100 percent interchangeability, determine what tolerance X should be applied to the two-inch dimension.

4–6. Repeat Problem 4–5 if a statistical interchangeability analysis is used.

4–7. Parts A and B are to be assembled by means of a bolt, as shown in Fig. P4–7, and the right side of the bolt is to *always* make contact with the hole in part A.
  (a) Based upon *average* dimensions, what is the average value of E that results under the stated restrictions?

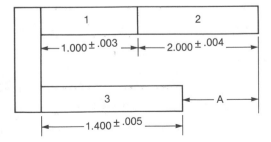

Figure P4–1

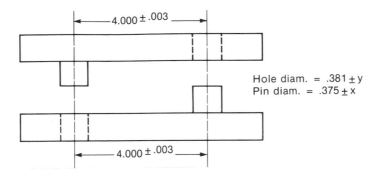

Hole diam. = .381 ± y
Pin diam. = .375 ± x

Figure P4–4

(b) Using your answer in part a, what tolerance X must be assigned to the four-in. dimension so that the tolerance of 0.010 on E is maintained? This is to consider 100 percent interchangeability.

(c) Repeat part b if statistical interchangeability is used.

4–8. A spacer S, bearing B, and a stepped shaft C are to be assembled in a housing as shown in Fig. P4–8. The minimum distance between surfaces A and D must not be *less* than 0.003, where D is always *below* A.

(a) For 100 percent interchangeability, what tolerance X should be assigned to the 0.766 dimension?

(b) Repeat a if statistical concepts are used.

4–9. Figure P4–9 shows original design dimensions. If for manufacturing purposes A is to be used as the reference surface for all operations, redimension the part accordingly so that all

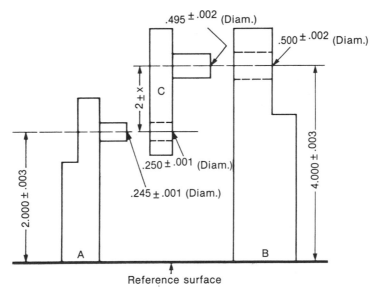

Figure P4–5

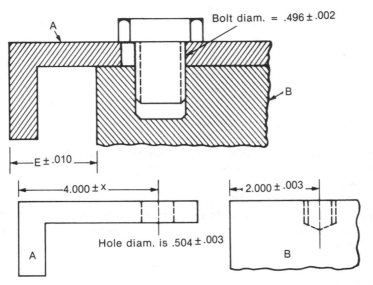

Bolt diam. = .496 ± .002

A

B

E ± .010

4.000 ± x

Hole diam. is .504 ± .003

A

2.000 ± .003

B

Figure P4–7

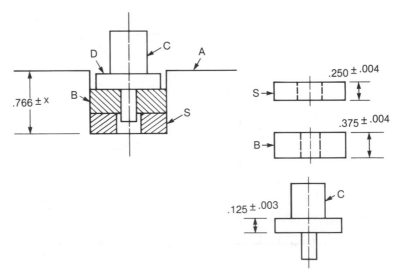

D

C

A

B

S

.766 ± x

.250 ± .004

S

.375 ± .004

B

.125 ± .003

C

Figure P4–8

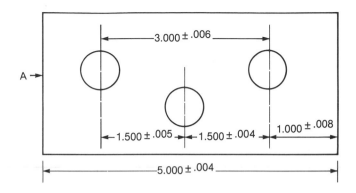

Figure P4–9

functional demands are satisfied and the maximum allowable tolerances are used. Keep the smallest single tolerance as large as possible.

4–10. Figure P4–10 is to be redimensioned by transferring dimensions with A as the reference surface.

4–11. Figure P4–11 shows original dimensioning. If A is to serve as a reference surface, transfer all dimensions so that all original specifications will be satisfied.

4–12. Redimension the part shown in Fig. P4–12, using A as the machining reference. In distrib-

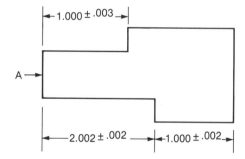

Figure P4–10

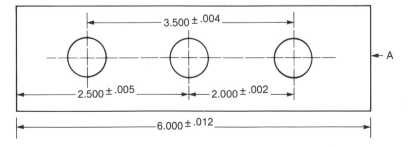

Figure P4–11

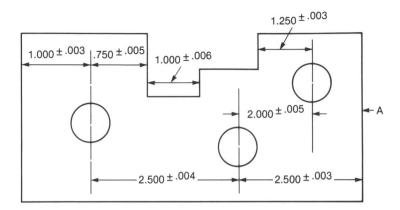

Figure P4–12

uting tolerances over any two new dimensions, use equal values for each individual dimension when possible. The maximum value of available tolerances must be used.

4–13. Consider the original dimensions of a part shown in Fig. P4–13. In transferring dimensions either surface *A* or *B* may be used.

    (a) Complete a transfer for each surface, using the maximum tolerance available for each dimension.

    (b) Which surface permits the greatest sum of individual tolerances?

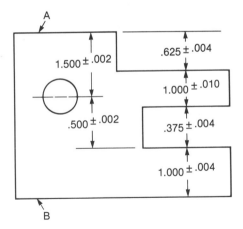

Figure P4–13

_5_ automation and computers _____

5.1 INTRODUCTION

One important consideration in a competitive industry is the degree to which manufacturing processes and human actions will be mechanized, automated, and computerized. There are clear benefits in doing so in many instances, but not in all. The deciding factor is whether a competitive advantage will be achieved by doing so. Recall that the economic survival of a manufacturer involves _delivering products, on time, with adequate quality, to sell in a well-informed market and gain a short-term profit._ A secondary objective is long-term economic survival, which is aided by offering new products, by making products good enough to assure repeat sales, by minimizing recall rates and warranty costs, and by a general reputation for fair commerce.

In the following sections, mechanization, automation, and computer technology will be defined and discussed. These topics have traditionally been linked in terms of mechanical progression; human controlled mechanical systems (machines) appeared first, after which the systems were made to _run themselves,_ and recently computers became available to manage the systems more efficiently.

In parallel with mechanical innovations, there has been mechanization and automation of _information_ transfer. At first there was paper and then the telephone, then photocopying and other conveniences. These systems speeded some aspects of information transfer, but they did not speed human comprehension of what all of the information means and where it should be going. Electronic devices are very helpful, particularly for transferring _appropriate intelligence._

Until small computers became readily available, there appeared to be little connection between mechanical automation and the automation of information transfer. Lately,

67

however, it has become very apparent that they are linked very strongly, and must be treated together. This may be seen in the following list of the tasks to be done in manufacturing.

5.2 THE TASKS TO BE DONE AND THE ASSOCIATED COSTS

The separate parts of the general industrial objective may be expressed under three headings, together with some of the major tasks in each, as follows. Many of these points were discussed in the previous chapters, and are listed here for emphasis.

1. The control of quality, by
   a. Inspecting incoming material.
   b. Routing the proper material to each process.
   c. Monitoring each process and machine, including handling, assembly, and packaging equipment, to ensure product quality.
   d. Keeping records of quality.

2. The control of scheduling or rate of production, by
   a. Allocation of machine use.
   b. Prediction and scheduling of downtime for failure, maintenance, and changeover.
   c. Effecting quick changeover of machines in order to process a variety of products on few machines.

3. The control of cost, by
   a. Maintaining minimum inventory.
   b. Balancing labor and machinery cost.
   c. Balancing production rate and tooling cost.
   d. Establishing a procedure to quantify the parasitic costs, such as
      (1) Defects in purchased material and the defects that develop in process, assembly, or handling. The cost of these defects increases if the defective material/product is allowed to continue through manufacturing, packaging, and marketing. *These costs may be minimized by complete product quality sensing, before processing and in process, but this costs money.*
      (2) Defects in machinery and tooling. These defects usually result in downtime of all or part of a product line, which is a loss of ability to recover investment. There may also be excess labor cost due to idle employees, and some loss of products that are damaged in a defective process. *These costs may be minimized by providing buffer storage between processes, by scheduled maintenance, and by adequate sensing of machine condition, all of which cost money.*

(3) Inefficient use of the skills and physical labor of employees due to inadequate attention to the needs of the manufacturing system and perhaps due to inadequate attention to human relations.

(4) Inadequate or inefficient communication of information. This is a problem throughout the entire factory; for example,

    (a) Design changes do not always reach the proper people to prevent the purchasing of the wrong materials or tooling.

    (b) Impending shutdown of a machine is not always known in time by the proper people to prevent severe unbalance in materials and labor in other parts of the plant.

    (c) Quick changes in production plans (in response to unanticipated market demands for more or fewer products, or for one product option over another) are not always communicated in sufficient time to purchase adequate materials and tooling to meet production needs, or to cancel orders for materials and tooling no longer needed, or to meet the changing needs in personnel or to change over machinery in proper time. *These costs can be minimized by a good organization, by thorough reporting procedures, and by the use of electronic communication, all of which cost money.*

The above list shows some of the challenges in industrial communication, or information flow. Some aspects of the challenge are very technical in nature, both in the content of the information and in the way information is transferred.

5.3 INFORMATION AND COMMUNICATION

It may be seen that for good management, information flow should be dynamic, responding to all changes in production rate, changes in machine condition, changes in the availability of labor, to name a few. Above all, specific people must be made accountable for every bit of communication, or for lack of it.

These are difficult requirements for a system of communication using paper or *word of mouth*. It is difficult for paper work to flow along unaccustomed channels, and it is even more difficult to grade recorded messages according to urgency and relevence. In the last three decades the telephone has been most helpful for some classes of information flow, but it lacks the ability to rank messages according to urgency. Telephones ring as loudly for trivial and bothersome messages as for urgent ones.

In the last decade the computer has been commercialized to the point where it provides a cost-effective aid to information flow. With computers it is easy to establish many different routes for information flow, beyond established mail routes or telephone extensions. It is possible to establish ways to rank the urgency of messages, and to establish accountability. Thus, if a production worker notifies the supervisor of an impending material shortage, the response of the supervisor can be recorded to establish responsibility, should there be an interruption of production. Or the supervisor could have previously sent messages on material needs to the material handlers or suppliers, and again responsibility could be fixed for any problems that develop. The importance of such an information network is that, in the short term, fewer material shortages are likely to occur, and in the longer term, deficiencies in the qualifications of personnel or organization will be highlighted rather quickly and definitively. It is apparent that information flow can strongly affect the cost of manufacturing. Thus the manufacturing engineer should be involved in the design of the communication system, both of the equipment and of the network of people involved.

Information transfer is a problem wherever there are people working together, whether that be in an airplane factory or in a fast-food restaurant. Using the latter example, we find that problems may begin when customers request the "product" in various degrees of processing in such terms as "rare, medium, or well done." They further request options in the product such as additions of catsup, pickles, mustard, and so on. The requests arrive at the "sales" department in random order, requiring complex communication, oral and nonoral, throughout the manufacturing facility (kitchen). Most often the system operates smoothly. But if there is a rush of orders for one or two products or options, and if the cook or other supplier of that product is either slow to receive the message or out of range of communication for a time, the system soon degrades into a state of confusion. Requests are shouted with increasing urgency, incomplete orders are set aside, which soon interfere with orderly traffic, and the number of errors increases. This sequence develops with varying frequency, hourly or daily, depending on many factors. The exact cause of the disorder is usually somewhat different each time, which taxes inexperienced workers and managers. And the search for the causes often is obscured by personal insecurities and blame shifting.

To prevent chaos it is vital to have good organization, which includes proper placement of people of the proper skills, with detailed assignment of tasks and responsibilities. In a restaurant there may be little need or opportunity to automate communication and systems of accountability. The need increases with the physical complexity of the system, with the sophistication of the product, and with the amount of competition in the market.

5.4 MECHANIZATION

We define mechanization as a state in which tools are applied to aid or replace animal power, but requiring human intervention for the purpose of control. Thus the hand shovel is a tool, since it allows the user to accomplish more than is possible without it, even though shoveling is hard work. The power shovel is also a tool. It moves large amounts of earth with little human effort, though it is guided in every motion by the human operator. The hand and power shovels are examples of the range over which tools may multiply human effort, and exemplify ranges or degrees of mechanization.

It is often surprising to see the great number of manual or unmechanized operations, even in a highly mechanized manufacturing facility. Assembly is often done manually, as is die-setting, machine loading and unloading, inspection, counting of inventory, and even communication. A factory therefore is usually only partly mechanized, and very likely none will ever be completely mechanized. First of all, the cost to mechanize completely is prohibitive. Second, there are some human capabilities involving aesthetics and judgment that will not be profitable to mechanize. After all, consumer products must appeal to the highly varied subjective sense of consumers, and human judgment is required to determine when products are ready to risk in the marketplace.

5.5 AUTOMATION AND CONTROL

We have seen that mechanization consists in the use of tools to multiply the physical efforts of humans. The next step is to relieve humans of the task of guiding or controlling the tool. Tools that control their own motions within prescribed limits are said to be automatic; or by virtue of having had control devices connected to them, tools are automated. Few if any power shovels have been automated, probably because of the wide range of tasks required of them. Certain components of power shovels are automatic, such as the speed control of the engine, and the lift-limiting feature to prevent tipping over. These automatic features prolong engine life and prevent accidents, but without significantly affecting the quality of the work done.

In many instances a tool or machine is automated to maintain product quality. A hamburger grill may become too hot or too cool to produce quality hamburgers, unless controlled. Likewise, the chemistry of an uncontrolled electroplating bath may vary from the acceptable range, necessitating the stopping of production while adjustments are made.

Automatic machines *can* often produce parts with greater accuracy, uniformity, and speed than can manually operated machines; they also relieve humans of some tedious, dangerous, and hard jobs. But automation is cost-effective only for simple tasks. The majority of humans can readily perform fairly complex tasks, and for moderate pay. Some industries use cost guidelines in deciding whether or not to automate an operation. If, for example, a machine that replaces a person costs more than three times the annual salary plus benefits for the person, then the person is retained. Such reasoning explains why the dispensing of newspapers is well automated but the setting and clearing of tables in restaurants is not.

In the latter example one can discern two aspects of the complexity of a task. One is manual dexterity, or the skill in making complicated movements. The other is in the use of the *senses* (sight, pressure, sound, feeling of vibration, etc.) in guiding action. Several human senses are used to monitor even the simpler tasks. To integrate only one or two senses with complex motions in robots or other automated machies is very expensive.

Automation involves control, and controllers range from the very simple to the very complex. One of the simpler controllers is used on railroads. Two rails effectively "control" the direction of the railroad vehicles, without sophistication or instrumentation. A more complicated system is the system that controls an airplane along its intended course.

An airplane or a train may at one instant actually proceed in the proper direction all by itself, but both will soon proceed in a different direction and will need corrective input from the control system; the rail exerts pressure against a wheel, and the rudder directs the airplane. Both control systems must be designed to meet certain requirements. These may include limits on the amount of deviation allowed from the intended path before correction is effected, or limits on the abruptness of correction in order to prevent high stresses on vehicle parts or discomfort and damage to the payload. These requirements and limits are called *constraints* in the technical terminology of the field of control.

The train and the airplane are both controlled, but in the conventional meaning of the terms, the rails are fixed guides rather than controllers or control devices. A *true* controller provides the flexibility to follow a different course each day.

The principles of control are simple. A controller reacts to a signal sent to it from one or more sensors. One type of sensor in an airplane is a compass. It *feeds* a signal to the controller on the direction the airplane is heading. If this is not the *preset* or prescribed heading, the controller changes the direction of the airplane, which action is sensed by the compass, which *feeds* a new signal to the controller. This type of control system with the sensor is referred to as a *feedback control* system. Since action follows a sensed variable and that action changes the second variable, the total action is referred to as *closed loop* control. These terms may be compared with what is often referred to as *open loop* control, which is a contradiction in terms. In so-called open loop systems a mechanized or electrical action is effected without any measurement or sensing of the effect of that action. For example, a mechanical cam and follower, or a servomotor and ball-screw may be actuated by a given amount in order to move the table of a milling machine some desired amount. At the end of the actuating signal, if there is no sensor, there is no *feedback* to indicate that the table has been moved the proper amount. In such cases the cam and the servomotor are merely actuators or *effectors* and are not a part of a control loop.

Controllers can further be classified according to the manner by which corrections are made. An airplane that is off course could have the rudder turned abruptly some fixed amount until the airplane is on course. Or the abruptness of this action can be reduced by providing some ramping or gentle transitions at the ends of the control events. This introduces the need to detect the points at which ramping must be actuated. Alternatively, a corrective action could be applied that is proportional to the amount the airplane is off course. The airplane would steer more sharply when it is far off course than when it is near course, subject to some constraints. This proportional control has some advantages in that simple analog electronic circuitry can be used in the controllers instead of digital devices. Analog circuits are readily characterized by linear differential equations, which makes their design easy. The same differential equations characterize the forced vibration of a series spring-mass-dashpot combination, connected in a U-frame, and are of the form:

$$m\ddot{x} + C\dot{x} + Kx = F(t) \tag{5-1}$$

where m is the mass, C is the damping coefficient, K is the spring constant, x is the displacement of the mass relative to the U-frame, and $F(t)$ is a time-varying force (forcing function) applied between the mass and the U-frame. The solution to the above equation depends on whether the quantity $\{C/m\}^2 - 4\{K/m\}$ is $+$ or $-$ or zero. (Beachley)

There is at least one mechanical system that behaves exactly like the example above. The cab of a truck is a mass, and for isolation of the cab from the frame of the truck the cab is mounted on springs with dampers. The frame of the truck is equivalent to the U-frame mentioned above, and it bounces or oscillates at frequencies that vary with road and load conditions. We will describe the problem as a one-dimensional, or single degree of freedom, problem, even though truck cab motion is much more complex.

Some of the frequencies of frame oscillation can be near to the natural frequency of oscillation of the body. The latter can be readily adjusted, since it is simply proportional to $\sqrt{(K/m)}$. For driver comfort the natural frequency should be low, requiring a large m. But a large m reduces the amount of payload the truck can carry under limits imposed by governmental regulations. A small m is therefore required for economic purposes, but a small m will oscillate at larger amplitudes as well. One solution is to place an accelerometer on the cab, and a hydraulic cylinder between the cab and frame. The hydraulic cylinder imposes a force, which acts to decrease the acceleration of the cab when the frame oscillates. Such systems can respond as quickly as 30 Hertz (cps) over an amplitude of 5 cm, which is sufficient for driver comfort.

In the example of the truck, the physical system was similar to the simple model given earlier. However, the important design problem is that of the control circuit. It receives a signal from a sensor, the accelerometer, indicating something about the movement of the cab. The circuit will then calculate and effect an appropriate response, by opening or closing a hydraulic valve, provided that its own components are direct electrical analogs of those of the physical system. This is a straightforward or standard control system.

A control system designed for one particular application cannot, in general, be used for another. Furthermore, it will function poorly if the physical device it is designed to control changes its properties in some way. For example, should the truck cab be occupied by three very heavy people instead of the usual one, the control circuit will not provide the appropriate response to truck frame vibration. It is a simple matter to provide a means for changing the circuitry of the controller. There might be a knob to turn somewhere in the cab which can be adjusted to compensate for weight of the cab. Or there could be a device that senses the "at rest" length of the springs on which the cab is suspended, which could damage the electrical characteristics of the controller. Any number of schemes could be designed into the controller to vary the nature of the ride in the cab. But all of these are done by altering a part of the electrical circuitry that represents a physical quantity. These changes are called *gain scheduling,* because the changes are usually made by adjustments in the gain loop of an electronic amplifier in the controller.

As noted above, gain scheduling can be done manually, or it can be done automatically from some input taken from the *static* condition of the device or machine to be controlled. It can also be done *dynamically.* That is, the control circuitry could include provision for comparing the dynamic response of the physical system against some model behavior. For example, it could monitor the amount of overshoot or undershoot of corrective action taken by the hydraulic cylinder. If such a deviation from critical damping is larger than a prescribed amount, the auxiliary circuitry could then adjust the gain of the primary controller to compensate for the change in cab weight. This is called *adaptive control.* Adaptive control is distinguished from other types of control in that it is a method of control in which the controller adjusts its characteristics according to changes in the dynamic behavior of the machine.

Control, particularly adaptive control, is done most often against one or two constraints. These may include some safety consideration, or noise limit, or the like. The computer has brought the possibility to operate with several constraints simultaneously, and in fact it is reasonable to optimize performance around some predetermined *index*. For example, a lathe could be operated within simultaneous constraints to prevent tool breakage, prevent vibration, and produce a good surface finish. At the same time the machine can be made to operate at the overall most economic feed and speed, when the cost of labor, the cost of cutting tools, and the cost of chip removal are taken into account. It is then necessary to have all of these variables either expressed in mathematical form or available in tabular form for use in approximate equations in order for the control system to be able to arrive at an optimum. There might even be provision for automatic updating of the cost of cutting and labor costs each day so that the controller can "find" a new optimum each day. These possibilities are yet several years in the future.

5.6 COMPUTERS

The increasing availability of digital computers alters the economics of automation and control. A computer and simple electronics now replace complex control systems, and provide virtually any desired degree of sophistication to the control routine. The problem of designing control systems has shifted from the design of complex electronics to the programming of computers so that they will respond to sensors and continually alter the gain setting of the controller.

A second useful capability of computers is electronic communication. Several computers hooked into a network become an instantaneous and precise medium for transferring information. The information ranges from merely words to the display of complex designs. The designs are those of the company's products or perhaps of a complicated mold for plastics or a die for sheet metal forming; they are entered into electronic storage in the *graphics* mode in computers of a wide range of sophistication. Several people in several locations can then connect simultaneously to the same source and discuss possible revisions in a design from identical displays, or viewers could connect serially, as the need arises; these might be the material purchasing department, the plant engineers, and the manufacturing engineers. The buyer of the part may also have access to the design in order to see how it fits with some other parts to be supplied from elsewhere. Someone responsible for production scheduling could have access to the plant schedules and predict when the viewed part may be finished. A small but important advantage of a system is that parts will not be delayed in manufacturing because the part print (on paper) has not yet made the rounds, or because someone has scribbled ambiguous notes and revisions on a print. The major disadvantage is that computers malfunction, but paper does not.

Designers will find the use of computer data bases to be mandatory in a few years. The current procedures in designing objects involves calculations of various types, for which most designers have references and text books. If these references are very familiar, it is a simple matter to find the desired equations for use in calculation. However, if the new problem involves some new boundary condition, for example, if available equations are for conditions of small strains whereas the new problem involves larger strains, some considerable study may be required to find appropriate equations. A well-developed communication network in a design group could include access to an *expert system,* which is an interactive program whereby an inquirer at a keyboard can be led through a series of questions. The answers given will be used by the program to direct the inquirer to the best source of information for solving the new problem.

A parallel problem for designers is to locate materials, components, and supplies. Usually catalogs are in hand for this purpose. However, some effort is required to update the stock of catalogs and the price lists that may be attached. If all of the cataloged information is in electronic storage, it is a simple matter for the supplier to initiate sending updated information daily or weekly. This avoids contacting suppliers by telephone for prices, availability, discontinued products, new products, and other vital information.

5.7 THE MACHINERY OF MANUFACTURING

A major responsibility of manufacturing engineers is to specify, purchase, and arrange installation of machinery. The broadest considerations are the degree of mechanization, automation, and control that will affect the costs over the long term. In this section we will focus on the hardware and not consider economics.

Machinery is available for most types of manufacturing operations, but it may be particularly useful to describe the equipment available for metal cutting. Metal cutting is an old process and is being done on a scale that encourages the development of a very wide range of sophistication in automation and control.

The simplest metal cutting machine is the lathe. The first lathe may have been a V-branch of a tree, in which another tree branch was manually turned while bark or twigs were being removed. At some point in history, bearings were made to aid in turning the wood more smoothly, and perhaps chisels and other cutting instruments were developed at the same time.

In the eighteenth century, lathes were used to cut metal. The cutting tools were plain carbon steel, so cutting could not be done at a high speed by modern standards. (This is covered in more detail in Chapter 7.) In early years the cutting tool was mounted on a small table that was moved manually by screw-nut pairs. Later, the tool was moved by an arrangement of shafts and gears connected between the screw-nut pairs and the spindle on which the workpiece was held. One could shift gears to effect a different amount of tool motion for each revolution of the workpiece, but when the workpiece stopped, the tool stopped moving. In a later development, the tool could be moved separately from the rotation of the spindle, but each motion was controlled by the human operator of the lathe.

As described, the old lathes were indeed automated, but lacked closed-loop electrical or mechanical control of tool motion. Actually, the human operator is the controller, a far more sophisticated one than any electrical or mechanical controller. But for many operations the human system is underutilized and therefore probably too expensive. Thus there have been many innovations in control devices to aid or replace some or all of the involvement of humans in some lathe operations. In those systems involving feedback from sensors, the controller uses only one or two sensors, thus implying that the several other senses resident in humans will not be needed. This is true in some cases, but it leads to very big surprises and failures in other instances. For example, a problem in metal cutting is the occasional, unexpected failure of a tool. This type of failure is immediately evident, even to the untrained eye. In fact, a machinist can give a continuous appraisal of the wear of the tool and can predict failure in time to prevent it. Automatic detection of impending tool failure has not so far been achieved, despite many efforts to do so. Cutting force and noise analysis (acoustic emission) have been used without great success.

A number of mechanical aids to the machinist have been developed over the years. In the cutting of parts requiring several operations, the machinist must exchange a sequence of tools. This takes time, partly to loosen and tighten tools, but more to accurately adjust the location of the tool tip so that it can be moved the proper amount to produce a part of the correct dimension. This problem was alleviated by mounting several tools on a precisely made turret. The turret is rotated to bring up any desired tool, and each tool is preset. Some of the tools on the turret can *feed* faster than others, so some machines are equipped to advance the turret at the proper rate for each tool that is in the cutting position. These developments increased the productivity of machines up to fivefold, and reduced the skill required to operate the machine. A skilled technician is still required to set up the machine, however.

One popular development in lathes and other machinery is "numerical control," or NC. Most NC lathes use a one-inch-wide tape of paper or Mylar plastic in which eight columns of holes may be punched with as many lines as are required to perform an operation. The tape progresses through a reader where an array of eight sensors (photocells or pneumatic sensors) reads which column has a punched hole and which does not. This constitutes eight bits of binary information, or up to 256 (2^8) different "numbers" or codes per line of holes. A series of relays, with no memory, is arranged to interpret each code as an instruction to effect some action. These might include:

1. Set the tool feed direction either left or right.
2. Set the feed rate to be any of ten different values.
3. Start the feed motor.
4. Advance the feed motor 600 pulses.
5. Set the spindle speed to any of five different values.
6. Start the spindle.
7. Stop all motors.
8. Rotate the tool holder to bring up a new tool.

These signals must be directed to proper locations, and part of the code includes such information. But there must also be the proper interpretation of that code by a receiver. The instruction to advance the feed motors is received by a servo controller which converts the instruction into a particular binary pulse stream to advance the servo 600 pulses. For this purpose the system contains many transistorized electronic elements called *chips*.

The tape reader and the associated electronics can accept code faster than most machine actions can take place. Thus the tape reader can sequentially read code for several separate servos of the machine. For example, both the feed and depth of cut can be varied at the same time, which is referred to as *two-axis control*. Some machines have even more motions actuated, perhaps as many as six. With multiaxis capability, a machine can cut complex contours and make very complex parts.

There are varying degrees of real control associated with NC. In the simplest case, servo motors that drive the screw-nut sets may receive a given number of pulses. There may be no method provided to assure that the motors have really moved. In this case the system should be called "numerically actuated." Better systems use servo motors that *feed* servo position signals *back* to its controller, and if the servo has not advanced in accord with the pulses sent to it, the controller may simply send some more pulses, or it may send a signal to the central controller, which may shut down the entire machine. In the older machines a human operator was then required to determine why the machine stopped. In newer machines a message is flashed, indicating the reason for stopping. This capability is a very elementary form of *artificial intelligence*.

The ultimate in NC is the system that uses sensors at various locations to verify that the tool or part has moved to its intended location. The sensor may be an interferometer, or a differential transformer, or other device. Interferometers are accurate to $300\overset{\circ}{A}$ (0.00003 mm), and the differential transformers are accurate to 0.01 mm. The advantage of sensing the position of the tool or workpiece is that greater accuracy of machining can be done. Most machines, workpieces, and tools deflect more than the desired accuracy when the forces of cutting are applied. An accurate and well-placed sensor will signal the servo controllers to compensate for this deflection. This is *feedback* of the most useful nature. An additional useful feature on NC controllers is a readout of the command from the controller and a readout of the actual position of the driven member.

One of the very tedious tasks connected with NC is the punching of the tape. This involves looking up code in a large table of codes, and it requires absolute precision in punching the holes as well. In the early days, after punching, the tape was tried out on the machine. Immediate success was rare. Failure required a lengthy and tedious search for errors.

Computers have been very helpful for punching tape. Computer programs of various sophistication are available, ranging from simple table lookup of NC code to programs that can receive information on the dimensions of the part to be machined, the size and shapes of tool to be used, and the particular machine to be used. In the latter case the computer then proceeds to punch the tape without human intervention. This system is variously referred to as computer-aided NC or computerized NC, or CNC.

Obviously, if a computer can be programmed to punch tape for the NC machine, all of the code and characteristics of the NC machine are part of that program. It is a small step, then, for a computer to control the machine without the intermediary punched tape. Some refer to this type of system as CNC, and others call it direct NC or DNC. (DNC is also an acronym for distributed numerical control, where several computers may be involved in controlling several machines.)

There is yet one type of machine that receives much attention, and that is the robot. It is built to the standards and by the methods of the machine tool industry, and thus it must be regarded as one extension of machine tool technology. Robots usually have a wider range of motion than most machine tools, with corresponding lower accuracy of motion. They are controlled by the same devices as are machine tools, but in some instances the robot may have one or two more movements or degrees of freedom than do machine tools. The movements of the machine tools may consist of four or five sliders, but each usually acts separately. Robots often consist of one actuator mounted upon another, which adds versatility to its motions but also "stacks" errors due to the inherent looseness of fit in joints, which is exacerbated by wear. The motion of the "end" of the system then is the result of the motion of several joints, and may be designed to move in rectangular, cylindrical, or spherical coordinates. The calculation of the location and velocity of the "end" involves a great number of trigonometric calculations, which must be done at great speed in order to effect smooth motion. Thus computers of larger capacity are required to control robots than those required to control most machines.

So far we have discussed single, stand-alone machines. These may be sufficient for some products in some industries. But most products are made in several operations, requiring several machines. A major economic difficulty arises when only 50 or 500 identical parts must be made, as is the case with some 70 percent of the products made in the United States. Again it is useful to imagine what takes place in familiar surroundings such as a restaurant; few survive if they produce one set menu. However, given the availability of equipment to produce one simple product, this equipment could probably be revised slightly so that it can expand the range of products. The system is now *flexible* as to product. With further revision the system could also become *flexible* as to production rate, the discontinuance of old product lines and the addition of new products.

The design of manufacturing systems usually requires considerable knowledge of logical groupings of manufacturing processes: groupings based on skills required to set up and maintain the machines, or based on the materials being processed, or even groupings based on the specific types of controllers on the machines. These groupings are called *manufacturing cells,* and factory automation usually starts by attaching simple microprocessors to machines in the cells.

Products usually must pass through a succession of *cells,* possibly including inspection stations and packaging machines. One major source of cost is the handling of materials as they move into and out of the cells. Material handling is readily automated, at some cost. But automated material handling requires coordinated control of all of the *manufacturing cells* in a system. Another dimension of flexibility must now be provided in terms of routing among the cells to provide for product mix and machine breakdown. Material handling can be done by robots, by linear conveyors, or by loop conveyors.

Coordinated control of *flexible manufacturing systems* (FMS) usually requires a special information transfer system. The reason is that within a manufacturing system there will almost inevitably be computers and peripheral equipment from several manufacturers, each of which uses a different communication method or protocol. This was not done capriciously in the industry. Rather, the several tasks to be done by computers, (for example, analysis, graphics, interactive communication, machine control, and the like) can best be done in special ways. A popular method for solving this problem is for all industries to purchase machine (computer) linkage or bus systems that operate with (nearly) the same data transfer rates, using the same system of priorities of access of computers to the bus, and the same limit on the amount of information that can be put into or taken from the bus when connected. Each manufacturer of computer/control equipment then supplies an interface module between his equipment and the bus, which then makes their proprietary communication method invisible to the bus system.

The useful range of information transfer extends well beyond the actual machines making parts. For example, the design of the part could be stored electronically (on magnetic media), and some of that stored information on dimensions and tolerances could be "accessed" directly to set up the measuring devices at inspection stations. Again, the measurements taken at the inspection station could be analyzed periodically, with results provided to management on the progression of product quality. When several functions are interconnected with manufacturing system control, then *computer integrated manufacturing* (CIM) has been achieved.

Total CIM is a massive undertaking which few industries can afford. Furthermore, unless there is some major change in societal thinking, total CIM will be limited as much by the human dimension as by dollar cost. Engineers are often advocates of increased automation, and find societal and money constraints irksome.

REFERENCES

Beachley, N. H., and H. L. Harrison, *Introduction to Dynamic System Analysis*. (New York: Harper and Row, 1978) or equivalent.

PROBLEMS

5–1. Assess the amount of automation in the following areas: discuss the cost of fully automating and the inflexibility that would result from full automation:
   (a) A restaurant.
   (b) A car wash facility.
   (c) A bank.

5–2. Describe the type of sensors used in the following, and classify the type of control in terms of unramped on/off, ramped on/off, or proportional control:
   (a) Trains moving along tracks.
   (b) Thermostats in home heating systems.
   (c) Thermostats in the coolant system of an auto engine.

(d) Power brakes of automobiles.
(e) Gasoline filler nozzles.
(f) Servos on lathes, with and without sensors.
(g) Strain gages on cutting tools for force measurements.
(h) Sensors for control of surface finish in grinding.
(i) Lawn mower engine speed.

_____ FOREWORD TO PART B

In the broadest and most traditional sense, there are four major processing categories. They may be called by different names; our preference is to use the terms _cutting_ (machining), _forming, joining,_ and _casting_. It would be a rare individual who would possess equal, in-depth knowledge and experience in all four areas. For example, one who has devoted his professional career to the subject of casting may possess little, if any, understanding about forming and vice versa. Yet is is not necessary to be an expert in all processes to understand and teach the most important aspects of each. Since entire texts have been written on an in-depth approach to each of these four processes, it must be realized that single chapters cannot cover, nor are they intended to, all that a specialist might desire. What we present here are the essential features and fundamental concepts of each process. If this much is understood then the reader should be able to continue with those specialized texts that go into greater depth and detail.

Chapter 6 is a minimal, but we think necessary, review of stress, strain, and mechanical behavior of solids. Of special importance is the coverage of strain hardening of metals and the material on nonmetals. This usually receives little if any coverage in other texts of this type, yet it is of decided importance if any quantitative analyses of certain processes are to be undertaken. This background finds major application in Chapter 8. Chapter 7 is a coverage of a number of the most important concepts of engineering materials; it is not a rehash of what is usually dealt with in a materials science course, nor is it intended to go into the many concepts of physical metallurgy. But with this background, the contents of Chapters 10 and 11 should be more meaningful. Chapter 12 introduces certain basics of the "newer" or "nontraditional" processing methods.

6

stress, strain, and mechanical properties _____

6.1 INTRODUCTION

Since the concepts of stress and strain are applied in later sections of this book, it seems sensible to discuss these terms in reasonable detail here, then use them directly whenever needed in chapters that follow. The coverage that follows is *sufficient* for the purposes of this text but is not put forth as an all-inclusive discussion of these two quantities. Both stress and strain are known as *tensors;* a tensor may be defined as a collection of components which may be *transformed* from one set of coordinate axes to another set by following certain mathematical rules. Because of the type of problems of major concern in this book, a background of tensor analysis is unncessary; however, it must be understood that stress and strain are not *vector* quantities in the usual meaning.

6.2 STRESS

This is usually defined as force per unit area. Although such a definition is adequate as far as it goes, neither the three-dimensional nature of stress nor the fact that it is a tensor quantity are made evident by this simple definition. Consider Fig. 6–1, where a force F acts normal to an area A as in *uniaxial tension,* with the x-y coordinate system shown. The stress S_y is simply F/A, and since no force acts *parallel* to A, no stress is associated with the x direction. Now consider a new coordinate system, x'-y', acting at some arbitrary angle Θ as indicated. What stresses act upon a plane at this angle due to force F? First, resolve this force into components F_x and F_y as shown; these act upon area A', which is simply $A/\cos \Theta$. Then since $F_{y'} = F \cos \Theta$ and $F_{x'} = F \sin \Theta$, we have

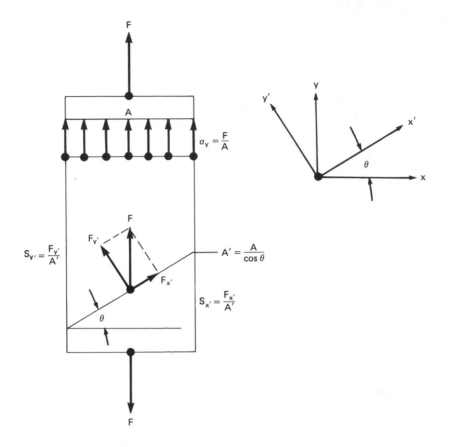

Figure 6–1 Forces and stresses related to different sets of axes.

$$S_{y'} = F_{y'}/A' = \frac{F \cos \Theta}{A/\cos \Theta} = \frac{F}{A} \cos^2 \Theta = S_y \cos^2 \Theta \qquad (6\text{–}1a)$$

and

$$S_{x'} = F_{x'}/A' = \frac{F \sin \Theta}{A/\cos \Theta} = S_y \sin \Theta \cos \Theta \qquad (6\text{–}1b)$$

This is the simplest illustration of stress transformation, and it is noted that to accomplish this requires the use of a *product* of two angular functions. Note that the equivalent force transformation required only one angular function; this illustrates one of the basic differences between vector (here force) and tensor (here stress) transformations.

By convention, *normal stresses* (such as S_y and $S_{y'}$) act perpendicular to planes, whereas *shear stresses* (such as $S_{x'}$) act parallel to planes. When a normal stress tends to cause extension of a body, it is considered *tensile* (or positive), whereas contraction is due to compressive (or negative) normal stresses. Shear stresses tend to distort the shape of the body. We shall not be concerned here as to how positive and negative shear stresses are defined, but it is worth mentioning that in problems where a detailed stress analysis is demanded, such a definition is essential. Note that from this point on, the symbol τ will be used to denote a shear stress, whereas either S or σ will denote normal stresses.

Example 6–1
  ━━━━━

A uniaxial tensile load of 10,000 lbf (44.5 kN) is applied to the ends of a solid round bar whose diameter is one inch (25.4 mm). Determine the stresses acting on a plane oriented 70 deg from the axis of loading.

Solution   With reference to Fig. 6–1, where Θ is 20 deg, from Eqs. (6–1a) and (6–1b),

$$S_{y'} = \frac{F}{A} \cos^2 \Theta = \frac{10{,}000}{\pi/4} \cos^2 20 = 11.24 \text{ ksi (77.5 MPa)}$$

$$\tau_{x'} = \frac{F}{A} \sin \Theta \cos \Theta = \frac{10{,}000}{\pi/4} \sin 20 \cos 20 = 4.09 \text{ ksi (2.82 MPa)}$$
━━━━━

Two points are noted here:

1. These answers do not describe the complete stress state of an element oriented as indicated, since, as shown shortly, there is a different normal stress acting 90 deg from $S_{y'}$. This does point out the danger in considering that stress is simply force divided by area.

2. To determine the full state of stress, the complete stress transformation equations should be used. Alternatively, this may be found by a graphical plot called *Mohr's circle*. Traditionally, normal stresses that cause extension are called *tensile* and carry a positive sign, whereas those causing contraction carry a negative sign and are called *compressive*.

6.3 MOHR'S CIRCLE AND PRINCIPAL STRESSES

For the majority of engineering problems, the magnitudes of the three so-called principal stresses are of greatest interest. Principal stresses are normal stresses acting on planes where shear stresses are zero. For *every* applied stress state there are three and *only* three principal stresses; they are mutually perpendicular. A circle plot, due to Mohr (Ref. [1]) is one way to determine the values of the two unknown principal stresses. For a uniaxial case, the applied stress is one of the principal stresses while the remaining two are both zero; thus there would be no need to plot a circle since all three of these stresses are noted directly. Thus, our main concern here is with so-called biaxial stress states.* Figure 6–2

* For those rare "three-dimensional" situations, the roots of a cubic equation must be found. These would be the three principal stresses. See, for example, reference [2].

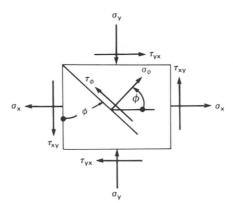

Figure 6–2 Stress element, in the physical plane, for a biaxial stress state.

illustrates such a situation, the known principal stress being zero. It is noted here that if this known normal stress were not *zero,* the stress state would still be *equivalent* to a biaxial condition in using Mohr's circle for determining the other two principal stresses.

Numerous *conventions* are used for plotting such a circle, and the one used here is the most sensible in our opinion. The convention employed is as follows:

1. Normal stresses are *plotted to scale* along the abscissa, tensile being positive and compression being negative.
2. Shear stresses are *plotted to the same scale* along the ordinate. If a shear stress would tend to cause clockwise rotation of the stress element, it is plotted as a positive value, whereas a tendency to cause counterclockwise rotation is considered negative.
3. Angles between two planes or directions on the circle plot are twice the corresponding angle on the physical plane containing the stress element. Figure 6–3 shows the correct circle plot for the general case given in Fig. 6–2, which indicates what is

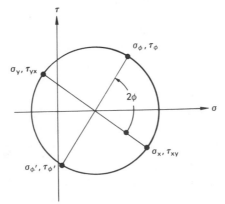

Figure 6–3 Plot of Mohr's circle for the stress state in Fig. 6–2.

meant by the physical plane. The *only* use of Mohr's circle in this text will be to define the *magnitudes* of the unknown principal stresses.

The equations that produce the values of the two unknown principal stresses and the maximum shear stress acting *in that plane* are

$$\sigma_{1,2} = \tfrac{1}{2}(\sigma_x + \sigma_y) \pm \tfrac{1}{2}\{(\sigma_x - \sigma_y)^2 + 4\tau_{xy}^2\}^{\tfrac{1}{2}} \tag{6-2}$$

note that the third principal stress, σ_3, must be known at the outset. Then

$$\tau_{max} = \tfrac{1}{2}\{(\sigma_x - \sigma_y)^2 + 4\tau_{xy}^2\}^{\tfrac{1}{2}} \tag{6-3}$$

gives the largest shear stress *in the x - y plane*. Note that the use of Eqs. (6–2) and (6–3) negates the need for using a plot of Mohr's circle if only the principal and maximum shear stresses *in the plane of concern* are required.

Example 6–2

1. A stress state is given by $\sigma_x = 10$, $\sigma_y = -5$, $\tau_{xy} = -6$ and $\sigma_z = \tau_{xz} = \tau_{zy} = 0$. Find the principal stresses from a plot of Mohr's circle.
2. A body is subjected to pure shear such that $\tau_{xy} = \tau_{yx} = 5$. Find the principal stresses. Pure shear implies that no normal stresses are applied.

Solution

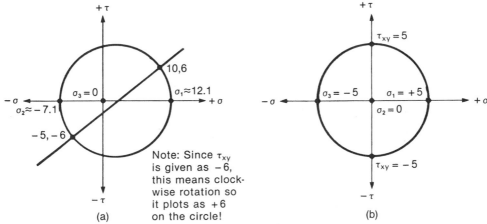

Figure E6-2

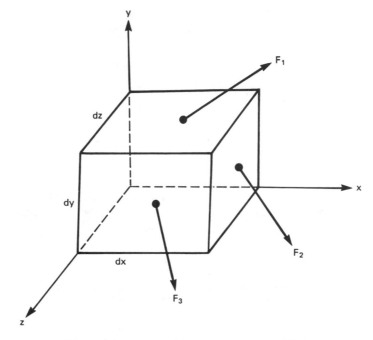

Figure 6–4   Generalized forces acting on a small body.

To portray the complete nature of stress, consider Fig. 6–4, where a number of forces act upon a small element which is under force equilibrium. All forces acting upon an individual face can be resolved into a single force; that is what is indicated in that figure. For clarity, only three such resultants are shown, but the reader should realize that equilibrating forces must act on the other three planes of the element. Now the forces F_1, F_2, and F_3 can be resolved into components acting parallel to the coordinate system $(x - y - z)$; if these components are then divided by the area of the face upon which they act, the total state of stress shown by Fig. 6–5 results. This collection of stresses is called the *stress tensor,* but in most real-life problems, and certainly in the majority discussed in later chapters, many of the stresses shown in Fig. 6–5 disappear (that is, they are zero). For this reason, further discussion of the stress tensor is unnecessary.

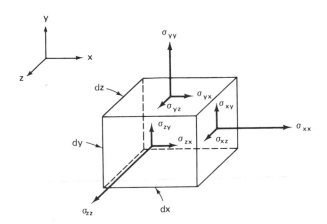

Figure 6–5 Stress element for a homogeneous state of stress. By convention, all stresses are considered positive as shown.

6.4 ALTERNATIVE DEFINITIONS OF NORMAL STRESS

Nominal (or engineering) normal stress carries the symbol S and is *defined* as

$$S \equiv F/A_0 \tag{6–4a}$$

where A_0 is the area of the body prior to the application of F.

True (or natural) normal stress carries the symbol σ and is defined as

$$\sigma \equiv F/A \tag{6–4b}$$

where A is the instantaneous or current area coincident with the load F. Shear stresses are not so distinguished. Although they are discussed in some detail later in this chpater, it is sensible here to indicate why these two definitions have been introduced. Students inevitably take courses in statics, strength of materials, and mechanical design *before* delving into manufacturing processes. Practically without exception, those earlier subjects are geared exclusively to elastic (small deformation) behavior of solids. Within such a constraint, any area *change* under a load F is so small that irrespective of whether one uses either A_0 or A, as in Eqs. (6–4a) and (6–4b), the computed values of S and σ are, for all *practical* purposes, identical. However, when large deformations of a body or workpiece are involved, as in plastic deformation, A becomes significantly less than A_0, and nominal stresses are no longer adequate to truly describe such a situation. Because *plasticity* is rarely ever discussed in traditional undergraduate courses, most students are never introduced to the concepts of true stress and true strain. In Sec. 6.6 we discuss their relationship.

6.5 STRAIN

When a solid is deformed under the application of external forces, points in the solid are displaced from their original positions or locations. Such displacements are used to *define* strain in such a way that excludes rigid body movements such as pure rotation or pure translation. For example, if a block rests upon a table and it is simply rotated or simply moved in some linear direction, none of the original dimensions has been altered; therefore, no strain has been induced in the block. If, however, one end of the block is held in place and external forces or, equivalently, stresses cause a change in any dimensions of the block, then strain has been induced. Like stress, strain is a tensor quantity, but nothing further need be said about that. The concepts of normal and shear strains are directly analogous to their stress counterparts if correct definitions are used. *Nominal* (or engineering) normal *strain,* which carries the symbol e, is defined as*

$$e \equiv \Delta \ell / \ell_o = (\ell - \ell_o)/\ell_o \qquad (6\text{--}5a)$$

where $\Delta \ell$ represents some change in length due to loading, ℓ is the length under some applied load, and ℓ_0 is the original length of concern *prior* to load application. The corresponding *true* (also called logarithmic or natural) *strain* is defined, in an incremental way, as

$$d\epsilon \equiv d\ell / \ell \qquad (6\text{--}5b)$$

which upon integrating between proper limits can be expressed as

$$\epsilon = \ell n(\ell / \ell_o) \qquad (6\text{--}5c)$$

As with stress definitions, tensile strains are considered positive, while compressive are negative. In Sec. 6–6, we show how ϵ can be expressed by parameters other than length. The use of nominal values is again appropriate where small (e.g., elastic) deformations are involved, but the use of true strain proves of greater use and convenience where large (plastic) deformations are encountered. Note that both e and ϵ are dimensionless quantities.

Example 6–3

_____

1. A bar of initial length of six inches (about 152 mm) is stretched to a length of eight inches (203 mm). Calculate the nominal and true strains induced in the direction of stretching.
2. If the eight-inch length is then further extended to twelve inches (304 mm), find the additional nominal and true strains induced during this step.
3. Compare the *sum* of both types of strain induced during the individual steps and then compare it with the strains that would have been computed if the six-inch dimension had been extended to 12 in. in a single extension. Assume that this uniform extension is possible for this problem.

Solution

1. $e_1 = 2/6$ [Eq. (6–5a)] $= 0.333$
   $\epsilon_1 = \ell n \ (8/6)$ [Eq. (6–5c)] $= 0.2877$

2. $e_2 = 4/8 = 0.500$
   $\epsilon_2 = \ell n \ (12/8) = 0.4055$

_____

* Whenever e is used as the base of natural logarithms, it will be so stated to avoid confusion. The use of exp will also clarify matters.

3. $e_1 + e_2 = 0.833$ and $e_{total} = 6/6 = 1.0$
   $\epsilon_1 + \epsilon_2 = 0.6932$ and $\epsilon_t = \ell n(12/6) = 0.6932$.

Note that summation of nominal strains is not equivalent to the total strain whereas the summation of true strains is. This indicates what is called the *additive property* of true strains. In most real-life engineering problems, four-place decimal accuracy is to be questioned, even though calculators are capable of providing this. Young as well as older engineers must realize that such accuracy is seldom warranted in practice.

6.6 USEFUL RELATIONSHIPS AND OBSERVATIONS

As mentioned earlier, the concepts of nominal stress and strain are used extensively in undergraduate courses, since elastic deformation is of major concern. In situations where plastic deformation dominates, true stress and strain are more useful. Assuming that nominal values are available, the corresponding true values can be determined quite readily. This is now shown.

Although changes in volume occur during elastic deformation of most solids, these are very small. Volume changes during plastic deformation, especially with *metals,* are negligible. What this implies is that for a fixed *mass* of metal, the *density* remains essentially constant during such deformation. Since metallic structures have very high packing factors (that is, the atoms are very closely packed), it is difficult to pack them any closer; thus, the structure retains its density. This is not true for other solids such as *polymers,* since they have structural configurations much different than metals. Because of these different conditions, what follows is related to metals subjected to plastic deformation. If volume constancy prevails (that is, the solid is practically incompressible), then

$$V_0 = A_0 \ell_0 = A\ell = V \qquad (6\text{–}6)$$

where the subscript zero refers to initial conditions under no load and the other terms refer to instantaneous values under load. Considering Eqs. (6–4a) and (6–4b) and assuming some force F, we have

$$F = SA_0 = \sigma A \quad \text{or} \quad \sigma = S(A_0/A) \qquad (6\text{–}7)$$

From Eq. (6–6), (A_0/A) equals (ℓ/ℓ_0) and with Eq. (6–5a), (ℓ/ℓ_0) equals $(1 + e)$; thus

$$\sigma = S\,(1 + e), \qquad \text{so } \sigma \text{ is always greater than } S. \qquad (6\text{–}8)$$

Now by combining Eqs. (6–5a) and (6–5c), we obtain

$$e = (\ell/\ell_0) - 1 \quad \text{or} \quad \ell/\ell_0 = 1 + e \qquad (6\text{–}9)$$

which leads to

$$\epsilon = \ell n\,(1 + e) \qquad (6\text{–}10)$$

Hence, for uniform deformation, e is always $> \epsilon$.

Thus, if nominal values of $S - e$ coordinate points are available, they can be converted to equivalent $\sigma - \epsilon$ values from Eqs. (6–8) and (6–10). Several points are worth noting here:

1. To use Eqs. (6–8) and (6–10) in a meaningful way, the nominal strain e must be *uniform*.
2. From Eq. (6–6), it is a simple matter to express Eq. (6–5c) as

$$\epsilon = \ell n(A_0/A) \qquad (6\text{–}11\text{a})$$

which for round cross sections gives

$$\epsilon = 2\ell n(D_0/D) \qquad \text{in terms of diameters.} \qquad (6\text{–}11\text{b})$$

3. If strains are small, say elastic strains where $e < 0.010$, then $e \approx \epsilon$ and $\sigma \approx S$. As an example, consider a nominal strain of 0.005; then

$$\epsilon = \ell n(1 + 0.005) = 0.004987 \approx 0.005 \qquad (6\text{–}12)$$

which is practically identical to e. The corresponding σ would only be 0.5 percent greater than S. This is illustrates why the use of true stress and strain is never introduced when elastic deformations are involved; there is no need to do so.

4. Next consider a specimen of length ℓ_0 that is doubled in length. The comparable strains are

$$e = (2\ell_0 - \ell_0)/\ell_0 = 1 \qquad (6\text{–}13)$$

and

$$\epsilon = \ell n(2\ell_0/\ell_0) = 0.693 \qquad (6\text{–}14)$$

both being tensile. What is required to induce equivalent compressive strains by decreasing ℓ_0 to a necessary level?
There,

$$e = -1 = (\ell - \ell_0)/\ell_0 \qquad (6\text{–}15)$$

so ℓ must approach zero; that is, the value of ℓ_0 must be reduced to *zero* thickness, which is physically impossible. Also, with this definition a value of e of -1.1 is impossible to attain. Now

$$\epsilon = 0.693 = \ell n(\ell/\ell_0) \qquad (6\text{–}16)$$

so ℓ must be equal to $\ell_0/2$ to provide this compressive true strain. It is noted that doubling the length or halving it (that is, the true strains are identical except for sign) does produce quite similar changes in the deformed structure of the metal as indicated by property measurements. The same is not true if equivalent nominal strains are involved (that is, $e = \pm 1.0$).

5. For volume constancy,

$$V_0 = t_0 w_0 \ell_0 = V = tw\ell \qquad (6\text{–}17)$$

where the symbols indicate thickness, width, and length, respectively, and any instantaneous volume V under load is equal to the original volume V_0. Then

$$dV = dV_0 = 0 = twd\ell + w\ell \, dt + \ell t \, dw \qquad (6\text{--}18)$$

or

$$dw/w + d\ell/\ell + dt/t = 0 \qquad (6\text{--}19)$$

so, using Eq. (6–5b) in a general way, we obtain

$$d\epsilon_1 + d\epsilon_2 + d\epsilon_3 = 0 \qquad (6\text{--}20)$$

Thus, for volume constancy, the sum of the three incremental normal true strains is zero.* This is a useful relationship in plasticity calculations, and it is noted that if nominal strains are used, the resulting expression is not so simple. If Eq. (6–20) is integrated by using proper limits, then

$$\epsilon_1 + \epsilon_2 + \epsilon_3 = 0 \qquad (6\text{--}21)$$

where these are total strains. Whenever this relation is used in plasticity analyses, the *elastic* portion of the total strain will be neglected; that is, the plastic portion is taken as being equal to the total. In large deformation processes, this is a reasonable assumption and leads to great simplification.

Example 6–4

1. A force of 2000 lb (8.96 kN) is applied to the ends of a solid rod of 0.250 in. diam. (6.35 mm); under load, the diameter reduces to 0.200 in. (5.08 mm).
Assuming uniform deformation and volume constancy,
a. Find the nominal stress and strain.
b. Find the true stress and strain.
2. If the original bar had been subjected to a true stress of 50,000 psi (345 MPa) and the diameter was 0.220 in. (5.59 mm) under that stress, what is the nominal stress and strain for these conditions?

Solution

1. a. With Eq. (6–4a), $S = F/A_0 = (2000)/(\pi/4)(1/4)^2 = 40.74$ ksi (280.9 MPa)
   With Eq. (6–5a), $e = (\ell - \ell_0)/\ell_0 = (\ell/\ell_0) - 1$ and with volume constancy, $A_0\ell_0 = A\ell$ or $\ell/\ell_0 = A_0/A$, so $e = (A_0/A) - 1 = [(0.25)^2/(0.20)^2] - 1 = 1.56 - 1 = 0.56$.
   b. Since $\sigma = F/A$, $\sigma = (2000)/(\pi/4)(0.2)^2 = 63.66$ ksi (439 MPa). Check with Eq. (6–8). $\sigma = S(1 + e) = 40.74 (1.56) = 63.6$ ksi.
   Now $\epsilon = 2\ell n \, (D_0/D)$ from Eq. (6–11), so
   $\epsilon = 2\ell n \, (0.25/0.20) = 0.446$
   Check with Eq. (6-10). $\epsilon = \ell n \, (1 + e) = 0.446$.
2. The true strain is $\epsilon = 2\ell n \, (0.25/0.22) = 0.256$, so
   $\epsilon = \ell n \, (1 + e) = 0.256$; therefore, $e = 0.292$. *Note* that
   $e^{0.256} = (1 + e)$ can cause confusion since e on the left is the *base of natural*

---

* Only normal strains relate to volume changes. Shear strains cause a shape change only.

logarithms, whereas e on the right signifies nominal strain. Using $\exp(0.256) = (1 + e)$ is far better.

Now $\sigma = S(1 + e)$, so $S = 50{,}000/1.292 = 38{,}700$ psi or 38.7 ksi.

Example 6–5

A metal specimen of 0.357 in. diameter (9.07 mm) is loaded in tension. When the applied force reaches 3000 lb (13.44 kN), elastic behavior ends; careful measurements show that at that instant, the diameter is 0.3566 in. Compare the nominal and true stresses.

Solution

$$S = (3000)/(\pi/4)(0.357)^2 = 29.97 \text{ ksi } (206.6 \text{ MPa})$$
$$\sigma = (3000)/(\pi/4)(0.356)^2 = 30.04 \text{ ksi } (207.1 \text{ MPa})$$

Note that the strain here is about 0.001 and that for all practical purposes, $S = \sigma$.

Example 6–6

A small circle of 0.500 in. diameter (12.7 mm) is printed on the face of a thin, flat sheet of steel. Under applied forces in the plane of the sheet, the circle changes into an ellipse having major and minor diameters of 0.600 and 0.520 in., respectively (call these the 1 and 2 directions). If direction 3 is normal to the surface, find ϵ_3 at this instant.

Solution

$$\epsilon_1 = 2\ell n\,(0.6/0.5) = 0.182 \text{ (positive)}$$
$$\epsilon_2 = 2\ell n(0.52/0.5) = 0.078 \text{ (positive)}$$

With volume constancy, Eq. (6–21) gives

$$\epsilon_3 = -(\epsilon_1 + \epsilon_2) = -0.260 \text{ (negative)}$$

6.7 MECHANICAL BEHAVIOR AND PROPERTIES OF SOLIDS

When any body or structure is subjected to external loads or forces, the body will elongate, shorten, distort, or undergo a combination of such shape changes; the overall effect will depend upon the type of loading. If the magnitudes of the induced stresses are less than some critical value, the deformation or behavior is called *elastic*, since upon removal of all loads, the body recovers its original shape or dimensions, much like the removal of a load from a spring.

If the induced stresses are large enough, then upon load removal, the body will display dimensions that differ from their original values. In addition a definite change in *shape* can occur. Such behavior is called *plastic,* and the body is said to have *yielded.* With many solids, elastic deformations are small, whereas plastic deformations can be much larger.

In many design considerations, plastic deformation is to be avoided, and *elastic properties* are used in calculations. However, in producing components such as an automobile fender or oil pan, it is essential to induce large plastic deformation in an initially flat sheet in order to *form* the desired part. There, the plastic properties of the material are of primary concern. What follows is a coverage of the major properties of interest; it discusses how they are determined and, in certain instances, points out some precautions that should be considered when one intends to use such properties selected from sources such as handbooks.

6.8 THE TENSILE TEST— ELASTIC PROPERTIES FOR ISOTROPIC SOLIDS*

Beyond question, this is the most widely used test in studies of mechanical behavior and, in all likelihood, most readers have been exposed to such a test. Although it is relatively straightforward, there are certain subtleties and restrictions that are often either over-looked or not pointed out to students when such tests are conducted. The measurements are generally obtained under controlled conditions, these being

1. The rate at which deformation is induced is quite low. In terms of *strain rates* (that is, the strain achieved divided by the time to achieve it), these are of the order of 10^{-3} per second. More is said about strain rate in Chap. 8.
2. Temperature of the surroundings is of the order of 25°C.
3. All measurements are restricted to a *gage section* that experiences a state of uniform, uniaxial stress during deformation.

For many materials, properties determined from a standard tensile test may be quite different if the magnitudes of strain rate and temperature are at noticeable variance with 1 and 2 above. Thus, care must be exercised if so-called *basic* property values, as usually listed in standard handbooks, are used in analyses where the actual conditions under which deformation will take place are greatly different from those under which that basic property was determined. A careful literature search will often provide more useful property values where, for example, strain rates and temperature are quite different from those listed above.

The basic information obtained from a tensile test is the magnitude of the force F required to cause a certain extension $\Delta \ell$. These findings are then converted to some type of stress-strain data; the definitions given earlier are used. Figure 6–6 is a schematic of a load-extension diagram typical of most *ductile metals*. Note that on the strain scale, which includes the total extension, the elastic region encompasses only a small portion of the full diagram. As such, it appears as a nearly *vertical* line if all data are plotted against a common scale. Figure 6–6(b) displays the elastic region where the strain scale is expanded; note that on Fig. 6–6 linear or cartesian coordinate scales are used. If the load-extension data are converted to nominal stress-strain data, by using Eqs. (6–4a) and (6–5a), the shape of the *S-e* curve is the same as the F-$\Delta \ell$ curve, since the constants A_0

---

* *Isotropy* means that a particular property is not affected by the *direction* in which the property is measured. *Homogeneity* means that a property does not vary with *location* in the body.

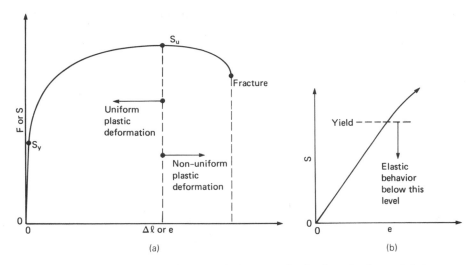

Figure 6–6 (a) Load-extension or nominal stress-strain plot for a ductile metal during a tension test, and (b) the expansion of the strain axis to clearly show the elastic region.

and ℓ_0 are used in the conversions; thus, Fig. 6–6(a) carries dual notations. Note that the load reaches some maximum or ultimate value and then proceeds to decrease.

One important elastic property is related to the initial part of the S-e curve, the slope of which is *linear* for almost all metals. This is the *elastic modulus E* (also called Young's modulus), which is

$$E = S/e \qquad (6\text{–}22)$$

This relation, which describes elastic behavior, is often called Hooke's law.* Physically, E is a measure of *stiffness* or rigidity and carries units of stress, since strain is dimensionless.

A second elastic property that can be evaluated from a tensile test is called Poisson's ratio, v. This requires the simultaneous measure of length and lateral (i.e., *across* the section) dimensional changes. Calling the loading direction 1 and the transverse direction 2 or 3, where these are perpendicular to each other, then

$$v = -e_2/e_1 = -e_3/e_1 \qquad (6\text{–}23)\dagger$$

and for *metals* the usual range of v is from about 0.2 to 0.4.

We note that a distinction is sometimes made among the elastic limit, proportional limit (neither of which will be discussed here), and the yield strength but such a distinction is unnecessary for our purposes. Sole emphasis will be placed upon the *yield strength,* which is considered to be the level of stress that coincides with the onset of plastic

---

* Strictly speaking, Hooke discussed the relation between load and extension, since stress and strain were not defined at that time; see Ref. [3] for a discussion. Later in this chapter, "Hooke's law" will be cast in the more general, three-dimensional form.

† $\epsilon_2 = \epsilon_3$ due to the assumption of isotropy, and both are compressive here. Note that since e_2 and e_3 are compressive, v is a positive number.

deformation. Technically this is not an elastic property, yet it is useful to define it here, using the symbol Y, so that

$$Y \equiv F_y/A_0 \qquad (6\text{--}24)$$

For most ductile metals, the actual onset of yielding is difficult to measure, and the simplest way to determine this property is by some definition, as is discussed shortly.

As with Y, the property called *tensile* (or ultimate) *strength*, which is certainly not an elastic property, is also included here; its symbol is S_u and it is defined as

$$S_u \equiv F_u/A_0 \qquad (6\text{--}25)$$

where the F_u implies the ultimate or maximum load experienced by the specimen. Note that S_u is the largest *nominal* stress measured during a test. What Eqs. (6–24) and (6–25) indicate are that *strength* properties of the test material are particular stress levels of importance. Do not *confuse* strength and stress. Depending upon the applied load, stress can have many values, whereas strength indicates a particular property of the material.

Attention is called to the fact that from the onset of yielding Y up to the tensile strength S_u the tensile specimen undergoes essentially *uniform* plastic deformation such that changes in area along the gage section decrease uniformly during loading.

Other plastic properties, of which little use is made in this text, are simply defined here. The *bulk modulus B* is

$$B = E/3(1 - 2\,v) \qquad (6\text{--}26)$$

while the *shear modulus G* is

$$G = E/2(1 + v) \qquad (6\text{--}27)$$

and the full derivation of these relations may be found elsewhere [2]. What is implied by Eqs. (6–26) and (6–27) is that of the four elastic *constants, E, v, B,* and *G*, only two are independent. Thus, if E and v are determined experimentally, then G and B can be calculated directly. If a solid exhibits anisotropic behavior, then these *constants* vary with direction and the simplified form of, say, Eq. (6–22) is no longer valid. Anisotropic elasticity is beyond the intent of this text, but coverage can be found elsewhere [4].

Returning to Eqs. (6–22) and (6–23), consider the case where three normal stresses are applied simultaneously to an isotropic body. Each not only induces a normal strain but also causes "Poisson strains" in the other two directions. Thus, in say the one direction, the induced strains are

$$e_1 = S_1/E, \qquad e_1 = -vS_2/E, \qquad e_1 = -vS_3/E \qquad (6\text{--}28)$$

Since the elastic equations are all linear, the concept of superposition can be applied so that the total strain in that direction is

$$e_1 = (1/E)[S_1 - v\,(S_2 + S_3)] \qquad (6\text{--}29)*$$

---

* If the three normal strains e_1, e_2, e_3 are added together, they equal zero *only* if $v = 0.5$; this would indicate constancy of volume.

where the signs of stresses [that is, positive (tensile) or negative (compression)] must be adhered to. This is the three-dimensional form of Hooke's law and includes the special case for uniaxial tension where $S_2 = S_3 = 0$. Strains in the other two directions result by altering the subscripts in Eq. (6–29).

Since $v < 0.5$ for most solids, there is a volume *increase* during tensile deformation or a *decrease* during compression. Since, however, these volume changes are very small, and since the primary interest in this book is not concerned with elasticity theory, we shall always *assume* volume constancy as a simplification.

Example 6–7

A specimen of steel, having properties of $E = 30 \times 10^6$ psi (207 GPa) and $v = 0.3$, is subjected to stresses $S_1 = 10$ ksi, $S_2 = 5$ ksi, and $S_3 = -7$ ksi. What is the strain in direction 3?

Solution   Noting that $E = 30 \times 10^3$ ksi, recast Eq. (6–29) as

$$e_3 = (1/E)\,[S_3 - v(S_1 + S_2)] =$$
$$1/(30 \times 10^3)\,\{-7 - 0.3(10 + 5)\} = -0.383 \times 10^{-3}$$

6.9 THE TENSILE TEST—
PLASTIC PROPERTIES FOR ISOTROPIC METALS

The ability of many metals to undergo a large degree of plastic deformation prior to fracture is referred to as *ductility*. To quantify this property, two commonly used measurements are developed from a tensile test; note that both are made *after* the specimen has fractured into two pieces. These properties are defined as

$$\% \text{ Reduction of area} = A_r = 100\,(A_0 - A_f)/A_0 \qquad (6\text{–}30)$$

and

$$\% \text{ Elongation} = \mathcal{E}\ell = 100\,(\ell_f - \ell_0)/\ell_0 \qquad (6\text{–}31)$$

where the subscript f refers to the *minimum* specimen area and the gage length after fracture. Figure 6–7 illustrates these points, noting that the $\mathcal{E}\ell$ is simply the nominal strain at fracture times 100. As seen on that figure, as extension is continued beyond the maximum load F_u, deformation becomes highly localized, and the phenomenon called *necking* begins. Most of the further plastic deformation, from this point until fracture, occurs in the region of the neck; as a consequence, the overall deformation between the initial gage marks becomes more and more nonuniform, so any measure of nominal strain is really a meaningless average. In situations where necking does occur, $\mathcal{E}\ell$ in Eq. (6–31) does not imply uniform extension. With regard to A_r, which is based upon the *minimum* area of the neck after fracture, this is indicative of the single *largest* deformation or strain induced in the specimen.

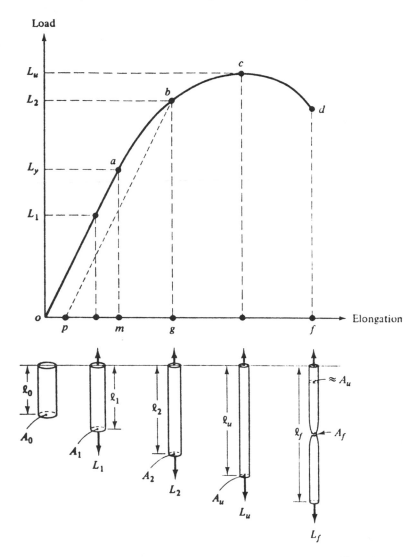

Figure 6–7 Tensile test results indicating changes in gage length at different loads. Note necking after L_u.

Although this fact is often overlooked, it should be realized that these indicators of ductility are influenced by whatever values of ℓ_0 and A_0 are chosen initially. For instance, with necking occurring, smaller values of ℓ_0 will lead to larger values of $\%\ell$ and, in a similar way, different starting values of A_0 can affect A_r. (The affect here is much less pronounced than is the effect of ℓ_0 on $\%\ell$.) Because of these possibilities, *standard* values of ℓ_0 and A_0 are used so that comparisons of these two properties, when obtained from standard specimens, will have some rational meaning. Yet, because the numerical values of both are geometry dependent, they are not truly basic in that context.

Example 6–8

Under tensile deformation, a specimen fractures when the *maximum* load is reached; the gage length was 60 mm and the area was 83.33 mm^2. Prior to loading, $\ell_0 = 50$ mm and $A_0 = 100$ mm^2. Find the true strain at fracture, ϵ_f, using both length and area changes, and explain why these results occur. Also find A_r and $\%\ell$ and show that the volume did not change during this deformation.

Solution

$$\epsilon_f = \ell n(60/50) = 0.182 \text{ and } \epsilon_f = \ell n(100/83.33) = 0.182$$

Since the deformation was *uniform* up to F_u and fracture occurred at maximum load, the fracture strains must be the same since $(\ell/\ell_0) = (A_0/A)$.

$$A_r = (100)(100 - 83.33)/100 = 16.67 \text{ percent from Eq. (6–30)}$$
$$\%\ell = (100)(60 - 50)/50 = 20 \text{ percent from Eq. (6–31)}.$$

Note that $V_f = A_f \ell_f = (1.2)\ell_0(0.833)A_0$, which agrees with the statement of uniform deformation (and constancy of volume) above.

Example 6–9

1. A tensile specimen, with $\ell_0 = 2.000$ in. and $D_0 = 0.505$ in., is pulled to fracture; necking preceded fracture. The final gage length was 2.780 in. and the *minimum* neck diameter was 0.321 in. Repeat Ex. 6–8.
2. A second value of $\ell_0 = 1.400$ in., placed on the same specimen, showed a length at fracture of 2.05 in. Find the value of $\%\ell$ for these numbers, compare with 1, and comment on the reason for any difference.

Solution

1. $\epsilon_f = \ell n(2.78/2) = 0.329$ (length changes)
    $\epsilon_f = 2\ell n(0.505/0.321) = 0.906$ (area changes)

Equivalence *does not* result here due to necking. The strain based upon length changes is a meaningless value, since the strain *along* the length is highly nonuniform. The value based upon the minimum neck diameter does indicate the *maximum* true strain induced; as such it is often more useful.

$$A_r = (100)(A_0 - A_f)/A_0 = 100\ (0.2 - 0.081)/0.2 = 59.5 \text{ percent}$$
$$\%\ell = (100)(2.78 - 2)/\ 2 = 39 \text{ percent}$$

Here, $V_f = (1.39)\ell_0(0.405)A_0 = 0.563A_0\ell_0$. This does *not* mean that volume constancy did not occur; rather, the nonuniform deformation does not permit this type of calculation as in Ex. 6–8, where deformation was uniform.

2. $\%\ell = 100(2.05 - 1.4)/1.4 = 46.4$ percent. After necking, further extension is concentrated in the region of the neck, and shorter starting gage lengths are more proportionally affected; this leads to a larger $\%\ell$.

The end of fully elastic behavior and the onset of yielding is not usually displayed by an abrupt or obvious change in the $F - \Delta\ell$ plot; rather, there exists a gradual region of transition. The most likely cause of this behavior is the variation of orientation of grains within the specimen. Due to differences in the alignment of crystallographic planes with respect to the applied load, slip occurs more readily in some grains than others.* In succession, as the load gradually increases, other less *favorably oriented* grains then slip until finally the entire structure has been plastically deformed. Beyond this point, as loading continues to increase, the entire specimen displays macroscopic uniform plastic deformation. This overall behavior is shown schematically in Fig. 6–8 (note that the elastic region is greatly exaggerated here to assist this explanation) for a specimen of aluminum and is *typical* of most ductile metals. Because of the uncertainty of exactly where yielding begins, it is common practice to *define the yield strength* by using an *offset* method. For a 0.2 percent offset, which is often used, a line is drawn parallel to the initial elastic region but offset by a *strain of 0.002* as indicated. The intersection of this line with the curve gives a stress that is taken as the yield strength (this is also called the *proof* stress), and it is obvious that this stress level will induce a small amount of permanent strain (0.002) in the specimen. A few metals, notably annealed, low-carbon steel, display a true yield point; see Fig. 6–8. In fact, one usually sees an upper A and lower B yield point for this material.† Since B is much less influenced than A by variations in loading rate, specimen alignment, and the stiffness of the testing machine, the yield strength in such cases is based upon B, and there is no need to employ an offset method. Note also that Fig. 6–8 shows plots on cartesian coordinates, and the higher elastic modulus of steel is reflected by a steeper slope.

With reference to Fig. 6–6(a) it is seen that as elongation continues beyond initial yield, an increase in load is demanded; this occurs at a continuously decreasing *rate*. As mentioned earlier, the deformation from the onset of yielding up to this maximum load is uniform. The reason for this $F - \Delta\ell$ behavior is attributed to the phenomenon called *work hardening* (or *strain hardening*). As explained by dislocation theory,‡ initial plastic deformation requires increased loading for the next increment of deformation to occur. As a consequence, plastic deformation increases the current yield strength of the specimen to such an extent that it offsets the accompanying decrease in cross-sectional area; therefore,

---

* See Chapter 7 for further discussion.

† At strains beyond B a horizontal sawtooth region follows. These small oscillations are caused by micro-neck formation that produces localized necks along the gage section; they are called Lüders bands, lines, or strains. See Chapter 9 regarding stretcher strains.

‡ See Ref. [2].

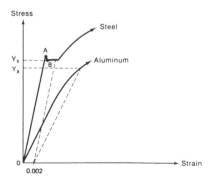

Figure 6–8 Offset method used to define yield strength.

the *load-carrying* capacity increases (that is, an increase in load is necessary to cause further elongation). Once F_u is reached, a condition of *tensile instability* occurs; note that $dF = 0$ at this point, and *necking* begins somewhere along the specimen. Although the metal in the neck *continues* to strengthen, it can no longer compensate for the faster-paced reduction of the minimum neck area, which now governs the load necessary to cause further extension. Thus, although the neck possesses the *current* highest yield strength throughout the specimen, it has the *lowest* load-carrying capacity, and, as a consequence, a drop in load results as extension continues. Testing machines are designed to apply only that load needed to cause further extension, and this is why the load drop from necking to fracture is observed. If static loads (for example, dead weights) were applied, the specimen would fracture at F_u, since the necessary load reduction would not be accomplished. Thus the major significance of the *tensile strength* S_u is that it indicates the maximum load-carrying capacity that a part of starting area A_0 can withstand.

The principal *plastic properties* determined from a tensile test are most readily found by plotting true stress versus true strain; that is, σ versus ϵ as defined by Eqs. (6–4b) and (6–5c). When plotted on cartesian coordinates along with their nominal counterparts, such results are shown schematically in Fig. 6–9.

From that figure it is seen that at initial yielding, S_y and σ_y are equivalent (see Example 6–5), so no distinction is made between the nominal and true yield strength; this is simply referred to as Y. Using the definitions from Eqs. (6–8) and (6–10) we see that it is obvious that as extension proceeds beyond Y, the $\sigma - \epsilon$ curve must be displaced above the $S - e$ curve, where, up to ultimate, for any $S - e$ combination the resulting σ is $>S$, whereas $\epsilon < e$. Beyond necking, σ continues to rise until fracture and, if careful

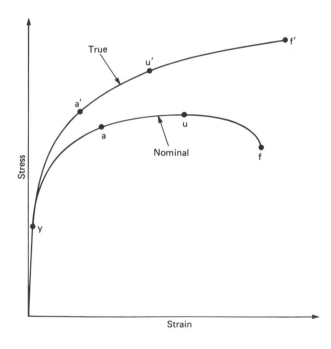

Figure 6–9 Comparison of nominal and true stress-strain curves.

measurements are made, and necking is relatively large, the values of ϵ based upon the minimum neck diameter can exceed e. For this reason, ϵ_f is shown to be larger than e_f on Fig. 6–9.

With many ductile metals that contain no effects of prior work hardening (that is, they are fully annealed) *prior* to tensile testing, the $\sigma - \epsilon$ behavior can be well described by a power-law expression,

$$\sigma = K\epsilon^n \qquad\qquad (6\text{–}32)$$

where K is called the *strength coefficient*, having the same units of stress as σ and n is the *strain hardening exponent*. Equation (6–32) is called the *strain hardening equation*. As will be shown shortly, n indicates the extent of uniform true strain a metal can withstand under uniaxial tension. Other forms of strain-hardening equations have been proposed (see, for example, Ref. [5]), but Eq. (6–32) will be used in this text. Manipulations of Eq. (6–32) as given below can be applied to other forms of $\sigma - \epsilon$ behavior.

If we assume that Eq. (6–32) is applicable, the most direct way to determine K and n is to plot the $\sigma - \epsilon$ data on logarithmic coordinates, since a straight line must result. Figures 6–10 through 6–12 show actual test results using commercially pure aluminum, a plain low-carbon steel, and an alpha brass.

To determine $K,$ each line is extrapolated to intersect a strain of unity (note that although ϵ_f is <1 in each case, we are only interested in the *equation* of the line itself) so that the value of K is equivalent to the stress corresponding with $\epsilon = 1$. By measuring the *slope* of the line, or by selecting two pairs of $\sigma - \epsilon$ values that lie *on the line* and using Eq. (6–32) with K defined numerically, the value of n is determined.

With reference to Fig. 6–10, the following are noted:

1. Up to a strain of about 0.0005 (that is, Zone 1) elastic behavior results. Here $\sigma = E\epsilon$ (or $S = Ee$) results, and if that line is extrapolated to a strain of unity, the corresponding stress level is the value of E. This provides a most practical way to determine E for solids, if, of course, Hooke's law prevails, since once the points are plotted they *must* lie on a line at 45 deg to either axis, so a best fit is easy to obtain without recourse to mathematical curve fitting.

2. For strains between 0.0005 and slightly less than 0.01 (Zone 2) the transition behavior can be seen. Some workers have fitted a straight line to such points and consider the material to show a *double n* value (that is, one at low plastic strains and another for higher levels). What seems to be overlooked here, and would be made perfectly clear if a *full* plot of $\sigma - \epsilon$ data were produced on logarithmic coordinates, is that seemingly *all* ductile metals would show such a double n value. Consider both Figs. 6–11 and 6–12 in this context where, for the plain carbon steel, the *initial* n is zero (because of the pronounced yield point). Use of double n values can cause confusion, and its practice is discouraged. In most forming operations where Eq. (6–32) finds greatest application, the strains are generally well beyond the transition region (that is, in Zone 3) and the behavior in Zone 2 is irrelevant. The use of n by the vast majority of persons who apply it in forming operations is understood to be related to the behavior in Zone 3.

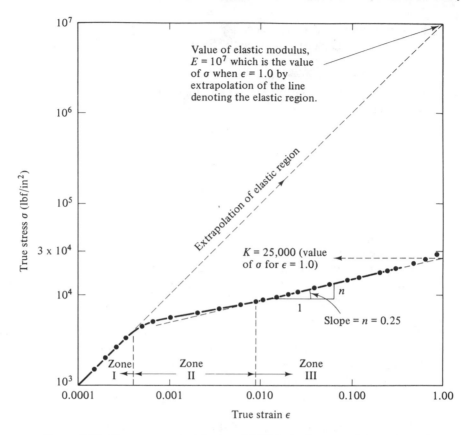

Figure 6–10   True stress-strain behavior of 1100-0 aluminum tested in tension and plotted on logarithmic coordinates.

3. Once the transition region is exceeded, the behavior can be viewed as being fully plastic, and it is those data points that should be used to define K and n; note that one should *not* give equal weight to any test points at strain levels less than those in Zone 3. We have seen instances where such an approach is overlooked and, in essence, points in Zone 2 are handled together with those in Zone 3, the result being that *incorrect* values of K and n are defined.

It is crucial, therefore, to note several points with regard to Eq. (6–32); these are

1. It is based upon true stress-true strain data beyond the initial transition region; usually this will be at strains of the order of 0.04 to 0.08.
2. To compute an original yield strength via $K(0.002)^n$ can be quite inaccurate; instead, a definition such as one based upon offset is more correct.
3. After necking, that region is subjected to a state of triaxial rather than uniaxial stress, and calculations of $\sigma = F/A$ and $\epsilon = \ell n(A_0/A)$, as based upon the minimum

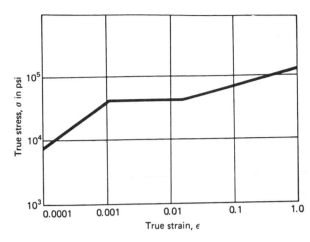

Figure 6–11   True stress-strain behavior of an annealed, low-carbon steel tested in uniaxial tension and plotted on logarithmic coordinates.

neck area, will produce points that begin to fall above the line described by Eq. (6–32). If careful corrections [6] are made, these corrected values will show quite a reasonable fit with Eq. (6–32).

4. With isotropic materials, the $\sigma - \epsilon$ data obtained from carefully conducted direct compression tests (other tests are also used) often display behavior equivalent to tensile results and can be conducted out to strains well in excess of that which coincides with necking in tensile testing. As a consequence, we do not *limit* the range of applicability of Eq. (6–32) to strains equivalent to the necking strain in tension. Indeed, the major stresses induced in many forming operations, where the use of a form such as Eq. (6–32) finds major application, are compressive and necking does not occur.

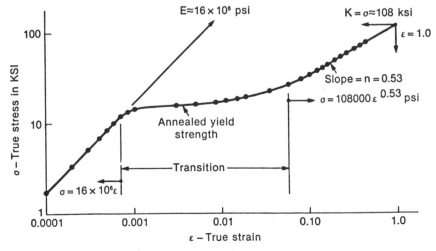

Figure 6–12   Same as Fig. 6–11 for an annealed alpha brass.

Example 6–10

Refer to Figs. 6–10 and 6–12. Using the strain hardening equations indicated, calculate σ for a strain of 0.002 in each case, and compare this with the probable yield strengths that would be found by using the offset method.

Solution   From Fig. 6–10, $\sigma = 25{,}000(0.002)^{0.25} = 5290$ psi. Note that this is slightly below the stresses in the transition region for a strain of 0.002, where an offset yield would be about 6000 psi. (This *cannot* be determined accurately from this plot.) Thus, this method for finding Y (annealed) might seem reasonable. Now, however, using Fig. 6–12, we obtain

$$\sigma = 108{,}000(0.002)^{0.53} = 4010 \text{ psi}$$

whereas the actual yield strength must be about 14,000 psi, as shown on the plot. This illustrates the error that can result if Y annealed is computed from the strain hardening equation. As indicated on both figures, K has units of psi.

Here, although *not* a property, an expression for the degree of cold work is introduced; symbolically

$$r \equiv (A_o - A)/A_o \tag{6-33}$$

where r is the cold work induced during a reduction of area from A_0 to some smaller A. If multiplied by 100 it is often called percent cold work. Note that this is *not* the same as the *property A_r* defined by Eq. (6–30). In essence, A_r is indicative of the maximum r that can be induced at fracture under *uniaxial tension*. We note that it is not always possible to define strain as a measure of r (such as in hardness testing), but whenever work hardening can be related to an area change, then the *homogeneous** strain induced will be calculated by using Eqs. (6–11) and (6–33) as,

$$\epsilon = \ell n(1/[1 - r]) \tag{6-34}$$

Thus, if a specimen is subjected to, say, 10 percent reduction of area, called *cold working* below certain temperature levels, then the *uniform* strain, from Eq. (6–34), is 0.105, since $r = 0.1$. Now, if Eq. (6–32) is defined for this material (that is, K and n have been previously determined), inserting $\epsilon = 0.105$ into the equation produces a value of σ that is equivalent to the new *yield strength* resulting from 10 percent cold working. In other words,

$$Y = \sigma = K\epsilon^n \tag{6-35}$$

provides a means for calculating the yield strength of a metal as a function of plastic strain induced due to work hardening. One major limitation to this concept is that most deformation processes do not induce uniform strains, and even in those cases where a value of r can be determined from area changes, the predictions of Y from Eq. (6–35) should be viewed as a *reasonable first approximation;* if anything, they provide a lower bound. With all of its qualifications, however, using Eq. (6–35) does at least *attempt* to include

---

* In most operations, the induced strain is not homogenous or uniform across the section; this is discussed in greater detail in Chapter 8.

the actual effects of work hardening into particular analyses rather than assuming that Y is constant as has often been done.

Example 6–11

The work (strain) hardening behavior of a metal is described by $\sigma = 690\ \epsilon^{0.2}$ (MPa). If a piece of the metal were plastically deformed in a uniform manner by 40 percent reduction of area, calculate the expected yield strength of the cold-worked metal.

Solution   Since $r = 0.4$, using Eq. (6–34) gives $\epsilon = \ell n(1/[1 - 0.4]) = 0.51$; thus, from Eq. (6–35), $Y = \sigma = 690(0.51)^{0.2} = 603$ MPa.

As discussed in relation to Eq. (6–40), the induced strain here (0.51) is much greater than n (that is, 0.2). Although a maximum uniform strain of only 0.2 (that is, n) can be induced due to *pure tension*, much larger strains can be induced by many deformation processes. Thus, the strain hardening equation is *not restricted* to use where the induced strains are limited by n or ϵ_u in a tensile test.

Two useful relationships are now developed. Recalling that $F = \sigma A$ and that at ultimate load F_u the slope of the $F - \Delta \ell$ curve goes to zero, it follows that

$$dF = \sigma\ dA + A\ d\sigma = 0 \tag{6-36}$$

or

$$d\sigma/\sigma = -dA/A \tag{6-37}$$

Considering that $-dA/A$, when integrated between limits of A_0 (lower) and A (upper), becomes $-\ell n A/A_0) = \ell n(A_0/A)$, it follows from Eq. (6–11) that $-dA/A = d\epsilon$. Thus,

$$d\sigma/\sigma = d\epsilon \quad \text{or} \quad d\sigma/d\epsilon = \sigma \tag{6-38}$$

with $\sigma = K\epsilon^n$,

$$d\sigma/d\epsilon = nK\epsilon^{n-1} = \sigma = K\epsilon^n \tag{6-39}$$

therefore,

$$\epsilon_u = n \tag{6-40}$$

because the analysis referred to the condition at ultimate load, hence ϵ_u. In words, the numerical value of the strain-hardening exponent equals the true strain at ultimate load, especially for most annealed metals. In practice, because of testing machine insensitivity it is truly impossible to decide the *exact* value of F_u (as seen in Fig. 6–6(a), the load curve is quite flat over a broad change in extension at the upper load levels) and hence the exact value of ϵ_u from which n could then be deduced. The most accurate method yet found to define n is from a plot such as that shown on Fig. 6–10.

Now the true stress at ultimate load is σ_u (*don't* call this the true tensile strength). Historically, tensile strength is defined as the largest *nominal* stress S_u, whereas the largest true stress coincides with fracture. Referring to σ_u in that way could cause confusion and is to be avoided for this reason. Now from Eq. (6–32) we have

$$\sigma_u = K\epsilon_u^n = K(n)^n \tag{6-41}$$

using Eq. (6–40). Considering the condition at ultimate load, we have

$$F_u = S_u A_0 = \sigma_u A_u = K(n)^n A_u \tag{6-42}$$

using Eq. (6–41). So

$$S_u = K(n)^n A_u / A_o \tag{6-43}$$

and from Eq. (6–11)

$$\epsilon_u = n = \ln(A_o/A_u) \quad \text{or} \quad A_u/A_o = e^{-n} \tag{6-44}$$

so that

$$S_u = K(n/e)^n \tag{6-45}$$

where the symbol e in this derivation is the base of *natural logarithms* and *not nominal* strain. The value of S_u in this equation is the value for the metal containing *no prior cold work* before tested in tension.

There is one very practical use of Eq. (6–45), and this is in regard to the comments made earlier about properly determining K and n from experimental data. Suppose a number of $\sigma - \epsilon$ values have been determined along with a measured value of S_u (this is quite accurately determined, since any errors arising from an initial measure of A_0 and the value chosen as F_u are truly minimal), and the data are plotted as in Fig. 6–10. If improper weight is given to any points in Zone 2, the resulting line would describe incorrect values of K and n. But if those values were used in Eq. (6–45) and the *calculated* value of S_u was different from the correctly measured value (say 2 percent or so), this should indicate to the experimenter that an adjustment of the initial line is needed and that Zone 2 points should be ignored.

Example 6–12

From a tensile test, the measured value of the tensile strength S_u is 28 ksi (193 MPa). When the $\sigma - \epsilon$ data are plotted on logarithmic coordinates, an experimenter fits a straight line to the large-strain data and concludes that $\sigma = 50\epsilon^{0.25}$ (where $K = 50$ has units of ksi). Comment on the appropriateness of the K and n values.

Solution   Using Eq. (6–45), we find that the *computed* value of S_u is

$$S_u = 50(0.25/e)^{0.25} = 27.54 \text{ ksi}$$

Since this is less than two percent from the measured value of 28 ksi, the values of $K = 50$ ksi and $n = 0.25$ are quite reasonable.

It is noted that cold working of metals increases the tensile strength as well as the yield strength. Details on this point can be found in Ref. [2].

6.10 MECHANICAL BEHAVIOR OF NONMETALS

6.10.1 Introduction

Although many of the basic definitions and concepts covered earlier in this chapter are applicable where ceramics or polymers are involved, there are certain differences that may be observed, and it is these that are discussed briefly here.

6.10.2 Ceramics

One of the characteristics of most ceramics is their inherent brittleness, which leads to higher compressive strength prior to fracture as compared to tensile strength. By and large, the stress-strain behavior to fracture is primarily elastic, so in applications where ductility is needed, these solids are not used. Yet the so-called traditional ceramics, and many newer ones that are man-made, do possess special properties that satisfy certain engineering applications. The making or building of chinaware, glasses, and structures such as bridges and cathedrals (which primarily support compressive loads) have utilized traditional ceramics for centuries. Applications of newer ceramics include refractory solids used in industrial furnaces, catalytic converters on cars, gas turbines, jet engines, and the like. In such high-temperature applications, ceramics retain their strength and thermal stability (that is, they do not *creep** very much) at temperature levels for which most metals and polymers would *not* be satisfactory.† It is noted that ceramics are also used as cutting tools for particular machining operations. In essence, traditional ceramics find use where the loading is primarily compressive,‡ whereas the more recent man-made types are used where the retention of strength and limited creep at elevated temperatures are needed.

6.10.3 Polymers

Some of the long-standing and traditional definitions of certain mechanical properties associated with metallic behavior have not been followed exclusively by those measuring and reporting the properties of polymers. This causes confusion and may lead to problems where designers are concerned.

One obvious difference between the behavior of ductile metals and ductile polymers is seen in a typical tensile test. Consider Fig. 6–13, which shows an annealed low-carbon steel specimen taken to fracture under tensile loading. As indicated by Eq. (6–40), the maximum *uniform* strain occurs at the ultimate load after which necking occurs. This is followed by a localized region in which most further plastic deformation occurs in the

---

* See Chapter 7 for a discussion on creep.

† Only so-called *superalloys,* which utilize, for example, cobalt, nickel, and chromium, approach the behavior of ceramics at temperatures of the order of 1000°C; however, ceramics are used at even higher temperatures.

‡ See the excellent texts by J. E. Gordon entitled *The New Science of Strong Materials—or Why You Don't Fall Through the Floor*, 1968, and *Structures—or Why Things Don't Fall Down*, 1978 (London: Penguin Books.)

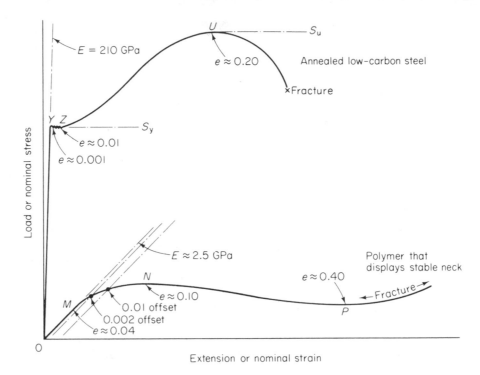

Figure 6–13 Nominal stress-strain curve for a polymer that displays stable neck propagation and an annealed low-carbon steel that displays a pronounced yield point.

vicinity of the neck. Further strain, as evidenced by a continued reduction of the diameter of the neck, follows until fracture occurs.

Compare that behavior with the specimen of high-density polyethylene (HDPE) shown in Fig. 6–13. The behavior up to the onset of necking is similar; that is, the area of the gage section decreases as the load increases. As a local neck begins to form, the area in that region continues to decrease under a decreasing load (just as with metals), but fracture does *not* occur. Instead, the neck reaches some value characteristic of the polymer, and a *stable* neck begins to propagate along the specimen towards the shoulders at the ends. This is called *cold drawing* in polymer terminology. During this stable neck growth, the *maximum strain* induced remains constant, since the minimum *area of the necked region does not essentially change during this* process. Note too that the period of stable neck growth proceeds under *constant* load. In essence the magnitudes of true stress and true strain remain fixed. This behavior is quite different from that observed with common ductile metals and probably accounts for some of the confusion found in the literature. For example, instead of considering the true strains involved, terms such as *draw ratio* (that is, ℓ/ℓ_0) or the use of nominal strain $(\ell - \ell_0)/\ell_0$ are most often used in the polymer literature. Both of these are practically meaningless, since they are a function of the initial gage length ℓ_0, which can take on numerous values. As shown by Ex. 6–9, that is why such values must be standardized. In addition, using nominal strains after necking occurs simply gives a type of average value due to the nonuniform strains occurring over the full gage length. If the definitions of the true stress and strain were used, confusion would be avoided, since, as mentioned above, those values are basically constant during stable neck propagation. It's unfortunate that this has been so misunderstood. Another misconception has also been apparent in the literature on the behavior of polymers; that has to do with what is called the *fracture stress*. We note that the necked region, probably due to alignment of backbone chains, must possess a greater *load*-carrying capacity than the regions adjacent to the neck. As the stable neck grows along the specimen, it eventually runs into the shoulders of the specimen. The much larger area of the shoulders requires the load to increase as they come into play, and eventually fracture occurs somewhere in the stable neck. To call this fracture *load* is pure nonsense. A more fundamental approach would be to produce a new test specimen from the region containing the stable neck and *continue* a test using that specimen. If measurements of true stress and strain were made for the entire test, a sounder description of such behavior would result. Figures 6–14 and 6–15 show such results for several polymers. Note that the test points at the higher strains under uniaxial tension were obtained with a specimen made from the region of stable neck growth.

Figures 6–13 and 6–16 show *nominal* stress-strain curves; the first concerns a polymer that displays stable neck growth and an annealed plain carbon steel showing

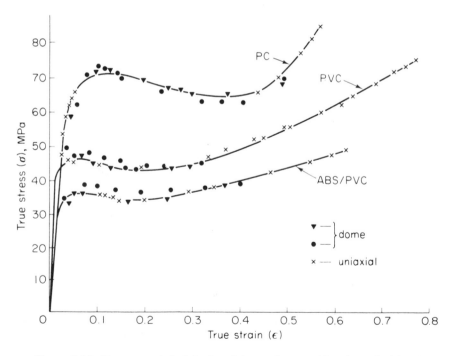

Figure 6–14 True stress-strain behavior of three polymers subjected to uniaxial and balanced biaxial tension.

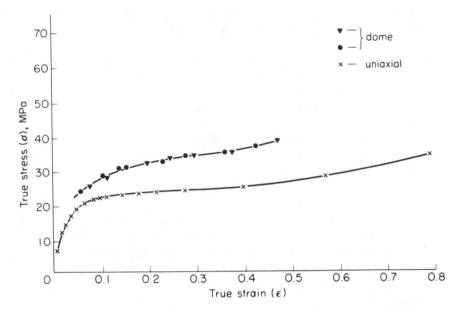

Figure 6–15 True stress-strain of high-density polyethylene subjected to uniaxial and balanced biaxial tension.

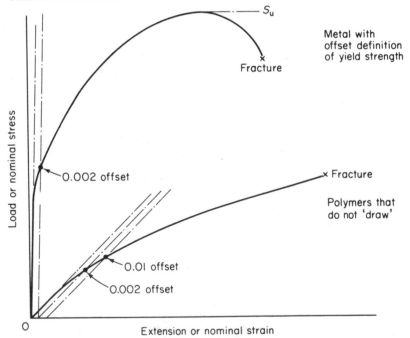

Figure 6–16 Nominal stress-strain curves for a polymer that does not display localized necking and a ductile metal that does not display a pronounced yield point.

a pronounced yield point, whereas the second involves a polymer that does *not* display a stable neck and a metal having no pronounced yield point. One reason for including these figures is to indicate why confusion often exists in defining the yield strength for polymers. Those that show stable neck growth display a load drop as necking begins, and the load associated with the start of necking is often taken as the load at yielding. The nominal stress at point N on Fig. 6–13 is called the *yield stress*. It is obvious that such a definition cannot be followed in Fig. 6–16, since no similar load drop occurs. We prefer to use an offset definition in any case; this at least provides consistency although the choice of percentage of offset is certainly arbitrary.

From these brief comments it is obvious that certain traditional methods used to define particular properties of ductile metals have not really carried over in a one-to-one correspondence for polymers. Thus, those who must select polymer properties in design or manufacturing studies should be aware of this. Reference [7] contains a more thorough discussion on this topic. Details of the structure of materials alluded to in this chapter are discussed in greater detail in Chapter 7 of this text.

REFERENCES

1. O. Mohr, Zivilingeneur, 1882, p. 113.
2. R. M. Caddell, *Deformation and Fracture of Solids* (Englewood Cliffs, N.J.: Prentice-Hall, Inc., 1980), pp. 5–9, 50–52, Chap. 7, 105, 113.
3. J. E. Gordon, *Structures* (New York and London: Plenum Press, 1978), Chapters 2 and 3.
4. N. H. Polakowski and E. J. Ripling, *Strength and Structure of Engineering Materials* (Englewood Cliffs, N.J.: Prentice-Hall, Inc., 1966), pp. 116–18.
5. W. Johnson and P. B. Mellor, *Engineering Plasticity* (New York: Van Nostrand Reinhold, 1973), pp. 15–17.
6. P. W. Bridgman, *Transactions,* American Society of Metals, **32** (1944), pp. 553–74.
7. D. K. Felbeck and A. G. Atkins, *Strength and Fracture of Engineering Solids* (Englewood Cliffs, N.J.: Prentice-Hall, Inc., 1984), pp. 276–80, 295–20.

PROBLEMS

6–1. A bar of area $A_0 = 0.2$ in.^{2} is subjected to tensile loading with the following results:

Force (lbf):   1000   2000   3000   4000
Area (in.^{2}):   0.19   0.185   0.181   0.17

Find the true and nominal stresses at each load and comment on the difference in magnitudes as F increases.

6–2. Calculate the ratio of true to engineering strains (ϵ/e) for values of e of 0.001, 0.01, 0.05, 0.10, and 0.50. Comment on the trend that is shown.

6–3. With the data in Problem 6–1, find the engineering strain e at each load, then compute σ using Eq. (6–8). If Problem 6–1 has been completed first, compare the values of σ for both approaches.

6–4. With the data in Problem 6–1 calculate e and ϵ for each load independently. Then use Eq. (6–10) to find ϵ from e and compare the results.

6–5. A metal bar of dimensions $\ell_0 = 76$ mm, $w_0 = 12.7$ mm, and $t_0 = 7.6$ mm is loaded in tension to 22 kN. Using a ruler (not the most accurate device), you eyeball the length under the load to be 89 mm, the width to be 11.9 mm, and the thickness to be 7.1 mm. Find the true strains in the three directions and comment on the accuracy of your measurements.

6–6. Find S and σ in Problem 6–5 and discuss any calculation that might be questionable.

6–7. A rectangle of dimensions one inch by 0.75 in. is printed on the surface of a metal sheet. In-plane loading causes these dimensions to change to 1.25 in. and 0.68 in. respectively. If the unloaded *thickness* of the sheet was 0.125 in., what is the thickness after the loads are applied?

6–8. A metal has an elastic modulus of 68.9 GPa and Poisson's ratio of 0.3. If elastic deformation results due to three normal stresses,
$$\sigma_1 = +20.7, \ \sigma_2 = +10.3, \text{ and } \sigma_3 = -6.89 \text{ (all in MPa)},$$
(a) Find the three normal strains.
(b) Determine if there is any volume change.

6–9. Repeat Problem 6–8 if $\nu = 0.5$ instead of 0.3.

6–10. Handbook data for a certain metal gives $\%\ell = 30$ percent and $A_r = 45$ percent. Did this metal neck when tested in tension? Explain.

6–11. The strain-hardening behavior of commercially pure aluminum is given by $\sigma = 22,000^{0.23}$. If a bar of this metal were cold-worked 20 percent, calculate the resulting yield strength. Note that K has units of psi.

6–12. During a tensile test of a brass specimen ($d_0 = 12.8$ mm) a *maximum* load of 53.4 kN is reached; at this point the area is 60 percent of the starting value. If a second identical specimen were loaded until the induced strain was $n/2$, what *load* is required to cause this condition?

6–13. A metal follows $\sigma = K\epsilon^n$ and displays a tensile strength of 300 MPa. To reach the maximum load required an elongation of 35 percent. From this information find K and n.

6–14. A tensile specimen having $\ell_0 = 2.000$ in. and $d_0 = 0.357$ in. is subjected to loading, and the following are found:
yield load = 2500 lbf.
diameter at ultimate load = 0.305 in.
diameter at fracture = 0.280 in.
elastic modulus = 15×10^6 psi
Upon completing this test you are informed that n for this material is 0.45 and in the annealed condition it obeys $\sigma = K\epsilon^n$. In addition, the specimen you started with contained some cold work *before* you pulled it in tension.
(a) How much strain had been induced *before* the specimen was tested?
(b) What maximum *load* was reached during the tensile test?

6–15. When an annealed metal is pulled to fracture, the maximum tensile strain is found to be 0.8. Suppose a specimen of this annealed metal has a $d_0 = 10.2$ mm and is pulled until the diameter is 8.9 mm (assume uniform reduction at this point). *Starting* with the 8.9-mm specimen that has been cold worked, what reduction of area at fracture would you expect?

6–16. At the end of this chapter it was stated that the *tensile strength, S_u,* as well as yield strength increases with cold working. Develop a relationship between the tensile strength due to cold working (call this S_w), the annealed tensile strength S_u and cold working r. *Hint:* Consider a general load-extension curve and think about what happens if you induce a load greater than initial yielding but less than the maximum load and then unload to zero. Now, starting in this condition, what happens if you carry the test to fracture?

7
engineering
materials _____

7.1 INTRODUCTION

The word *materials* covers a spectrum of different atomic and molecular structures which find many applications. In some cases, ready availability is the reason why certain materials are used. For example, eskimos build igloos out of ice, whereas Indians of the southwest used clay to build shelters. Fabrics are used in large quantities in the furniture and automotive industries; wood finds extensive use in the housing industry, while paint is used in cases too numerous to mention. Depending upon one's viewpoint, materials can be categorized in various ways. Solid, liquid, and gaseous are one such category whereas organic and nonorganic are another. Composites are often considered to be natural or man-made, whereas metals are often considered to be ferrous or nonferrous.

In this text, *engineering materials* are intended to include metals, polymers, and ceramics, since, by and large, it is these three categories that are most often subjected to the major *manufacturing processes* discussed in chapters that follow. With much of today's publicity being devoted to topics such as robotics, computers, and processes that find important but rather limited application, the traditional and most widely used industrial processes seem to have been relegated to a position of limited importance. Product design, inventory control, scheduling, and the like are all important aspects in the broad field of manufacturing, yet materials must be *selected* and *processed* or parts would never be produced. It is our contention that a reasonable knowledge of the interaction among the material being processed, the design, and the processing method itself should be of definite concern to manufacturing engineers. They should not be expected to be experts in materials science, but an understanding of certain broad, basic principles, and a reasonable awareness and vocabulary connected with materials will permit a better interaction with specialists, which is often necessary.

Many other texts that address manufacturing processes begin the coverage of engineering materials with sketches of unit cells, dislocation models, polymeric chain structure, Miller indices, and so forth, but little if *any* use is then made of aspects of materials science. Except for certain terms or definitions that are needed for explanatory purposes, we see no need to present a detailed review of such topics since the reader (as mentioned in the preface) is expected to have covered that material in an introductory course in materials science.

One last point pertains to the depth of coverage to be presented. Consider the topic of equilibrium diagrams of metals. Since the success of various *strengthening mechanisms* is limited by phase changes, we shall discuss a few in detail in order to demonstrate their importance. The reader can then extend these ideas to the myriad of other equilibrium diagrams without the need of specific discussion.

7.2 METALS

The basic differences among the three groups of engineering materials, which leads to their different properties, arise from the manner in which atoms are arranged and held together by bonding forces. For our purposes, metals are considered to form a *crystalline* structure.* A small number of atoms form a single *unit cell,* where the combination of the type and size of cell varies from one element to the next. Each cell attaches to adjacent cells, and this repetition in three dimensions is called the *space lattice.* The most common cell structures are the body-centered cubic (BCC), face-centered cubic (FCC), and hexagonal close-packed (HCP), and the balance of forces between positively charged nuclei and negative electrons is referred to as the *metallic bond.* One important characteristic of metallic structures is their relatively high *packing factor;* that is, the individual atoms are packed tightly together because of high bonding forces. This leads, for example, to the high density and, generally, high strength of metals among other properties. To cause metals to deform under applied loads or forces, one of two major mechanisms is required.†
They are called *slip,* where a relative sliding displacement between atoms on adjacent planes is caused by *shear* stresses, and *cleavage,* where adjacent planes of atoms are literally split due to *tensile* stresses. In either case, the applied forces must overcome the bonding forces for such results to occur.

From a crystallographic point of view, a *perfect* structure is one in which every atomic site is occupied by an atom, and the space lattice would range undisturbed from one free surface to another. Any type of interruption of such a perfect structure is deemed a *defect* by the crystallographer and would appear to be undesirable. To the engineer, however, such defects can have important positive benefits, as we will discuss shortly.

---

* Metals can be produced in an amorphous form, but the total industrial application of such metals is highly limited at the present time.

† Twinning can also occur but is not discussed here; see, e.g., R. A. Flinn and P. K. Trojan, *Engineering Materials and Their Applications* (New York: Houghton Mifflin Co., 1981), p. 64.

The major defects are called *point, line,* or *surface* (area). A point defect is called a *vacancy,* which is simply a missing atom; a line defect is called a *dislocation,* which may be a row of missing atoms, while an area defect is a *grain boundary,* where a mismatch of atoms occurs between adjacent grains. In the main, the latter two types of defects are of the greatest importance to us as they provide certain benefits. First, except for very small metallic *whiskers,* all metals contain dislocations in exceedingly large numbers. The presence of dislocations explains why real metals do not possess the theoretical shear strengths that should prevail if the structure were perfect.* Also, any aspect of the structure that acts to inhibit the *easy glide* or movement of dislocations is considered to be a *strengthening mechanism.* The only way for slip to occur in a perfect metal would be for entire planes of adjacent atoms to displace simultaneously; of course, if that were to happen, immediate fracture would most likely occur. Although extremely high stresses would be needed to cause fracture, no significant degree of *plastic* deformation would result, and those processes that require such deformation to produce a part would be basically nonexistent. With the presence of dislocations, the stress needed to cause slip is about 0.1 percent of the theoretical value based upon perfect structure analyses, but because of dislocations, plastic deformation can occur before fracture is reached. This ability of *ductile* metals permits many parts to be formed to shape by the application of appropriate forces. Thus, the crystallographic defects can be a benefit to the engineer concerned with plastic deformation.

Practically all commercially used metals are *polycrystalline;* that is, a large number of separate crystals or grains make up the overall structure. The interface between adjacent grains is called a *grain boundary.* Figure 7–1 is a photomicrograph of a steel containing 0.02 percent carbon, showing distinct grain boundaries. Why and how grains form can be explained with the schematics shown in Fig. 7–2.

Initially, solidification will begin in regions where the rate of heat transfer is greatest. If it is assumed that the molten metal has been poured into some kind of container, this rate is usually highest at the interface between the melt and the walls. Solid nuclei begin to form, much like the freezing of water in an ice-cube tray, as in Fig. 7–2(a). Solidification towards the center continues and is shown as rows of atoms in Fig. 7–2(b), where the atomic spacing is uniform. Eventually, rows growing from different directions reach a point where a mismatch or interference occurs, as shown in Fig. 7–2(c). The atom shown solid cannot meet the spacing demands of the two competing rows, so it is forced away from the equilibrium position of either and is now considered to be at a position of *higher energy level.* The continuation of this process finds a number of adjacent atoms undergoing this same process, and it is these atoms that form the grain boundary. For this reason, the final grain boundary displays a *higher* energy level as compared with the grain interiors; this is shown in Fig. 7–2(d). As will be discussed later, the average *grain size* can be altered. Again, this is considered to be a defect, even though the strength of the structure can be altered with grain size. The engineering viewpoint is to take advantage of grain size to modify the strength or ductility accordingly; this is discussed in some

---

* See R. M. Caddell, *Deformation and Fracture of Solids* (Englewood Cliffs, N.J.: Prentice-Hall Inc., 1980), Chapter 7.

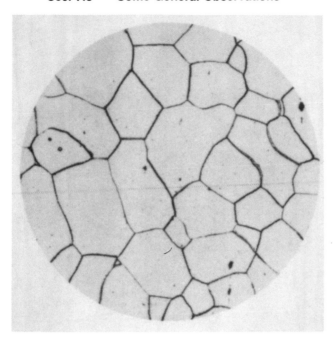

Figure 7–1 Photomicrograph of a low-carbon steel showing distinct grain boundaries.

detail later. With this brief introduction we have indicated many terms and a few basic concepts that will be used in much of this chapter.

7.3 SOME GENERAL OBSERVATIONS

The word *equilibrium* requires understanding, and we introduce a simple but useful explanation that will serve to explain various phenomena. Essentially, the concept of energy levels is involved, and we can refer to a book lying on a desk as an illustration. Since the book is several feet above the floor, it possesses potential energy; yet unless a force is applied to move the book to the edge of the desk and cause it to fall to the floor, it will remain in place. One might consider the book to be in a state of equilibrium, yet it is not at its lowest energy level with respect to the floor. To bring this about requires an input of energy, such as pushing the book to the table's edge, thereby causing it to fall; the energy supplied can be called the *activation energy* and is needed to cause a lowering of the potential energy that the book possessed prior to falling. Next, consider Fig. 7–3, which shows a marble in a trough. The datum plane shown is assumed to be at a position of lower energy level; let us consider this to be loosely the level of equilibrium. To reach this level, the marble must first be *lifted* to the top of the trough, then under the force of gravity it will fall to the reference plane. In order to cause this lowering of the energy of the marble, an input of energy must be first induced; again this is viewed as activation energy. One other observation is important here. If the marble rested in a trough that was higher from the datum, it would be further from equilibrium than in the first case. Then to drive the marble to the datum requires *less* activation energy than before. Thus, the

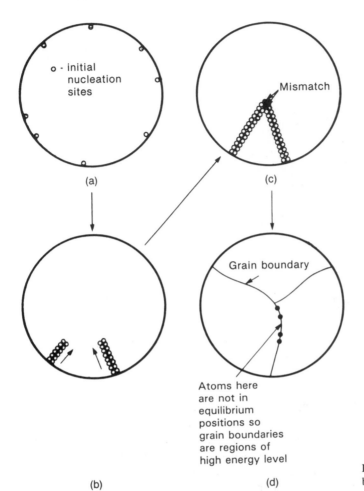

o - initial nucleation sites

(a)

Mismatch

(c)

Grain boundary

Atoms here
are not in
equilibrium
positions so
grain boundaries
are regions of
high energy level

(b)   (d)

Figure 7–2   Schematic showing how grain boundaries form.

further from equilibrium is the condition of a structure (we might say the more unstable is the condition), the lower is the activation energy needed to drive the structure towards a lower energy level. Although a rigorous analysis of this topic involves certain concepts of thermodynamics, this simple physical explanation will prove adequate for our needs.

Bulk metals are initially cast to shape; that is, molten metal is poured into a container of some sort and permitted to solidify. Subsequent operations are then used to change the initial shape into the many types that find industrial use, for example, plates, sheets, tubes, and the like.* It is important to realize that no piece of bulk metal exists in a fully unflawed condition. Small voids or holes are *always* present; sound processing techniques can usually keep such flaws to a tiny and acceptable level, *but* they cannot be

---

* See Chapters 8 and 11 for greater detail.

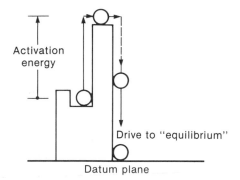

Figure 7–3 Conceptual model to illustrate activation energy and energy levels.

wholly eliminated. Note that these are in addition to the atomic level defects discussed earlier.

7.4 EQUILIBRIUM DIAGRAMS

To understand how and why the properties of metallic alloys can be altered by various *heat treatments* it is essential to have at least a minimal understanding of *equilibrium diagrams* as a starting point. What follows is not intended to cover the many details of physical metallurgy; rather, a few of the more commonly used types of alloys will be addressed to illustrate some of the *strengthening mechanisms* that can have a significant influence on the mechanical properties of these alloys. An *alloy* can be considered to be a mixture of elements that displays various levels of solubility; for example, nickel and copper will mix together in any percentage (by mass) of either element, and thus they are said to be completely soluble and can form an alloy of any composition. This is an exception rather than the general rule, as we shall see.

Certain words require definition, as they will be used extensively in discussing metals. A *phase,* which may be liquid or solid, is a homogeneous mass of matter. For particular alloys and levels of temperature, a *phase transformation* may occur. Such transformations may be from liquid to solid (such as the freezing of water) or from one solid phase to a different solid phase. Often the term *solid solution* is used; this is nothing more than a solid phase.

In general, the mechanical properties of alloys cannot be fully explained by considering the amounts of phases alone; rather it is the size, shape, and distribution of phases that are important. Here the words *microconstituent* and *microstructure* are introduced. A microconstituent may be defined as a repeatable portion of the overall structure as viewed with an *optical* microscope; it may be a single phase or may be composed of two phases that are insoluble in each other.* The microstructure is the combination of all microconstituents. Discussing a few equilibrium diagrams should make these definitions clear.

---

* This restricts considerations to features larger than the resolution of light microscopes, i.e. of the order of about 100 μm. Electron microscopes are capable of much greater magnifications, but the use of only traditional optical microscopes is implied here.

7.4.1 The Iron-Iron Carbide Diagram

Because they are one of the most widely used engineering materials, steels will be discussed first. Although considered to be alloys of iron and carbon, in practice they *always* include other elements, some of which are introduced intentionally to produce *alloy steels* whose particular properties are superior to *plain-carbon steels*. Other elements, called impurities, are controlled within allowable limits. To remove them completely is uneconomical and not really essential.

  Ignoring the minor effects of impurity elements, the iron-iron carbide (or simply the iron-carbon) diagram shown in Fig. 7–4 can be used effectively to discuss the various equilibrium structures of *plain-carbon steels*. Figure 7–5 is an enlarged view of the left-hand end of the full diagram; since few steels contain more than 1 percent carbon, this figure will receive major attention. To simplify this discussion we have rounded off

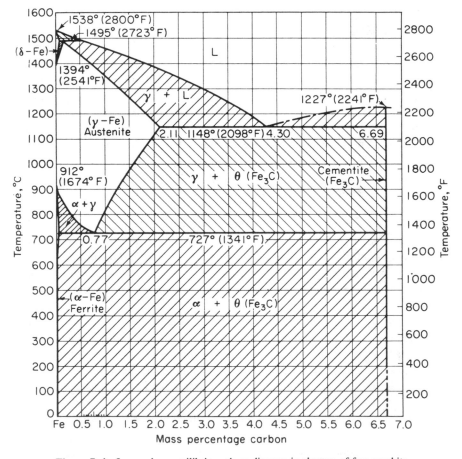

Figure 7–4 Iron-carbon equilibrium phase diagram in absence of free graphite.

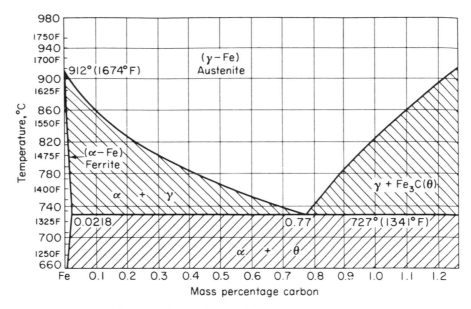

Figure 7–5   Iron-carbon equilibrium phase diagram to 1.2 percent carbon.

certain compositions and temperatures. For example, what is called the *eutectoid* is indicated as having a composition of 0.8 percent carbon, whereas the most recent measurements indicate 0.77 percent. The approximations we make have no real affect on the conclusions we shall draw. It is important to remember that Figs. 7–4 and 7–5 are equilibrium phase diagrams, and the various microstructures that result are due to a cooling rate that approximates equilibrium cooling. In practice this does not mean an infinitely slow cooling rate. Of the commonly available means of cooling, *furnace* cooling is the slowest and, therefore, closest to equilibrium. There, the part is heated to an appropriate temperature and held for an adequate time period to produce a homogeneous structure (that is, all grains are identical as to chemical composition); then the furnace is turned off. With well-insulated furnaces it may take up to half a day before the inside reaches room temperature. *Air* cooling is the next slowest method used; there, after homogenization, the part is removed from the furnace and simply cooled by the surrounding air. In many cases, but not in all, removing the heated part and plunging it into special *quenching oils* rather than air will also produce the phases shown on the diagrams. Thus, as long as such phases result, the cooling rate is considered *close* enough to equilibrium that the diagram can be used for making *reasonable* predictions of the final structure. Now there are three distinct *phases* indicated on these diagrams. Their symbols and other comments are given below.

1. Austenite, γ, is an elevated temperature phase. It is *extremely* soft and cannot exist below the *lower critical* temperature of 727°C (1341°F) if true equilibrium cooling

prevailed.* Note that the carbon content of γ can vary over a wide range, with the maximum being about 2 percent.

2. Ferrite, α, results when γ of *less* than 0.8 percent C transforms as the temperature is lowered. The γ is a face-centered cubic structure (FCC), whereas the α is a body-centered cubic (BCC). Although harder than γ, the α is also soft, having a Brinell Hardness Number (BHN) of 80 at room temperature.† Note the α is almost pure iron, containing a maximum amount of carbon of a little over 0.02 percent at 1341°F (727°C) and decreasing to practically zero at room temperature.

3. Cementite, Θ, results when γ of *more* than 0.8 percent C transforms as the temperature is lowered. It is iron carbide, Fe_3C, and has a fixed composition of about 6.7 percent carbon. The unit cell of Θ is more complex than either α or γ (actually orthorhombic) and, unlike the solid structure of those two, is an *intermetallic compound*. One characteristic of such structures is their extreme hardness as compared with solid solutions; the BHN of Θ is of the order of 1000. Because of this high hardness, such compounds are quite brittle; however, microstructures containing free Θ display good wear resistance and are most effective when the Θ exists as relatively small particles dispersed uniformly throughout the microstructure.

To this point, equilibrium phases have been discussed. Now consider what results if a steel containing 0.2 percent C is heated to 1650°F (899°C), homogenized, and then *furnace* cooled (FC) to room temperature. The resulting microstructure is shown in Fig. 7–6. The light grains are ferrite, and those that show a dark, lamellar pattern (alternating plates of α and Θ) are called *pearlite* (P). Although the phases are α and Θ, the *microconstituents* are α and P. Pearlite forms whenever γ of 0.8 percent carbon transforms at the lower critical temperature (727°C), so the average carbon content of P is also 0.8 percent. To estimate the relative *amounts* of α and P, the *inverse lever rule* can be used. The first step is to locate the fulcrum point of the lever. Here, it is the average carbon content of 0.2 percent, as seen on Fig. 7–5. Since our interest is with microconstituents and not phases per se, the length of the lever of concern is determined from the compositions of the microconstituents at the temperature of transformation (1341°F). To *simplify* this type of calculation (little error results) consider the α to have zero carbon while the P has 0.8 percent carbon; thus, the *length* of the line of concern is 0.8 unit, and this is to be divided into two portions to give the relative amounts of α and P. Physically, the fulcrum point is much closer to the α composition than that of P, so a greater amount of α results in the end structure. The distance from the fulcrum (0.2 percent C) to the two ends of the line (from 0 to 0.8 percent C) gives ratios of ¼ and ¾. Therefore, the final structure contains 75 percent α and 25 percent P. Note that to arrive at the 75 percent value for α, the distance from the fulcrum to the P composition must be used and vice versa to arrive at 25 percent P. This demonstrates the *inverse* nature of the lever rule.

---

* This γ is for plain-carbon steels. *Austenitic stainless steels* are substantially harder than this type of austenite.

† BHN is one of various hardness designations used.

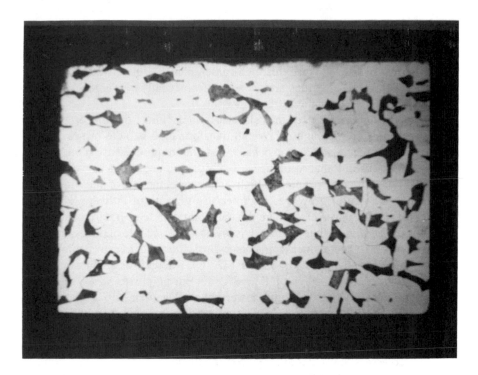

Figure 7–6 Photomicrograph of 1020 steel furnace cooled from the austenite region.

If oil quenching rather than furnace cooling had been used and if equilibrium phases resulted, a structure of about 75 percent α and 25 percent P would again result. However, the properties of the FC and OQ specimens would differ; this is readily indicated via hardness tests. How can this be when we have not only the same amounts of the two phases but also the same microconstituents? The answer is we do *not* have the same microconstituents in the grains of P. Careful observation of such grains made at the same magnification would indicate that the platelets of Θ in the furnace-cooled specimen are *thicker* and *less numerous* than those in the OQ specimen. In these cases the platelets of Θ in the FC specimen are about 10 times thicker than those in the OQ specimen. Because the *size* and *distribution* of the Θ phase differs, the pearlite in these specimens is considered to be a different microconstituent. With furnace cooling it will be called coarse pearlite, P_c, while the oil quench produces fine pearlite, P_f. Note that if air cooling had been used, the resulting structure is called medium pearlite, P_m, since the thickness and number of platelets of Θ would be between the others; this occurs because air cooling is faster than FC but slower than OQ.

To explain in a simplified manner why these differences occur, the terms *nucleation, growth,* and *diffusion* must be introduced and defined. Recall that for pearlite to form we must start with grains of γ that contain 0.8 percent C and that these are in the solid state. At the transformation temperature a group of iron and carbon atoms must find themselves positioned to form a unit cell of Θ. This will occur most likely at γ grain-boundaries, since they are at higher energy levels than the grain interiors and the structure desires a lower energy level. Thus the nucleation of Θ begins in those regions. Since Θ has a fixed composition, 6.7 percent carbon, atoms of carbon must diffuse from adjacent regions in order to produce this high carbon phase from γ, which contained 0.8 percent carbon. Thus those regions are depleted of practically all carbon, thereby becoming α. Figure 7–7 is a schematic of this process.

Diffusion is rapid at this elevated temperature, and the original nuclei begin a growth process. Although the overall transformation is quite fast (a matter of seconds), this *pearlite reaction* is *time dependent* and does require diffusion of carbon atoms. Now, the differences in P that result because of different cooling rates can be explained. When such a transformation occurs at a high temperature, the rate of nucleation is relatively low, whereas the rate of growth is high because the rate of diffusion is greater at elevated temperatures. Conversely, where the transformation temperature is lower, the nucleation rate is greater *but* the rate of growth is lower, since the rate of diffusion decreases at lower temperatures. Careful measurements indicate that the transformation of γ for furnace cooling occurs slightly below 1341°F (727°C), whereas for oil quenching it takes place at lower temperatures, say, 1000°F (538°C) for our purposes. So we should expect that the P for furnace cooling will have fewer but thicker plates of Θ than for oil quenching. Remember that *neither* of these are true equilibrium cooling and oil quenching is further from that condition than is furnace cooling. So, although both rates lead to *equilibrium* phases, the end microstructure differs.

To assess the actual difference in the harness of the *types* of P, three specimens of *eutectoid* composition (0.8 percent C) could be heated to, say, 1400°F (760°C). By using furnace, air, and oil quenching on one of each specimen, we would end up with pieces having 100 percent coarse, medium, and fine P respectively. The BHN of each have been found to be 240 (P_c), 280 (P_m), and 380 (P_f); these are reliable average values. Note that P_f is almost 60 percent harder than P_c; this could never be explained on the basis of phases alone and is an obvious example of why the use of microconstituents is more practical for explaining differences in properties. *Caution:* Oil quenching eutectoid steel of small size or cross section can produce a substantial amount of martensite (see Sec. 7.4.2). This

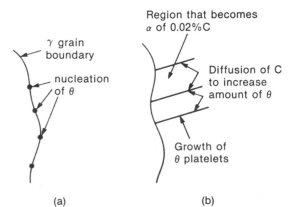

Region that becomes
α of 0.02%C

γ grain
boundary

nucleation
of θ

Diffusion of C
to increase
amount of θ

Growth of
θ platelets

(a)   (b)

Figure 7–7 The pearlite reaction illustrated schematically.

makes welding of high carbon steel pressure vessels difficult and that is why the carbon content in boiler plate is kept so low by most state codes.

The approximate hardnesses of the FC and OQ pieces of 0.2 percent C (this is a *1020* steel using the AISI designation) can be computed by using a rule-of-mixtures type of calculation. The relative fraction of each microconstituent is multiplied by its BHN; then they are added. For the FC specimen, we had $\frac{3}{4}$ α (80 BHN) and $\frac{1}{4}P_c$ (240 BHN), so the overall hardness is $\frac{3}{4}(80) + \frac{1}{4}(240) = 120$ BHN, while the OQ would give $\frac{3}{4}(80) + \frac{1}{4}(380) = 155$ BHN.

A useful relationship between the BHN and the *tensile* strength S_u of many *steels* is

$$S_u = 500 \text{ BHN, psi} \quad \text{or} \quad S_u = 3.45 \text{ BHN, MP}_a$$

Based upon the predicted hardnesses, S_u for the FC 1020 steel should be 60,000 psi (414 MPa), while that for the OQ is 77,500 psi (534 MPa). This is just about exactly what would be measured from a tensile test. Three points are worth noting here.

1. The method used above could be questioned on purely metallurgical grounds and may be due to a type of averaging affect in these polycrystalline structures. We note a similarity in connection with the elastic modulus of iron or steels. Tests on single crystals of iron indicate that the elastic modulus varies between 18 and 42×10^6 psi (124 to 289 GPa) depending upon the direction of loading. An average of 30×10^6 psi (207 GPa) results when these extreme values are used. Since bulk metals are almost always polycrystalline, it would seem that an averaging must occur, since these values typify the modulus of those metals.

2. The important point to see with such calculations is that it is the microconstituents and *not* the phases per se that govern properties. Our real concern is not to expect exact correlation between prediction and measurement with such a technique. Rather its use is to indicate the reason behind the noticeably different hardnesses and strengths of the two different structures that contain the same phases.

3. We know of no other alloy systems where the above averaging approach gives adequate predictions. Perhaps it is fortuitous that even one such system provides the means to indicate the importance of microconstituents as compared with phases.

For all plain-carbon steels up to say 1 percent carbon, the weighted averaging of the amounts of microconstituents and their appropriate hardnesses will give similar, reasonable predictions.* With low alloy, superalloys, and stainless steels, no such predictions should be attempted, *yet* the general affect on properties as a function of microstructure is still observed.†

---

* For steels with carbon less than 0.1 percent, lamellar pearlite may not form, so this averaging technique becomes questionable.

† In these metals, the alloying elements can have a great influence on the hardness of α, and the BHN of 80 is no longer appropriate. Also, the relative amounts of each microconstituent are not readily calculable.

Example 7–1

Estimate the tensile strength of a 10-mm-diameter rod of 1040 steel that is air-cooled from the γ region.

Solution   The structure would be about 50 percent α and 50 percent P_m, so the BHN would be (½)80 + (½) 280 = 180, or S_u = 180 (500) = 90,000 psi.

7.4.2 Nonequilibrium Microconstituents

Compared with practically all other commercial alloys, steels possess a unique ability. If they are quenched at a *fast* enough rate, a nonequilibrium microconstituent called *martensite** can form upon the transformation of γ. An entire text could be written by experts in the many subtleties of martensite, M. For our purposes it is sufficient to cover only the major aspects of the formation of M and its importance in relation to mechanical properties.

Martensite is a body-centered tetragonal (BCT) structure of iron that is supersaturated with carbon; that is, it contains much more carbon than equilibrium cooling would allow. It is useful to realize that if all of the excess carbon were removed (this is discussed shortly), the BCT structure would revert to the BCC structure of α. There is no single hardness of M and for plain-carbon and low-alloy steels, *maximum* hardness is solely a function of the carbon content. Figure 7–8 shows a relationship of hardness as a function of percentage of carbon. We emphasize that this plot shows the *maximum* hardness attainable; in practical situations, lower values result, and the shaded region illustrates this. Note that the divergence from maximum becomes greater as the carbon content is decreased. When M is first produced, it is called *primary* or *as-quenched* martensite and is often defined as a diffusionless, shear-induced transformation. Unlike the P transformation that is time-dependent (since diffusion is needed), the M transformation occurs in

---

* Martensitic transformations do occur in other alloy systems, such as Cu-Mn and In-Th, but such alloys do not find *widespread* industrial use.

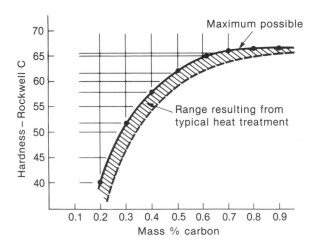

Figure 7–8   Hardness of martensite as a function of carbon content.

about 10^{-7} seconds, which is considered to be instantaneous. Although much harder than α and P, M is seldom used in the primary condition. The reason for this has to do with residual stresses that develop across the section of the quenched piece. The surface is the first region to transform from γ to M, and this requires a volume *expansion* of the γ; the center of the piece is still γ. As the center transforms to M, α, or P (depending upon the carbon content), a volume expansion of γ must occur. This causes one of two consequences, since the surface M is strained in tension due to the volume expansion of the center. First, the surface may crack due to the tensile strains induced, and cracked pieces are not too useful. If the piece does not crack, the surface is left in tension. Now as the center cools, say, to room temperature, it undergoes thermal contraction, thereby forcing the surface to also contract. This complex sequence of events locks in strains; the net result produces *residual* stresses which can reach a fairly high level. The expected tensile strength, based upon the hardness of the surface (via S_u = 500 BHN psi) would not be reached in a tensile test; rather a lower brittle fracture stress would be measured. Figure 7–9 illustrates this, noting that the applied tensile stress is added on to residual tensile stresses. Thus the measured tensile *strength* would fall below that which would result *if* no residual tensile stresses existed. To overcome this problem, the quenched part should, at the very least, be given a thermal stress-relief where diffusion effects tend to remove the locked-in strains. Heating to 200–300°F (93–149°C) for one hour per inch (25 mm) of cross section is usually adequate. Although there is little if any change in the hardness of the piece, the removal of the residual stresses leads to an increase in strength.* In fact, the relation between tensile strength and BHN used earlier is now appropriate. In the as-quenched condition, however, large deviations between measured hardness and predicted S_u can occur. Figure 7–10 illustrates this point. Even stress-relieved martensite (M_{sr}) has certain limitations in applications. Although it possesses high strength and wear resistance, its ductility and *impact resistance* are minimal. For that reason, it is usually heated

* Hardness involves plastic flow under high hydrostatic compressive stress. This inhibits fracture of primary M. Such is not the case under tension, as shown by the different values of S_u.

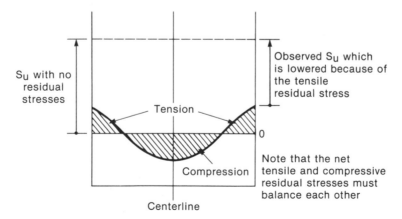

Figure 7–9   Illustration of a residual stress distribution resulting in primary martensite.

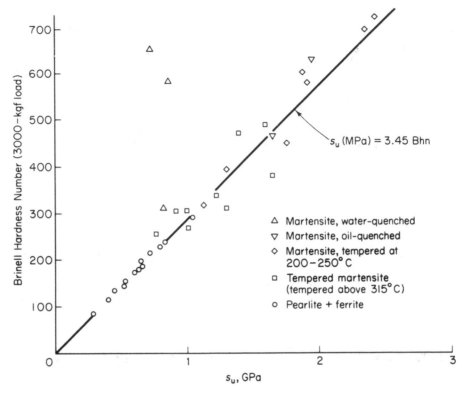

Figure 7–10   Relation between BHN and tensile strength for a number of steels. Note that the water-quenched steels, giving primary martensite, do not follow the usual trend.

(that is, *tempered*) to form what is called tempered martensite (M_T). There is no way to give a single simple description of the structure or hardness of M_T, since the end result is a function of both the temperature used and the time at which the piece is held at that temperature. Figure 7–11 shows how the hardness of primary M varies with temperature for some selected steels; note that the time is constant on that plot. Perhaps the most important point to understand is that the ability to alter the properties of M by tempering allows a trade-off of strength for an improvement of ductility and impact resistance thereby permitting a type of optimization of overall properties for specific applications. Figure 7–12 illustrates this feature in that one cannot attain maximum strength, ductility, impact resistance, and so on at the same time; some sacrifices must be made.

Example 7–2

Consider a ¾-inch round of 1050 steel that is water-quenched from just above 727°C. Estimate the probable hardness of the quenched piece.

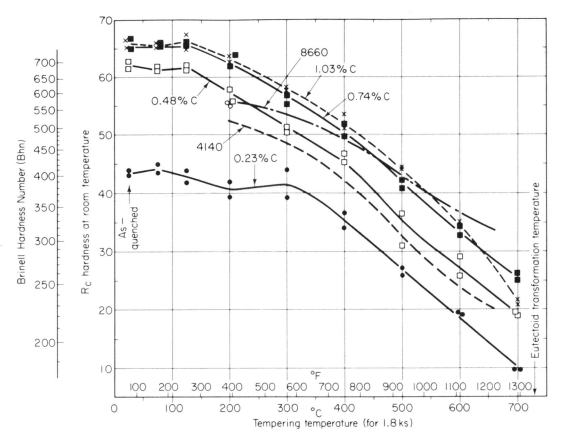

Figure 7–11 Hardness at 20°C as a function of tempering temperature for a number of steels, all tempered for 1.8 ks.

Solution   Since the structure before quenching is about ⅜ α and ⅝ γ, the structure after quenching will be ⅜ α and ⅝ M of 0.8 carbon. From Fig. 7–8, the M will have a hardness of about 65 R$_c$ or about 700 BHN. So the approximate hardness will be (⅜)80 + (⅝)700, or 467 BHN.

Example 7–3

If the quenched piece in Example 7–2 were heated to 400°C for 1.8 ks, then cooled to room temperature, estimate the tensile strength of the piece after this tempering operation.

Solution   Since only the martensitic portion of the structure will temper, use Fig. 7–11 to estimate the hardness of that portion as about 500 BHN. Note the plots for 0.74 and 1.03 percent carbon. Thus the BHN of the tempered structure is (⅜) 80 + (⅝) 500 = 342, so S_u = 500(342) = 171 ksi.

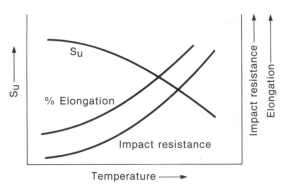

Figure 7–12   Schematic illustration of the effect of tempering temperature on certain mechanical properties.

7.4.3 Hardenability

An important property of many steels, both in understanding some of the rudiments of heat treating as well as in the process of welding (see Chapter 10), is called *hardenability*. We define this as the ease with which martensite forms to depth across a section. There is no quantitative method used to assess this property in the sense of saying that one steel possesses twice the hardenability of another. Instead, we say that one steel has better hardenability compared with another. The reader should *not* confuse hardness and hardenability. As discussed earlier, the hardness of M is basically a function of carbon content. Consider Fig. 7–13, where pieces of 1040 (plain-carbon) and 4340 (low-alloy) steels* have been quenched at rates fast enough to form M at the surface. The pieces have been sectioned and a hardness traverse made from surface to center. Since the carbon content is identical, both pieces show the same hardness at the surface, but a rapid decrease is seen with the 1040 steel as one moves towards the center. No such rapid drop-off occurs with the 4340 steel, so this steel has better hardenability; that is, it fully hardens at a slower cooling rate. As section sizes become larger, a requirement to form M across the section demands the use of a steel possessing good hardenability. This property is influenced by both higher carbon content and larger γ grain size prior to quenching, but the most effective means for improving hardenability is the addition of alloying elements such as molybdenum, chro-

* The numbers refer to the AISI designation for steels.

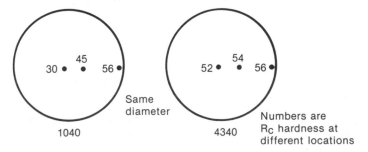

Figure 7–13   Surface to center hardness variation of 1040 and 4340 steeels to show the superior hardenability of the low-alloy steel.

mium, nickel, manganese, and boron. A useful concept to keep in mind can be illustrated by considering a eutectoid steel. If γ is to transform, it must choose between P and M, and any action which diminishes the chance of forming P will enhance the formation of M. Since P requires diffusion of carbon (M does not), the presence of alloying elements must retard such diffusion, thereby improving hardenability.

Two other points are worth mentioning.

1. The concept of hardenability pertains *only* if an alloy has the capability of forming M. This term is never used with the many commercial alloys that cannot form martensite.
2. With metals such as plain-carbon and low-alloy steels, both the section size of the part and the rate at which it is cooled from γ will influence the depth at which M can form. Because plain-carbon steels have poor hardenability, it is almost always essential to use water quenching to form any reasonable depth of M, but even with this fast quenching rate, it is difficult to form a significant amount of M with such steels if the carbon content is low (say less than 0.3 percent). The better hardenability of low-alloy steels permits them to be cooled at slower rates, and still form M to reasonable depth; oil quenching is often used. It is always sensible to use as slow a cooling rate as possible to form M, since that reduces any tendency to cause cracking. For that reason, water quenching low-alloy steels is usually avoided if at all possible.

In closing the discussion about hardenability, consider two additional figures. Figure 7–14(a) is a so-called time-temperature-transformation (TTT) curve (or IT curve, which is short for isothermal transformation) for a plain-carbon steel of eutectoid composition. The diagram is obtained in the following way. Consider a series of identical specimens of eutectoid composition, all homogenized at 1450°F (788°C); they all are composed of grains of γ containing 0.8 percent C. If the temperature of one specimen were suddenly reduced to 1200°F (649°C), held at that temperature for a time period denoted by point 1, and then rapidly quenched in water, a structure of martensite, M, would result. In essence, since M can only form from γ, there was *no transformation* of γ during time interval 1 at 1200°F. By repeating this procedure with longer and longer time intervals, we eventually determine the time at this temperature where a small amount of γ transforms to pearlite, P, before water quenching; the final structure in this case would be M plus a small amount of P. (Again recall that M forms from γ, and if some of the γ→P, we can't end up with 100 percent M.) Thus point 2 indicates the time needed to cause the *start* of transformation of γ at 1200°F. Continuing this procedure, we then find the time to produce complete transformation of γ→P at 1200°F; this is designated by point 3. Note that between points 2 and 3 the structure after water quenching would contain differing amounts of P and M; ratios of these two microconstituents would depend upon the amount of γ→P *prior* to water quenching, and the degree of γ→P itself depends upon the time interval. Repeating the procedure for a number of temperature levels produces points comparable to 1 and 3 with the 1200°F temperature; in essence, the time intervals to begin the γ→P transformation are determined. Connection of all points for

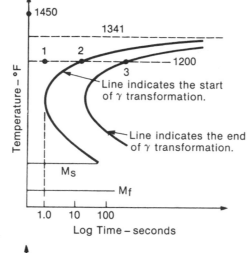

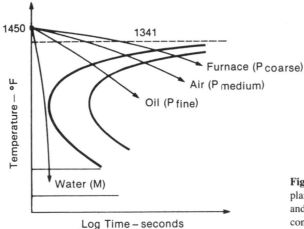

Figure 7–14 (a) T-T-T curve for a plain-carbon steel of eutectoid composition and (b) superposition of curves of different constant cooling rates on the curve in (a).

each of these conditions produces the boundary lines shown on Fig. 7–14(a). To complete such a diagram, it is necessary to determine the temperatures at which γ begins to transform to M and at which that transformation is completed; these are shown as the M_s and M_f lines on that same figure.

Of course, most practical heat treatments do not involve transformation of γ under isothermal conditions. Instead, parts are quenched in media such as water, oil, or air, or in a furnace, all of which produce continuous cooling transformations, or CT for short. To illustrate this point, and to avoid undue clutter on a single figure, Fig. 7–14(a) is reproduced as Fig. 7–14(b), and curves of *constant* cooling rates are superimposed to reflect the approximate rates of the four widely used quench media. We *emphasize* that this approach is taken to illustrate concepts and is not intended to provide wholly accurate predictions of the final microstructures. With water quenching, the cooling rate is considered to be high enough that no γ transforms until the M_s temperature is reached (that is, the *nose* of the curve has been missed, so no P forms). At the other extreme, furnace cooling (FC), all of the γ transforms at a high temperature and coarse P results. For the intermediate cooling rates, medium P results with air cooling, whereas fine P is obtained with oil quenching. The following comments are important here:

1. Since Fig. 7–14(a) was the result of isothermal transformations, it is not technically correct to superimpose cooling curves on such a diagram and expect to make accurate predictions of the resulting microstructures. In Sections 7.4.1 and 7.4.2 we have alluded to the influence of cooling rate on the resulting microstructure, and with Fig. 7–14(b) an attempt is made to tie in the effect of cooling rate on the structures likely to result.

2. Both the carbon content and the percentage and type of alloying elements lead to IT diagrams that look quite different from Fig. 7–14(a).* In addition, the size of the piece and the quenching medium used can lead to variations in microstructure from surface to center, since the cooling rates at these two regions can be quite different. Thus the use of appropriate CT diagrams and associated data are far more reliable in predicting resulting microstructures than one would expect by using figures such as Fig. 7–14(b).† Once again, the technical details and decisions related to hardenability considerations are best left to the specialists in metallurgy and not to the typical manufacturing engineer.

3. The most widely used manner for presenting hardenability data directly is with the use of *Jominy* curves. Such results are obtained by using the so-called Jominy test.‡ Figure 7–15 compares typical findings for 1040 and 4340 steels. Note that the maximum hardnesses, obtained at a high cooling rate, are the same, since they both possess the same carbon content. However, due to the presence of alloying elements, the 4340 steel *maintains* a much higher hardness as the cooling rate is decreased. In essence, thicker sections of 4340 could be hardened, so its hardenability is superior to the 1040.

7.4.4 The Copper-Aluminum System

The reasons for discussing these alloys in some detail are twofold. First, they find widespread commercial use, and second, they involve the important ability to undergo *precipitation hardening* (also called *age hardening*). Figure 7–16 shows the portion of the equilibrium diagram of interest. Other than the liquid phase, two equilibrium phases are indicated, these being an α solid solution and an intermetallic compound Θ. Note that these *symbols* were used with the diagram in Fig. 7–4, but they are not to be confused with

---

* See, for example, *Isothermal Transformation Diagrams*, 3rd ed. (United States Steel, Pittsburgh, Pa., 1963).

† See, for example, M. Atkins, *Atlas of Continuous Transformation Diagrams for Engineering Steels* (British Steel Corporation, Sheffield, England, 1980).

‡ See, for example, D. K. Felbeck and A. G. Atkins, *Strength and Fracture of Engineering Solids* (Englewood Cliffs, N.J.: Prentice-Hall, Inc., 1984), pp. 226–28.

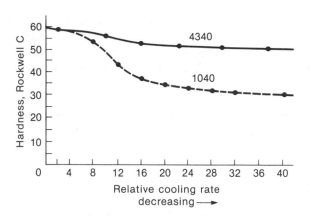

Figure 7–15 Typical Jominy curves for 1040 and 4340 steels.

ferrite and cementite. Here, the compound Θ is copper aluminide ($CuAl_2$), while the α is a solid solution of copper and aluminum where the maximum solubility of copper in aluminum is 5.65 percent.

First consider an alloy containing 4 percent copper that is heated to about 500°C, as shown by the dashed line on Fig. 7–16. When it is homogenized—that is, when Cu is uniformly dispersed in Al—we have a solid solution containing 96 percent Al and 4 percent Cu. Upon slow cooling, say by shutting off a furnace, the phase structure remains constant until a temperature just below point A is reached. From there down to about 200°C the diagram indicates that there are now two phases (α and Θ) that exist and that the amount of Θ increases as the temperature decreases. The phase boundary line beginning at 548°C and showing 5.65 percent copper drops downward to the left to about 05 percent copper at 200°C. This is an indication of *decreasing solid solubility* with tem-

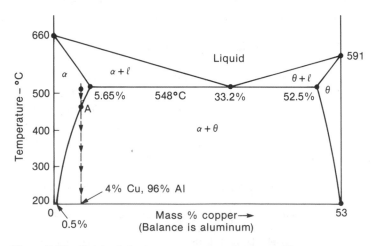

Figure 7–16 Sketch of aluminum-copper phase diagram to 53 percent copper.

perature; that is, the amount of copper that can be retained in solution with aluminum decreases with temperature under equilibrium cooling. As the copper content of the α changes from 4 percent at about 450°C to 0.5 percent at 200°C, the copper atoms that comprise this 3.5 percent difference combine with some aluminum atoms to form the Θ phase. During this temperature drop, the grain boundaries are at a higher energy level than are the grain interiors, so as the Θ forms, it will tend to precipitate principally in the grain boundaries (again tending to decrease the overall energy level of the structure). Figure 7–17 is a *schematic* microstructure indicating the end result. There are two practical disadvantages to such a structure. First, it would possess relatively low strength, since its major microconstituent α is almost pure aluminum. Second, the hard, brittle Θ phase, being confined to grain boundaries, would produce a more brittle structure than expected for this low strength and ductile matrix (α). Whenever intermetallic compounds (Θ here) are located primarily in grain boundaries, fracture often begins in those regions and proceeds as intergranular cracking, so the tougher, more ductile matrix does not contribute as much to fracture resistance; the end result is a lower ductility than would be expected. As mentioned regarding cementite in steels, hard phases are generally most effective if they are dispersed as small particles throughout the matrix. Precipitation hardening does that very thing, as now described.

Example 7–4

Refer to Fig. 7–16. If an alloy containing 15 percent copper were slowly cooled to room temperature, describe the resulting microstructure.

Solution   After full solidification at just under 548°C, a structure of α solid solution and eutectic exists. As further cooling occurs, Θ will precipitate out of the α solid, primarily in α grain boundaries; the end result will consist of grains of α solid solution, Θ precipitate, and grains of eutectic.

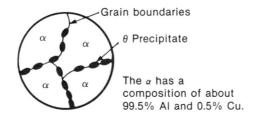

Grain boundaries

θ Precipitate

The α has a composition of about 99.5% Al and 0.5% Cu.

Figure 7–17 Schematic microstructure of a 96 percent aluminum-4 percent copper alloy furnace cooled from the α region to 200°C. Note that the Θ phase precipitates primarily in the grain boundaries.

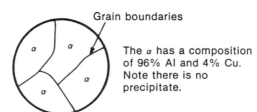

Grain boundaries

The α has a composition of 96% Al and 4% Cu. Note there is no precipitate.

Figure 7–18 Schematic of the alloy in Fig. 7–17 that has been water-quenched from the α region. A supersaturated α solid solution results.

If this 4 percent copper alloy were quenched in water from 550°C, there would be inadequate time to form the Θ phase as earlier, since diffusion could not occur, and that is essential if Θ is to form upon the lowering of temperature. A microstructure of supersaturated α would result as shown in Fig. 7–18 and in that condition, it would be stronger and usually more ductile than the structure indicated in Fig. 7–17; note that *no precipitate* exists after the initial water quench, and all the copper has been used to form the supersaturated solid solution of α. This α is not near equilibrium and would like to lower its copper content as it did when furnace cooled. What follows will not describe certain metallurgical subtleties, since it is the end results that are of major interest here. Both time and temperature are of importance, but for simplicity, consider that temperature is held constant for now. Suppose we conduct a series of tests where we heat the supersaturated α to, say, 400°C for a few minutes, quench it in water, measure its hardness, then repeat this procedure for ever-increasing time intervals. The sketch in Fig. 7–19 shows a typical result of hardness versus time, and we note the following:

1. The hardness begins to increase.
2. The hardness reaches some maximum value.
3. The hardness begins to decrease and eventually becomes *lower* than that of the *initial* α.

A series of schematic microstructures, shown in Fig. 7–20, will assist in explaining this sequence of events. First there occurs a precipitation of relatively *few, fine* particles.* These particles tend to interfere with the easy movement of dislocations, thereby increasing hardness and strength. To form the precipitate, some of the copper atoms from the supersaturated α combine with aluminum atoms; this *reduces* the degree of supersaturation of the α, thereby lowering the energy level and causing the α to be closer to equilibrium. As this process is continued at longer times, more and more precipitate forms and the hardness continues to increase. Note that two effects† are counteracting here. To form

* These will always be indicated by dots, although they have a small platelet-like form during the early time stages and are of the order of 10 nm in size.

† The degree of *coherency* of the structure is also important but is not included here. See D. K. Felbeck and A. G. Atkins, *Strength and Fracture of Engineering Solids,* p. 194 for a discussion.

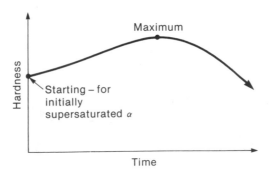

Figure 7–19 Typical hardness versus time plot as an initially supersaturated α solid solution undergoes precipitation hardening. Beyond the point of maximum hardness, overaging results and the hardness begins to decrease.

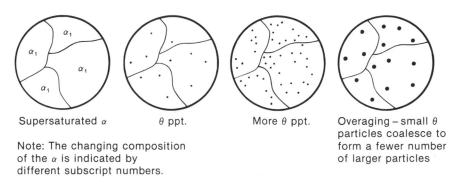

Supersaturated α θ ppt.   More θ ppt.   Overaging – small θ
                                                                                   particles coalesce to
Note: The changing composition   form a fewer number
of the α is indicated by   of larger particles
different subscript numbers.

Figure 7–20   Series of schematics to illustrate the effects of precipitation hardening and
subsequent overaging.

more Θ, a greater number of copper atoms must *come out* of the α solid solution, and the
degree of supersaturation is continually lessened. The overall hardness at any time is,
therefore, the result of two competing factors. If Θ is to form, it must do so at the expense
of the supersaturated α, so at longer times, the α itself becomes softer, since the *solution
hardening* effect of the excess copper diminishes. When the influence of the precipitate
more than offsets the diminishing solution hardening effect, the *net* result is an increase
of hardness and illustrates what is meant by precipitation hardening. This important
strengthening mechanism permits one to attain a variety of combinations of strength and
ductility for the many aluminum alloys used commercially. Note that some of these alloys
will age-harden fairly quickly at room temperature (called *natural aging*), but most
require the use of an elevated temperature for specific time periods (called *artificial aging*)
to attain the desired structure in a reasonable time period. It is important to consider the
possibility of *overaging* when such alloys are used in practice. If they are subjected to
elevated temperatures for adequate time, they will begin to soften, and the resulting
decrease in strength could lead to serious consequences. We note that various time-
temperature combinations can produce similar results where higher temperatures lead to
an increase in hardness in shorter time.* Designers must be aware of this and other
potential problems when they specify the use of such alloys. For example, to produce
what is known as a T-6 (artificially aged) condition can lead to residual stresses, while
welding such a structure can cause a serious decrease in the original strength.

7.4.5 Cold Work, Recrystallization, and Grain Size

The effect of *cold working* (also called work hardening or strain hardening) was discussed
in a quantitative way in Chapter 6. Here it is covered qualitatively as a strengthening
mechanism and may be defined as plastic deformation carried out at temperatures below

---

* An excellent reference on this topic is John E. Hatch, ed., *Aluminum: Properties and Physical
Metallurgy* (Metals Park, Ohio: American Society for Metals, 1984).

which *recrystallization* would occur. To explain these words, a series of schematic microstructures will help. For simplicity, consider a *single phase* material such as commercially pure copper or aluminum, or an alpha brass. In Fig. 7–21(a), the structure is considered to contain *equiaxed* grains of a particular average size. Due to cold working, the grains become highly distorted and although there is a definite change in the *shape* of the grains, their average size is not altered. During cold working, the number of initial dislocations increases tremendously,* and the energy of the grain interiors thereby increases. If the structure is then heated at an elevated temperature, nuclei of new strain-free grains will begin to form at *both* grain boundaries and interiors, and this begins to lower the energy level of the overall structure. With time, these initial grains begin to enlarge as other new nuclei form. During this time, many dislocations disappear and the energy level continues to decrease. When this process is first completed, we have a strain-free structure of relatively small grain size and an overall dislocation density about the same as existed prior to cold working. Figures 7–21(b) and (c) illustrate these comments. If the process is not stopped at this point, the phenomenon called *grain growth* will occur wherein smaller grains of severe boundary curvature (that is, smaller radius) will tend to disappear and a smaller number of larger grains results; see Fig. 7–21(d). In essence, the coarse grain structure is closer to equilibrium than the finer grain structure, since the total amount of grain boundary surface area is smaller. The above procedure is used if we wish to *refine* grain size, that is, cold work the metal then cause recrystallization to be completed throughout. It is noted here that the final recrystallized grain size, as in Fig.

---

* In the annealed condition, the dislocation density of bulk metals is about $10^6 cm^2$ whereas after severe cold working it increases to the order of $10^{12} cm^2$ (planar density). For the model used to explain this increase see, for example, R. M. Caddell, *Deformation and Fracture of Solids,* p. 172.

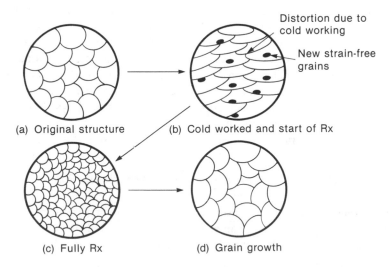

(a) Original structure   (b) Cold worked and start of Rx

(c) Fully Rx   (d) Grain growth

Figure 7–21 Series of schematics to illustrate the sequence of cold-working and annealing a single-phase structure. Note that a strain-free, fine-grained structure undergoes grain growth if heating is continued.

7–21(c), will be influenced by the degree of initial cold work and the time-temperature combination used in this process.

With single-phase metals, whose microstructure cannot be influenced by methods discussed in previous sections, their properties can only be altered by cold working or by changing grain size. Taken individually, this occurs as follows:

1. Cold working by itself causes a drastic increase in the *total number* of dislocations, but the number capable of *easy glide decreases*. As a general consequence, strength increases and ductility decreases.

2. If we compare a coarse versus a fine-grained structure, where grain boundaries provide the major impediment to dislocation motion, we see that the *average* glide distance is smaller with the finer grain size. That is, as dislocations move under the application of stress, they travel a shorter distance before reaching a grain boundary in the fine-grained structure; this ties in with an increase in strength.*

Figure 7–22 illustrates the typical behavior of the cold work-annealing cycle that is observed with single-phase structures.

In forming a part to its final desired shape, it is sometimes necessary to induce strains that would cause fracture *before* the end shape is reached. To avoid this, cold working is carried out to a level short of causing fracture, and the piece is then annealed to remove the effects of this initial cold work. Now that ductility is restored, the piece is further cold worked to its final shape.

7.5 POLYMERS

7.5.1 Introduction

Words such as *plastics* or, more recently, *engineering* plastics are now commonly used to denote those types of *polymers* that are usually man-made and used to make engineering components. As such, they are often called *synthetic* polymers to distinguish them from *natural* polymers such as wood, bone, and the like. The polymers of major importance in this context are comprised of organic molecules that are bonded together by various means to form a solid structure. Just as the unit cell formed the basic building block for crystalline metals, what is called the *mer* or monomer is the smallest repetitive unit from which polymers (that is, many mers) are developed. However, the fundamental forces that hold a polymer together are very different from metallic bonding, and this is why the properties of polymers are so different from metals. It is this difference in properties that usually leads to the use of polymers in numerous industrial applications, since they often fulfill a need that other solids cannot provide, or they do so more economically. To give a reasonable understanding of the behavior of polymers, it is necessary to delve into a brief coverage of a few of the principal aspects of structure as was done with metals.

---

* At temperatures above what is called the *equicohesive* temperature, the coarser-grained structure can be viewed as being stronger. This is where the phenomenon called creep becomes important.

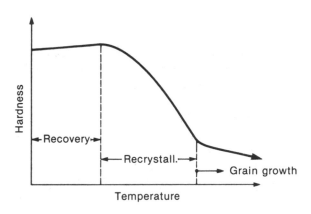

Figure 7–22 Sketch of the change in hardness of an initially cold-worked single phase metal that is heated to various temperature levels; time at temperature is constant here.

Again we stress that this coverage pretends no attempt of completeness; in fact, the numerous aspects involved with altering such structures are in many ways far more complicated than was the case with metals. Consequently, detailed coverage is best left to specialized books in the field of polymer science or polymer chemistry.*

7.5.2 Structure of Polymers

Carbon has been called the backbone of organic materials, and is the element of greatest single importance in synthetic polymers. Because of its chemical valency, carbon can provide anywhere from one to three bonds with other atoms, thereby forming molecules. As the length of the molecules increases, the *total* bonding force that holds the molecules together also increases. In most cases, the polymers of major interest to us consist of molecules that contain carbon plus elements such as hydrogen, nitrogen, fluorine, oxygen, sulfur, and chlorine. This rather large number of possible combinations leads to the variety of properties displayed by such polymers.

The atoms which comprise the molecules that constitute the backbone chain are held together by covalent bonding; this is the primary bond. It is a useful analogy to view a chain of such molecules as resembling a string of beads. Secondary bonds, such as Van der Waals forces, hold adjacent strings together.

It is important to realize that the molecular chains usually occur in a random orientation and the *packing* of atoms is nowhere near as close as in metals. As a result, the overall bonding forces are lower; thus, polymers are, on the whole, not as strong as metals nor as dense. When the backbone chains are held together only by secondary bonds, this is called a *linear* polymer, but this does not mean that all chains are parallel to each other. A loose analogy is to view the chains as the individual strands in a plate of spaghetti; in such a condition they are quite flexible to alter their positions under loading. There are several actions that can be taken to alter the properties of linear polymers. These include:

* An excellent *introduction* to this topic is the text by G. M. Moore and D. E. Kline, *Properties and Processing of Polymers for Engineers* (Englewood Cliffs, N.J.: Prentice-Hall, Inc., 1984).

1. Forming a *copolymer*. Here the backbone chain contains more than one type of repeating mer.

2. *Crosslinking*, wherein covalent bonding between chains is used. This leads to an increase in strength and rigidity.

3. The addition of *plasticizers*. These are solvents that act between the chains and tend to make the end product softer. Note that, depending upon the relative *amount* of such additives, end properties can be quite different.

4. The addition of *fibers* that do not dissolve in the polymer. Such materials are usually called either reinforced plastics or, preferably, *composites*, and they display strength and stiffness much higher than the polymer itself.

7.5.3 Manner of Classification

7.5.3.1   Thermoplastic and Thermoset   Of the several ways to categorize polymers as to groups, the use of *thermoplastics* and *thermosets* is an important one and is distinguished chemically by the degree of crosslinking. *Thermoplastics* utilize secondary bonds to hold the backbone chains together; upon being heated, such bonds are easily broken, and the polymer becomes quite pliable. It can then be formed to a desired shape and cooled to retain that shape. Upon reheating, this process can be repeated time after time. Note, however, that although the desired shape can be produced quite easily, if it is subjected to even modest temperature levels, a shape reversion can occur. We also note that thermoplastics can be recycled in a manner somewhat like that used with aluminum cans. With *thermosets*, the backbone chains are crosslinked in their final form. Upon initial heating and shaping to form, the crosslinks occur; upon cooling to a final shape, the end result is, essentially, one giant molecule. If reheating takes place, there can be no reshaping; instead, such a polymer simply chars and degrades. These polymers are more rigid and stronger than the thermoplastics but are incapable of recycling. In essence the heating, shaping, and cooling of thermoplastics is reversible, since the process can be repeated, whereas the heating, forming, and cooling of a thermoset is irreversible. Commonly used polymers such as polyethylene (PE), polyvinyl chloride (PVC), polymethyl methacrylate (PMMA), and polycarbonate (PC) are thermoplastics, while epoxies and phenolics are thermosets.

7.5.3.2   Amorphous and Crystalline   No polymer has yet been produced commercially with a structure that is fully *crystalline* where the ordered repetition of the basic building block is as complete as in metals. Nevertheless, in a general sense, the simpler the mer, the greater the possibility of obtaining some crystallinity. The rest of the structure is *amorphous*, where the atomic arrangement is random in nature. The highest degree of crystallinity (about 90 percent) attainable with any of the commercially used polymers occurs in high-density polyethylene (HDPE); the remainder of the structure is amorphous. Polymers such as polycarbonate (PC) and polymethylmethacrylate (PMMA) are completely amorphous, while others can be produced with varying degrees of crystallinity. Various models* have been proposed to explain how the large molecules form crystals, but we shall not be concerned with such detail. It is sufficient to state that in the amorphous structure the atoms are less densely packed than in the crystalline structure. Let's consider a polymer that is *capable* of being produced in at least a partially crystalline form. Reference to Fig. 7–23 will assist here, where T_m refers to the melting temperature. If the cooling rate at T_m is *slow,* the tendency for crystallization is enhanced, tight packing of atoms occurs, and the specific volume decreases abruptly. Further cooling to room temperature shows a further drop in the specific volume. If, however, the rate of cooling at T_m is much faster, the time to produce crystallization is inadequate, and the structure that results is basically amorphous. Owing to the looser packing of atoms, the drop in specific volume is not as great compared with the crystalline structure. Further cooling shows a break in the curve, after which the decrease in specific volume occurs at a lesser rate. The temperature at this break is called the *glass transition temperature* T_g and has important consequences. Above T_g, the amorphous structure is usually called *rubbery,* since it is pliable and lacks any degree of stiffness. Below T_g the structure is much stiffer and is called *glassy,* probably because glass, which is amorphous, behaves in a similar manner. The strength and elastic modulus of such solids are greater when below rather than above T_g. Figure 7–24 illustrates the variation of modulus with temperature.

Another important property is the damping of capacity. This can be explained with the use of a rubber ball, where we desire the ball to bounce after it is dropped. Initially, the ball possesses potential energy; as it drops and strikes a surface such as a floor, some energy will be absorbed by the ball and the remainder will cause the ball to bounce. If the ball has a large damping capacity, the rebound height will be small, whereas the reverse occurs if the damping capacity is low. A schematic of damping capacity versus temperature is shown in Fig. 7–25; we note that at or near T_g the damping capacity is greatest. Returning to the rubber ball, we address its purpose. Assuming a child is to use the ball, we would want a low damping capacity (that is, high bounce) and a soft (rubbery) rather than a hard (glassy) structure. From Fig. 7–25, a polymer (or rubber) whose T_g is well

* See, for example, G. M. Moore and D. Kline, *Properties and Processing of Polymers for Engineers.*

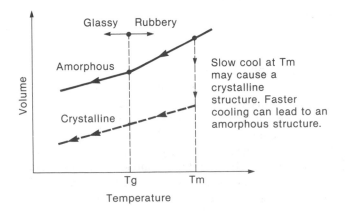

Slow cool at Tm may cause a crystalline structure. Faster cooling can lead to an amorphous structure.

Figure 7–23 Formation of an amorphous or crystalline structure of a polymer as influenced by the cooling rate at T_m. Note that the atoms would be packed more tightly in the crystalline form.

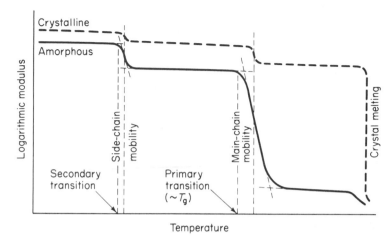

Figure 7–24 Schematic showing the influence of temperature on the modulus of a thermoplastic.

below the temperature of application, say, room temperature, would satisfy both requirements. This simple example illustrates a few reasons why T_g of polymers must be given consideration before an individual polymer is specified for an application.

Two points are worth stressing. First, T_g and its effect on certain properties is applicable, only to the amorphous portion of the structure. Second, care must be exercised in comparing the influence of the degree of crystallinity on properties. One should not conclude that polymers of high crystallinity such as HDPE are stiffer (that is, higher modulus) than polymers that are amorphous. The modulus of HDPE (highly crystalline) is decidedly lower than PC (completely amorphous). A correct conclusion results if the *same* polymer, having different percentages of crystallinity, is considered. For example, high- versus low-density PE (90 versus, say, 60 percent crystallinity) shows a higher tensile strength and elastic modulus for the HDPE.

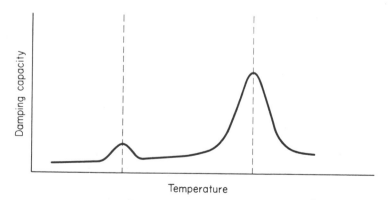

Figure 7–25 Schematic showing the influence of temperature on the damping capacity of a thermoplastic.

Example 7–5
━━━━━━

With reference to Fig. 7–23, how would the ductility of the amorphous structure compare with the crystalline structure?

Solution   Due to the tighter packing of atoms in the crystalline condition, that structure will be stronger, stiffer, and more likely to exhibit lower ductility.

━━━━━━

7.5.4 Viscoelastic Behavior

The discussion in this section pertains to the *thermoplastic* type of polymers discussed earlier, and it is useful to compare their behavior with those metals most commonly used for load-carrying purposes. Parameters such as temperature, strain-rate, and time are important as to their influence on macroscopic behavior of metals and polymers. Although we shall not discuss why and how such factors are influential at the atomic or microscopic level, it is noted that similar behavior, at least phenomenologically, results for different reasons, since the atomic structure of these classes of solids is so different. For example, the phenomenon of *creep*, which is an increase in strain with time under constant load or stress, can occur with either type of solid. Although the observed macroscopic behavior appears similar, the fundamental *causes*, in terms of structural arguments, are quite different. To discuss them is beyond the intent of this book.

Consider a bar of steel subjected to a tensile load which causes elastic deformation at room temperature. Regardless of how much time elapses, any additional deformation would be minimal. If under similar conditions, a specimen of polyethylene had been used, we would note a continual extension as time passes; that is, creep would occur. What this really means is that under *equivalent* conditions, polymers can display a behavior not as readily exhibited by metals. This *does not* mean that metals don't creep, but to cause such an occurrence, both the temperature and magnitude of stress must be increased to much higher levels than those needed to cause such behavior with polymers.

At least four discrete types of behavior enter here. These are:

1. *Creep,* which is an *increase* of *strain* with time under constant load or stress.*
2. *Recovery,* which is a *decrease* of *strain* with time as the load is reduced or fully removed.
3. *Stress relaxation,* which is a *decrease* of stress with time as the strain or displacement of the specimen is held constant.
4. *Strain-rate effects* on the stress-strain behavior. In general, as the strain rate is increased, the stress-strain curve is raised.

These four results are shown schematically in Fig. 7–26. The simplest way to describe such behavior, at least qualitatively, is with mechanical analogs involving springs

---

* Creep tests are usually run under constant load for simplicity. To maintain a constant *true* stress would require a decreasing load, since the area of the test specimen decreases as elongation occurs. If nominal stress is used, then this stress is constant with constant load.

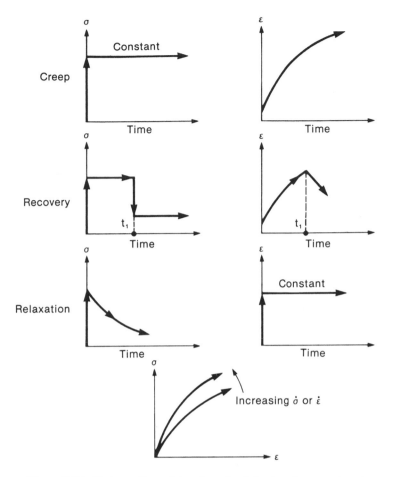

Figure 7–26 Various effects of time-dependent behavior of stress and strain.

and dashpots. These can then be expressed by equations which may then be used for predictive purposes. Shown in Fig. 7–27 are a single spring and dashpot. Under stress, the behavior of the spring is given by

$$\sigma = E\epsilon \qquad (7\text{–}1)$$

while the dashpot behaves according to

$$\sigma = \eta\dot{\epsilon} = \eta \, d\epsilon/dt \qquad (7\text{–}2)$$

Here both E, the stiffness or modulus of the spring, and η, the viscosity coefficient of the dashpot, are assumed constant or time independent; $\dot{\epsilon}$ is the strain rate. Three points are worth noting. First, both equations are linear in the sense that E and η are not functions of time. Second, over a broad range of temperature, a single value for E in Eq. (7–1) would not be adequately descriptive, nor would a single value of η. Finally, both E and η *are* rate sensitive with most real polymers. If one attempts to account for this, nonlinear equations result with accompanying complexities. Yet, ignoring these points still permits a good physical feel for viscoelastic behavior, and that is the main point in this discussion.

Equation (7–1) implies the type of instantaneous elastic response observed with most solids; removal of the stress leads to recovery of elastic strain. Equation (7–2) implies a linear creep behavior since, if integrated, it shows that $\epsilon = \sigma t/\eta$, but this viscous strain is nonrecoverable. Individually, these components explain some of the behavior of polymers but by themselves are quite limited. When combined in series they are called a *Maxwell* model, whereas in parallel, the name *Kelvin* or *Voigt* model is used; Fig. 7–28 illustrates these. Even these models are limited as to their prediction of the four major types of behavior mentioned earlier, but when combined as a *four-element* model, as in Fig. 7–29, their combined effects provide a much better description of such behavior.

First consider creep, where Fig. 7–30 illustrates the most general type of such behavior seen. An initial elastic response occurs immediately as the load is applied and is accounted for by the spring in series whose modulus is E_1. Next, the *increasing* strain which occurs at a *decreasing rate* occurs from the *combined* effects of the dashpot in series (that is, η_1) and the Kelvin branch (E_2 and η_2). As will be shown shortly, the Kelvin strain will tend to saturate to the value of σ/E_2 as time approaches infinity (in a mathematical sense) or essentially at long times in practice. From that time on, the single dashpot continues to strain at a constant rate. We emphasize here that *nothing* is predicted regarding *fracture* with this or any of the other models mentioned here. Although not derived here, the strain of the Kelvin branch* is

$$\epsilon = \sigma/E_2 \, [1 - \exp{(-tE_2/\eta_2}] \tag{7-3}$$

Because Eqs. (7–1) through (7–3) are all linear, the principle of superposition can be applied; in essence the individual strains can be added together to give the total strain as

$$\epsilon_{\text{total}} = \epsilon_{\text{elastic}} + \epsilon_{\text{viscous}} + \epsilon_{\text{Kelvin}} \tag{7-4}$$

or

$$\epsilon_t = \sigma/E_1 + \sigma t/\eta_1 + \sigma/E_2[1 - \exp{(-tE_2/\eta_2}] \tag{7-5}$$

which in equation form basically describes Fig. 7–30 up to the tertiary region.

---

* See R. M. Caddell, *Deformation and Fracture of Solids,* Chapter 6 for a more detailed derivation of the various equations quoted in this section.

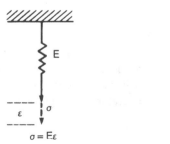

$$\sigma = E\varepsilon$$

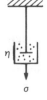

$$\sigma = \eta\dot{\varepsilon} = \eta\frac{d\varepsilon}{dt}$$

Figure 7–27   Schematic of a spring and a linear dashpot.

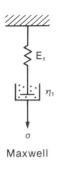

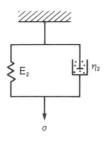

Maxwell   Kelvin or Voigt

Figure 7–28   Schematic of a Maxwell (left) and a Kelvin (right) model.

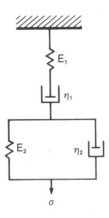

Figure 7–29   A four-element model.

Now consider recovery. Removal of the stress σ would cause an immediate elastic recovery of σ/E_1 followed by a time-dependent recovery of the Kelvin branch, which is exponential in form. Physically this occurs because the spring E_2 would tend to be restored to its initial, unstressed condition and would force the dashpot η_2 to return with it. Any strain due to the individual dashpot η_1 would be a nonrecoverable viscous strain. Figure 7–31 illustrates this behavior and is in qualitative agreement with actual behavior.

As to stress relaxation it is simplest to consider the behavior of the Maxwell and Kelvin parts individually. When the spring E_1 and dashpot η_1 are *strained* to some fixed level (which is then held *constant*), most of the strain must be accommodated by the spring, which is time-independent. As time passes, the dashpot must undergo strain (that is, $\sigma t/\eta_1$) and since the total strain is fixed, this can happen only if the spring contracts.

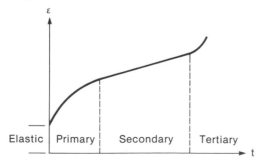

Elastic | Primary | Secondary | Tertiary

Figure 7–30   A generalized creep curve.

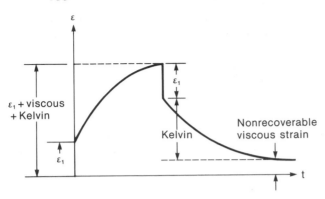

Figure 7–31 Strain-time behavior of a four-element model during loading and then unloading.

This decrease in the strain of the elastic component can result only if the stress σ decreases, hence stress relaxation. Using this same analysis of the Kelvin portion, any strain initially induced is common to both the spring E_2 and dashpot η_2 so there can be no relaxation with time, since both components are locked together. The conclusion is that any relaxation displayed by the four-component model must result because of the behavior predicted by the Maxwell portion; Fig. 7–32 displays this schematically.

As to rate effects, both the Maxwell and Kelvin models predict a raising of the stress-strain curve as the strain rate is increased, although the shapes of the curves differ. When these are combined into a four-element model, the combined effect is more realistic. Figure 7–33 illustrates this, and the interested reader can pursue the details elsewhere.*

The intent of these few pages is to provide a physical understanding of particular aspects of the behavior of certain solids and to impart the idea that for equivalent conditions of temperature, stress, strain rate, and time, polymers, as a *class* of solids, are far more susceptible to such time-dependent or viscoelastic behavior than are metals. It really comes back to the fact that the structure, as dictated by the bonding forces, is so different for these solids.

* R. M. Caddell, *Deformation and Fracture of Solids,* Chapter 6.

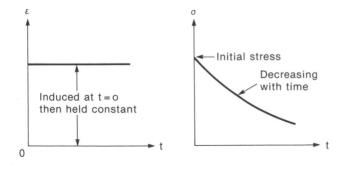

Figure 7–32 Illustration of stress relaxation.

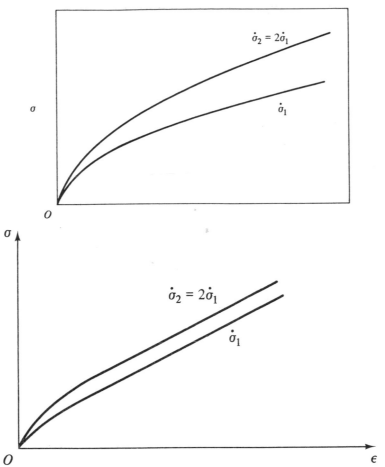

Figure 7–33 The influence of rate effects on the resulting stress-strain behavior as predicted from a Maxwell and Kelvin model.

Example 7–6

Explain why a Kelvin model by itself cannot adequately describe the long time creep behavior as shown in Fig. 7–30.

Solution Refer to Eq. (7–3) and note that as t approaches infinity (or at large values of time), the value of ϵ approaches or saturates to σ/E. Thus continued linear creep cannot be predicted by this model.

7.5.5 Composites or Reinforced Plastics

Natural composites such as wood are not considered here; rather, our attention is directed towards *man-made* composites. For our purposes, a composite is defined as a mixture of *fibers* (also called filaments) that are bonded with a *matrix* material such that a distinct interface results between the two ingredients. Matrices may be thermoset or thermoplastic polymers, metals, or even ceramics.

Consider Fig. 7–34, which illustrates a composite made from *aligned, continuous,* and *unidirectional* filaments embedded in a matrix material. It is assumed that the loading is to be parallel with the fibers as shown, and during loading there is *no slip* at the interface between fibers and matrix. This means that any strain in the fibers, matrix, and composite is equal. A force balance, involving the load supported by *all* of the fibers, P_f, all of the matrix, P_m, and the overall composite, P_c, is

$$P_c = P_m + P_f \tag{7–6}$$

and in terms of stresses and appropriate areas,

$$\sigma_c A_c = \sigma_m A_m + \sigma_f A_f \tag{7–7}$$

Since the fibers are continuous, for a given length, the areas in Eq. (7–7) must be proportional to their corresponding volumes; then

$$\sigma_c V_c = \sigma_m V_m + \sigma_f V_f \tag{7–8}$$

or

$$\sigma_c = \sigma_m (V_m/V_c) + \sigma_f (V_f/V_c) \tag{7–9}$$

If V_c is considered as unity, the *fractional* volumes of matrix and filaments are v_m and v_f, where $v_m + v_f = v_c = 1$. Then Eq. (7–9) can be written as

$$\sigma_c = \sigma_m (1 - v_f) + \sigma_f (v_f) \tag{7–10}$$

This is called the *rule of mixtures* (ROM) and is described by Fig. 7–35. This is similar to the ideas discussed in Sec. 7.4.1. In essence, if σ_m and σ_f are known, a particular value of v_f then defines the strength of the composite σ_c *in the loading direction* shown in Fig. 7–34. If such a composite were loaded at right angles to that shown, the strength of the composite would be dictated primarily by the strength

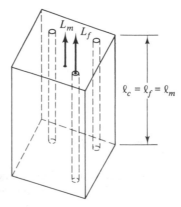

Figure 7–34 Section of a composite made with parallel and continuous fibers.

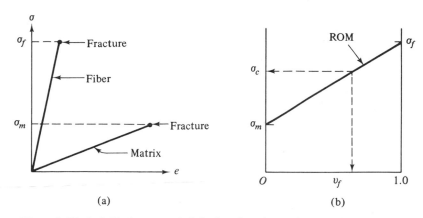

Figure 7–35 Individual stress-strain behavior of matrix and fiber materials (a) and the rule-of-mixtures approach for calculating composite strength as a function of volume fraction of fibers (b).

of the bonding *between* the two components. It would be, in general, considerably *lower* than that predicted by Eq. (7–10), indicating a decided condition of anisotropy.

Since the induced strain is common for the loading described by Fig. 7–34, the elastic *modulus* of the composite is, from Eq. (7–10),

$$E_c = E_m(1 - v_f) + E_f v_f \tag{7–11}$$

where each σ in Eq. (7–10) is given, in general terms, by $\sigma = E\epsilon$. Many experiments show that predictions from Eq. (7–10) overestimate the strength of such composites. To analyze *one* possible reason for these discrepancies, consider the *individual* stress-strain behavior of a matrix and fiber material as shown in Fig. 7–36. Because the fibers are almost always stronger and more rigid (that is, higher modulus) than the common matrix materials, the relative position of the $\sigma - \epsilon$ plots results. Under loading, the fibers will

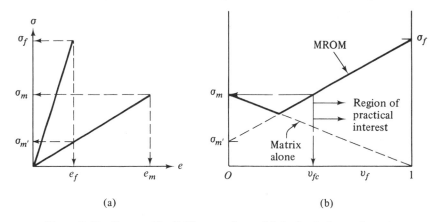

Figure 7–36 Same as Fig. 7–35 except the modified rule-of-mixtures is used.

reach their *fracture strains* before the matrix ever reaches its *fracture stress*. In many, but not all cases, fracturing of the fibers can lead to almost immediate fracture of the entire composite, since we lose some of the load-carrying capacity of the stronger component. Thus, Fig. 7–35 is modified by using the value of $\sigma_{m'}$, instead of σ_m, that is, the matrix stress which corresponds to the fracture strain of the fiber. Figure 7–37 shows this comparison and indicates why this *modified rule of mixtures* (MROM) predicts lower strengths than the ROM. We note two important points here.

1. The two approaches predict different values of σ_c for the same value of v_f, but the stiffness E_c is still governed by Eq. (7–11).
2. Unless a minimum or critical value of v_f is used (see v_{fc} on Fig. 7–36), the MROM predicts a composite strength that is *lower* than the matrix strength σ_m. In a practical vein, there is no sense in adding costly fibers unless we get a value of σ_c that reasonably exceeds σ_m so, as shown in Fig. 7–36, the practical values of v_f are indicated.

Example 7–7

Suppose you are told that the fracture strengths of a matrix and fiber material are 35 and 300 MPa, respectively, and you are to design a composite having parallel and continuous fibers acting parallel to the applied load. The strength of the composite is to be 120 MPa. What volume fraction of fibers would you specify?

Solution   Since nothing is known about the fracture strains of the two components, the rule of mixtures must be used, that is, Eq. (7–10). So

$$120 = 35(1 - v_f) + 300(v_f) \quad \text{or} \quad v_f = 0.32 \text{ or about 32 percent}$$

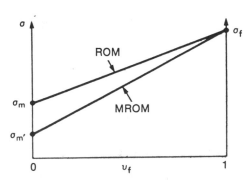

Figure 7–37   Comparison of the two approaches for predicting composite strength as a function of volume fraction of fibers.

7.6 CERAMICS

7.6.1 Introduction

In a general sense, ceramics may be defined as compounds that contain both metallic and nonmetallic elements, even though such a distinction is not as sharply defined as one might expect. However, distinguishing elements as being metallic or nonmetallic is useful, especially for the nonspecialist in materials science.

Just as we noted with polymers, there are different ways by which ceramics can be classified without going into great detail regarding structural configurations. *Traditional* ceramics would include items such as bricks, fine china, and pottery, whereas *modern* ceramics include those used as refractory solids in furnaces, catalytic converters in cars, components in jet engines, and cutting tools used in machining operations. In the discussion of the iron-carbon diagram, the phase called cementite, Fe_3C, is a ceramic even though it wasn't pointed out at that time. Other compounds such as titanium carbide and tungsten carbide are also examples of fairly simple ceramic compounds.

Regarding the primary bonding forces involved with ceramic structures, these may involve both the *ionic* and *covalent* types. As with metals, the unit cell can be viewed as the basic building block and the importance of microstructure, as it affects properties, is as crucial with ceramics as with the other materials discussed in previous sections. Although we shall not include any equilibrium diagrams here, it is noted that they are as important with ceramic materials as with metals in regard to phase transformations.

7.6.2 General Comparisons with Other Materials

As a class of materials, ceramics of certain types have been in use longer than either metals or the type of polymers that are man-made; brick and glass typify ceramics that have been used for centuries. Yet it is reasonable to state that most engineers who are not materials scientists probably possess a better understanding of metal structure and properties because ceramics are, in general, more complicated. For our purposes it is adequate to present a broad overview of comparative properties of these various classes of materials and to stress that it is the design engineer, more than the manufacturing engineer, who must contend with the selection of the most advantageous material in terms of functional as well as environmental requirements.

First, we shall compare ceramics with metals, bearing in mind that exceptions to such general comparisons can exist. If metals are considered as a class, they possess better electrical and thermal conductivity, have much greater ductility, possess greater tensile strength, and display greater resistance to fracture. Ceramics are better insulators, tend to have more stable properties at elevated temperatures, are harder, in most instances display a higher elastic modulus, but have much lower resistance to fracture. Although most bulk metals find widespread use regardless of whether the loading is tensile or compressive, ceramics possess much greater strengths in compression than in tension. Here we note one obvious exception for metals; gray cast iron, like ceramics, is much stronger in compression than it is in tension.*

Compared with polymers, ceramics are stronger and more rigid (greater elastic modulus), and retain their properties at elevated temperatures. Polymers are more ductile and can be formed to a desired shape by techniques that could not be used if ceramics were involved; in addition, polymers have lower densities, which could be important where weight considerations are of concern. Both classes are capable of varying degrees of crystallinity but *on the whole* cannot be considered to be as fully crystalline as are metals. One other concern with ceramics is that they often contain a greater degree of porosity than do either polymers or metals. Although specific processing techniques can reduce such porosity to minimal levels, the tendency for brittle fracture† is still much greater with ceramics than with either polymers or metals.

7.6.3 A Final Wrapup

In this chapter we have followed the philosophy that the manufacturing engineer is not the individual who selects or specifies a particular material for a given component. That task is left to a design engineer. It has been our experience that the great majority of design engineers have a relatively limited background in the field of materials and often select a material on the basis of strength after completing a stress analysis. But in many applications, properties related to corrosion resistance, chemical attack, temperature effects, resistance to crack propagation, and the like are of at least equal importance. Then, it seems to us, the design engineer must consult with those who possess more specialized backgrounds in metallurgy, ceramics, or polymer science for sound advice and help. It is in this same vein that the manufacturing engineer is often placed. Such a person is not expected to be a materials expert, *but* realizing and accepting one's limitations and then seeking proper assistance will usually produce a sensible and economic solution.

We also admit that the coverage of metals in this chapter has been the most extensive, whereas that for ceramics was least. Up to now the understanding of metals has been of longer standing, and their use in manufacturing has been much greater as compared with polymers and ceramics. However, this situation is rapidly changing and, due to weight considerations, for example, many applications today are seeing some type of polymeric material being substituted for metals. Again, the use of ceramics is increasing greatly and it is likely that the manufacturing engineer of the future will have to contend with a broader use of all engineering materials than did his predecessors. The brief coverage in this chapter should provide enough background in the fundamentals so that

---

* See Chapter 11.

† We have not discussed *fracture toughness* as a property in this text, but we mention here that this is of great importance in resisting crack propagation. Unfortunately, the broad subject of fracture is barely covered, if at all, in most undergraduate curricula today. Thus, though we are aware of this important property, we cannot in any meaningful sense cover it in this text without going into the type of detail that has been deliberately avoided on other topics. For an introduction to the subject of fracture and fracture toughness, the reader can consult D. K. Felbeck and A. G. Atkins, *Strength and Fracture of Engineering Solids,* Chapter 14, or R. M. Caddell, *Deformation and Fracture of Solids,* Chapter 8.

the reader gains a reasonable appreciation and awareness of the three classes of engineering materials.

PROBLEMS

7–1. Suppose you could control the dislocation density of a bulk metal. What advantages would result if the metal contained no dislocations? What would be disadvantageous about producing such a structure?

7–2. A piece of commercially pure copper can be made to contain a fine grain size (many small grains) or to have a much coarser grain size (fewer large grains). In which condition would the yield strength be greater? Explain why.

7–3. Two specimens of AISI 1040 steel are heated to 1600°F (871°C), homogenized, and then cooled to room temperature. One is furnace cooled, whereas the second is air cooled. Based upon the likely microstructure that results, determine the probable tensile strength of each specimen.

7–4. One-half-inch-diameter solid bars of 1040 and 4340 steel are heated to the austenite region for one hour; each is then quenched in water. Determine the maximum BHN you would expect at the surface of each bar.

7–5. Explain why a piece of steel containing a structure of as-quenched martensite might cause problems in service.

7–6. Two pieces of eutectoid, plain-carbon steel are heated to 760°C and 1050°C, respectively. Both are quenched in water. Which would display better hardenability? Explain why. Remember when γ transforms it has a choice to produce P or M.

7–7. Refer to Fig. 7–14. If a thin specimen of that metal were transferred quickly from 1400°F to 1100°F, held at that temperature for 20 seconds, and then quenched in water, what are the likely microconstituents that result after quenching?

7–8. You produce a *surface* hardness of 55 Rockwell C using a bar of 1040 steel. Unfortunately, the hardness at the *center* is about 25 R_c. If it were desired to obtain a hardness of 55 R_c throughout most of the section, what action would you recommend?

7–9. An aluminum alloy is heated to a single-phase (α) region for one hour; it is then quenched in water. The specimen is then heated to 400°C.
   (a) Draw a curve indicating how the hardness of the water-quenched specimen would change with time.
   (b) At time = 0, and for a few other time intervals, indicate how the microstructure changes by drawing schematics. On each schematic you should label the constituents to distinguish any differences in the structures.

7–10. Explain the procedure you would use to refine the grain size of a piece of pure aluminum whose current grain size is quite large.

7–11. For a certain application you want a polymer to be *glassy* and to display a *low* damping capacity. In service, the polymer will experience a temperature variation between 20°C and 40°C. Three possible choices have glass transition temperatures of $-40°C$, $+30°C$, and $+100°C$, respectively. Which would you select? Explain why.

7–12. Considering the overall general creep behavior displayed by many solids, where does the predicted creep behavior of a Maxwell model fit into the overall spectrum? Repeat this for a Kelvin model.

7–13. Two tensile specimens of polyethylene are individually subjected to a tensile test. With one, the strain rate induced is about 10^{-3}/s whereas the other experiences a strain rate of 10^{-1}/s. A measure of the elastic modulus shows a decidedly higher value with the higher strain rate. Three models are proposed to explain this behavior in a qualitative manner. They are

1. A spring.

2. A spring in series with a Maxwell unit.

3. A dashpot in series with a Kelvin unit.

Of the three models, which would provide best agreement with the test results? Why did you choose that particular model? No numerical calculations are required.

7–14. A glass-reinforced tape, such as that used in wrapping packages, consists of a matrix (much like Scotch tape) whose fracture strength is 35 MPa. The glass filaments possess a fracture strength of 4 GPa. By conducting a tensile test, the reinforced tape displays a strength of 420 MPa. Using the rule of mixtures, what volume fraction v_f of fibers would you expect was used?

7–15. Taking the results of Problem 7–14 one step further, the fibers are circular and have an average diameter of 0.076 mm while the *thickness* of the reinforced tape is 0.152 mm. If the width of the tape is 25 mm, and the fibers are spaced uniformly across that width, calculate the *number* of fibers involved. Note that the fibers must be continuous and parallel to the length of the tape.

7–16. A composite is to be made using a matrix that displays a fracture stress of 20 ksi and strain of 0.100. Fibers to be used show a fracture stress of 300 ksi and strain of 0.005. The σ-ε behavior for both of these components is linear to fracture. You want to produce a composite having continuous, parallel filaments where loading will be parallel to the filaments.

(a) Using the *modified rule of mixtures,* what volume fraction of fibers v_f should be used if the composite is to have a strength of 100 ksi?

(b) Determine the elastic modulus of the composite.

8
forming

All readers, whether they realize it or not, have been exposed to many formed products. Such diverse items as nails, automobile fenders, casings on washing machines, metal cabinets and chairs, and plastic milk bottles have all been subjected to some type of forming operation. The principal objective in any forming process is to change the original shape of a workpiece to some final desired shape by the application of adequate forces in combination with constraints applied by specific tooling. In forming, the original material is *moved about* to obtain the end result. A few operations, such as blanking and punching, involve separation by a shearing action. These are not typical of most forming operations.

Various requirements must be met to achieve successful forming. First, forces of adequate magnitude are essential if an operation is to be effective; therefore, the manner in which forces are applied and reasonable predictions of the magnitudes of necessary forces are important in understanding forming operations, *unless* one is content with a purely descriptive coverage of hardware. Second, any individual forming process is directly dependent upon the tooling used, since that governs the final shape of the semi-finished or finished workpiece. Finally, the solid being deformed must possess the ability to move or flow as forces are applied; that is, it must have adequate ductility. These interactions are considered in much of this chapter.

The field of forming is currently dominated by the manipulation of ductile metals, and much of this chapter deals with that topic. Even so, some discussion on the forming of polymers is presented. Brittle metals (for example, gray cast iron) and other brittle materials cannot be formed successfully by usual practices, so they are excluded from further concern. Because our initial emphasis covers the *cold forming* of ductile metals, secondary effects, other than the shape change and the forces needed to cause the metal flow, must be considered. These relate to the change in the properties of the metal as it is cold formed.

There is a multitude of different types of forming machines and a variety of ways in which the tooling acts in contact with the surface of the material. For example, the power supply devices used in forging operations are sometimes classified as *path or stroke* restricted, *load or force* restricted, or *energy* restricted. Mechanical presses often employ an eccentric drive to move a loading surface up and down over a fixed length as it contacts, deforms, and then moves away from the workpiece. Here, the length of the stroke is limited to some fixed value. Hydraulic devices are designed to provide some maximum force that can be applied to the workpiece; as such, they are considered to be force restricted (of course, any such individual device will also have some maximum travel). Finally, certain presses employ a flywheel as a source of energy and, depending upon the size of the flywheel, a certain maximum energy is available for deforming the workpiece. To attempt to apply consistently such definitions to the numerous types of machines used in forming would require the kind of excessive descriptive material we seek to avoid. Instead, we shall consider forming *processes* in terms of either of two broad categories, namely *bulk forming* or *sheet forming*. This places emphasis on the shapes to be produced and leads more directly to the kind of coverage we feel is of greater importance compared with a descriptive coverage of forming machines per se.

8.2 BULK FORMING

Operations that fall under this broad category cause shape changes, in general, by the application of compressive loads to the workpiece. Thus most of the deformation results from compressive stresses and large contact between the tooling and workpiece. We now cover a sample of processes that are widely used.

Plates, sheet, and foil result from *flat rolling* in which the initial material is reduced in thickness. Here, the cylindrical rolls are driven by a motor, and the entire unit is called a *rolling mill*. When the width-to-thickness ratio of the original metal is large (at least 10 to 1), the metal exhibits little tendency to widen as it passes through the deformation zone (see Fig. 8–1), owing to the elastic constraint of material on either side of the roll gap. Thus, the decrease in thickness is accompanied by a change in length so as to maintain constant volume. This is called a *plane-strain* operation, since there is almost no movement of metal across the width, so the strain is practically zero in that direction. If the rolls are shaped (that is, not merely parallel) more complex shapes, such as I-beams, rails, and the like can be produced.

Extrusion is the process whereby constrained metal, often called a *billet,* is forced to flow by the application of pressure. If the metal is forced to flow in the same direction as the piston movement, that is called *direct* or *forward* extrusion; see Fig. 8–2. The chamber, which surrounds the billet, induces compressive stresses and as the metal is forced to flow through a converging die, the die also exerts a compressive effect on the

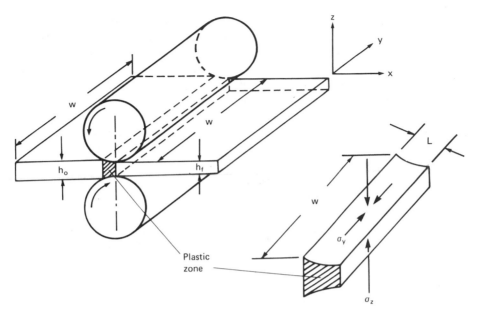

Figure 8–1   Schematic of the deformation zone in flat rolling.

metal.* Note that direct extrusion is similar to squeezing a tube of toothpaste which forces the contents through an opening. *Indirect* or *backward* extrusion is illustrated in Figure 8–3.

* For certain combinations of die angle and reduction, it is possible to develop hydrostatic tensile stresses along the centerline of the billet. (See Ref [1]). Possible consequences are discussed in Sec. 8.8.4.

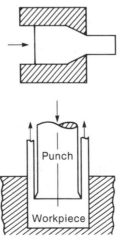

Figure 8–2   Schematic of direct or forward extrusion.

Figure 8–3   Schematic of indirect or backward extrusion.

Drawing is the process whereby rod or wire can be made; see Fig. 8–4. In contrast with extrusion, the metal is pulled through the die so that tensile stresses are developed in the outlet section. The reduction per pass must be limited; otherwise, the outlet material will neck and fracture. In wire drawing, successive reductions are made by using a series of dies of ever-decreasing outlet diameter, and quite large *total* reductions can be made without fracture or intermediate annealing.

Forging processes are usually classified as either *closed-die* or *open-die;* Figs. 8–5 and 8–6 illustrate these processes, both of which basically employ direct compression to the workpiece. With closed-die forging, the accuracy of the dies is high, and finished parts are often produced. Because dies are expensive, such processes are economical only if high production (that is, a large number of parts) is involved. Open-die forging often utilizes flat tooling surfaces, so smaller production rates can be tolerated. As shown in Fig. 8–7, *coining* is a type of forging process whereby the metal is moved to produce an embossed or raised surface according to some predetermined pattern; coins and medals typify this result. *Upsetting,* as shown in Fig. 8–8, is another forging-type process; the heads on bolts or nails are produced in this way.

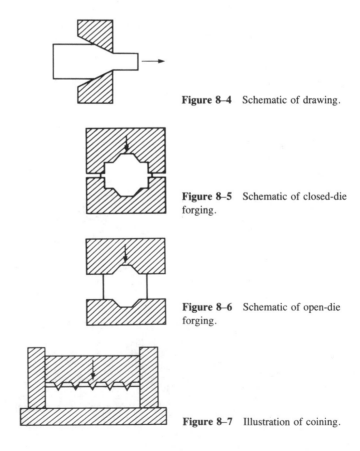

Figure 8–4   Schematic of drawing.

Figure 8–5   Schematic of closed-die forging.

Figure 8–6   Schematic of open-die forging.

Figure 8–7   Illustration of coining.

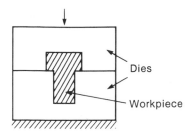

Figure 8–8 Schematic of an upsetting operation.

Other processes could be included here, such as swaging, thread rolling, and tube or pipe production, but those described above should provide an adequate introduction to the meaning of bulk forming.

8.3 SHEET FORMING

Sheet forming is distinguished from bulk forming, since the metal properties of importance, the tooling and the metal-to-tool contact conditions, are quite different. In most sheet-forming processes, the metal deforms in certain regions under tensile stresses, and the extent of deformation is often governed by the onset of necking, which is well short of actual fracture. In contrast, bulk-forming operations utilize compressive loads, and the limit of deformation (often referred to as *formability*) is the formation of cracks.

Bending, by which components such as brackets, cabinets, and the like are produced, is, perhaps, the simplest and most obvious type of sheet-forming process; it is shown in Fig. 8–9. If pure bending is used, the inner fibers experience compression while the outer experience tension. Release of the bending moment results in partial unbending or *springback* due to elastic recovery. This, by the way, is a real problem, since it is the shape of the workpiece *after* springback that is of major concern. If during bending, tensile forces are applied simultaneously, springback may be minimized, since the entire section can be subjected to net tension at both the outer and inner fibers. This is usually called *stretch-forming.*

Deep drawing (also called cupping) is used to produce items such as cartridge cases, beverage cans, flashlight cases, and the like; see Fig. 8–10. A round blank, initially clamped by means of a *hold-down* device, is forced to flow into a die by means of the punch force. The flange of the blank is subjected to *compressive* stresses, while the cup wall experiences tensile stresses. If the latter become excessive, the cup will fail, usually near the bottom, where the blank material has not been work hardened. The hold-down

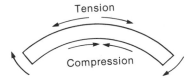

Figure 8–9 A bending operation; note the signs of the surface stresses.

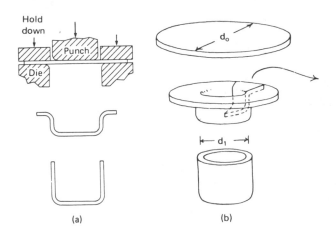

Figure 8–10 An illustration of deep drawing.

force must be of adequate magnitude, or else the compressive stresses will induce a type of buckling or *wrinkling* of the flange; see Fig. 8–11.

Often, a single drawing operation will not produce a cup of desired height, so *redrawing* is used, as shown in Fig. 8–12. As a final operation, *ironing* may be employed. Since the wall thickness from the top to bottom of the cup is not truly uniform, the cup is forced to flow through a final annular opening to produce a thinner and more uniform wall thickness; this is the ironing stage. Figure 8–13 illustrates a drawing, redrawing, ironing sequence using concentric dies and one stroke to finish the cup.

A number of sheet-forming operations is considerably more involved than those discussed above. During deformation, some regions of the sheet experience stretching

Figure 8–11 Wrinkling in a partially 5drawn cup due to insufficient hold-down force. From D. J. Meuleman, Ph.D. thesis, the University of Michigan (1980).

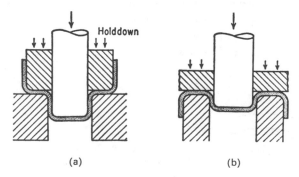

Figure 8–12 Examples of (a) direct and (b) reverse redrawing.

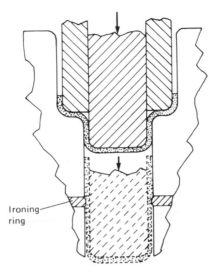

Figure 8-13 Schematic of deep drawing operation showing the final stage of ironing.

while other regions undergo bending. These can be called *complex stampings* of which car fenders, oil pans, and panels are typical. In a later section, the concept of *forming limit diagrams* will be introduced to show how limiting deformations of such complicated products are often analyzed.

8.4 YIELDING

To induce plastic deformations, adequate loads must be applied to the body so that the *applied* stresses will cause the body to yield. Note that to *continue* deformation, the stresses must be *increased* due to work hardening, as discussed in Chapter 6.

Except for the tensile test, true uniaxial loading is not encountered in forming operations, since combined states of stress are usually involved. In such cases, it is necessary to employ a *yield criterion;* this is a postulated mathematical expression involving the combination of stresses that will induce yielding. For our purposes, only one yield criterion will be used, although it is noted that others are available (see, for example, Ref. [2]). All such criteria are based upon certain assumptions, and if these are reasonably met, the more closely will predictions agree with experimental measurements. The assumptions involved are

1. The metal is isotropic and homogeneous.
2. The yield strength in tension and compression is equivalent; thus there is no Bauschinger effect.*
3. Constancy of volume prevails; thus the *plastic* equivalent of Poisson's ratio is 0.5.
4. The magnitude of the *mean normal stress,* σ_m does not influence yielding, where

$$\sigma_m = (\sigma_x + \sigma_y + \sigma_z)/3 = (\sigma_1 + \sigma_2 + \sigma_3)/3 \qquad (8-1)$$

Note that in Eq. (8-1) the stresses are added algebraically where tension is positive and

* Discussed later in this chapter.

compression is negative. Since plastic deformation of most metals occurs by *slip*, which is a shearing process, common yield criteria are all really functions of shear stresses and when particular combinations of such stresses reach a critical value, yielding is incipient.

Although referred to by different names, the criterion used here is called the von Mises (or simply Mises) yield criterion [3]. In its most useful form it is expressed as a function of *principal stresses*. This was discussed in Chapter 6, and they will *always* be designated as σ_1, σ_2, σ_3. The postulated criterion is then

$$(\sigma_1 - \sigma_2)^2 + (\sigma_2 - \sigma_3)^2 + (\sigma_3 - \sigma_1)^2 = C \quad \text{(a constant)} \tag{8-2}$$

Again, it is essential to invoke the proper signs for each stress (that is, plus for tension and minus for compression). If this is applicable for any state of stress, the constant C is most readily evaluated from a tensile test where $\sigma_2 = \sigma_3 = 0$ and σ_1 is positive. Yielding occurs when $\sigma_1 = Y$ (uniaxial yield strength), and introducing these values into Eq. (8–2) shows that $C = 2Y^2$. If a torsion test is used (this is a case of pure shear), the principal stresses are $\sigma_1 = -\sigma_3$, and $\sigma_2 = 0$.* At yielding, $\sigma_1 = k$ (shear yield strength) and using Eq. (8–2) gives $C = 6k^2$. So the Mises criterion becomes

$$(\sigma_1 - \sigma_2)^2 + (\sigma_2 - \sigma_3)^2 + (\sigma_3 - \sigma_1)^2 = 2Y^2 = 6k^2 \tag{8-3}$$

Note that Y and k are material properties *obtained from experiment. Any* yield criterion must be expressed as a function of such properties.

Although stated without proof, this criterion in its most general form is

$$(\sigma_x - \sigma_y)^2 + (\sigma_y - \sigma_z)^2 + (\sigma_z - \sigma_x)^2 + 6(\tau_{xy}^2 + \tau_{yz}^2 + \tau_{zx}^2) = 2Y^2 = 6k^2 \tag{8-4}$$

Since in many forming processes, the applied stresses are principal (or nearly so), Eq. (8–3) finds major use.

8.5 PLASTIC WORK

Consider a round bar of initial length ℓ_0 and area A_0 subjected to a force F that causes an extension $d\ell$. The work done is $F\,d\ell$ and on a *unit volume* basis, using the area and length under load, we obtain

$$dw = (F\,d\ell)/A\ell = \sigma\,d\epsilon \tag{8-5}$$

or

$$w = \int_{\epsilon_1}^{\epsilon_2} \sigma\,d\epsilon \tag{8-6}$$

so w is equal to the area under the true stress-true strain curve between appropriate strain limits. Using principal components, and in the most general case, we find that Eq. (8–5) becomes

$$dw = \sigma_1\,d\epsilon_1 + \sigma_2\,d\epsilon_2 + \sigma_3\,d\epsilon_3 \tag{8-7}$$

* See Chapter 6, where Mohr's circle can be used to show this result.

Example 8–1

The tensile yield strength Y of a metal is 175 MPa. A block of this metal is subjected to principal stresses $\sigma_1 = 70$ MPa and $\sigma_2 = -35$ MPa. What is the magnitude of the tensile stress in the third principal direction that would cause yielding?

Solution   With Eq. (8–3),

$$(70 - [-35])^2 + (-35 - \sigma_3)^2 + (\sigma_3 - 70)^2 = 2(175)^2$$

from which
$$\sigma_3 = (35 \pm 299)/2$$

Since a tensile stress is required,

$$\sigma_3 = (334)/2 = 167 \text{ MPa}.$$

Note if a *compressive* value of σ_3 were required, the negative root would be used and $\sigma_3 = -132$ MPa.

8.6 EFFECTIVE STRESS

In plasticity problems it is useful to define an *effective* stress $\overline{\sigma}$, which is a function of the applied stresses, and which may be thought of as the equivalent single stress that has the same effect as the actual combined stresses in regard to yielding. If the *magnitude* of $\overline{\sigma}$ reaches a certain level, yielding is predicted. Thus,

$$2\overline{\sigma}^2 = (\sigma_1 - \sigma_2)^2 + (\sigma_2 - \sigma_3)^2 + (\sigma_3 - \sigma_1)^2 \qquad (8\text{–}8)$$

for the Mises criterion. Note that $\overline{\sigma}$ is *always* positive by this relationship. When $\overline{\sigma} = Y$ or $\sqrt{3}\, k$, yielding is predicted.

8.7 EFFECTIVE STRAIN

This is *defined* so as to express the incremental work *per unit volume* as

$$dw = \overline{\sigma}\, d\overline{\epsilon} = \sigma_1\, d\epsilon_1 + \sigma_2\, d\epsilon_2 + \sigma_3\, d\epsilon_3 \qquad (8\text{–}9)$$

where the symbol $\overline{\epsilon}$ is taken as the total strain.* In conjunction with the effective stress defined by Eq. (8–8), the companion effective strain is expressed as

$$d\overline{\epsilon} = [2/3(d\epsilon_1^2 + d\epsilon_2^2 + d\epsilon_3^2)]^{1/2} \qquad (8\text{–}10)$$

or in terms of total strains

$$\overline{\epsilon} = [2/3(\epsilon_1^2 + \epsilon_2^2 + \epsilon_3^2)]^{1/2} \qquad (8\text{–}11)$$

where the incremental strains in Eq. (8–10) can be converted to total strains by proper integration. For the derivation of Eq. (8–11), see Ref. [1]. With Eqs. (8–8) and (8–11), Eq. (6–11) can be written as $\overline{\sigma} = K\overline{\epsilon}^n$; this is the form we will use in all further work involving a strain hardening equation.

* For most of the problems of concern in this chapter, the plastic strain is nearly equivalent to the total strain, and ignoring elastic effects causes little error.

In terms of effective stress and strain, the stress-strain relations (or *flow rules*) connected with Eqs. (8–8) and (8–10) can be expressed as

$$d\epsilon_1 = (d\overline{\epsilon}/\overline{\sigma})([\sigma_1 - (\tfrac{1}{2})(\sigma_2 + \sigma_3)]) \tag{8.12}$$

which shows certain similarities with Hooke's law in three-dimensional form, Eq. (6–29); note that $d\overline{\epsilon}/\overline{\sigma}$ is *not* a constant as is E; rather it is a positive ratio whose magnitude changes during deformation. Expressions for $d\epsilon_2$ and $d\epsilon_3$ result by interchanging subscripts.

Example 8–2

A plate is subjected to plane stress loading such that $\sigma_1 = 2\sigma_2$, $\sigma_3 = 0$, and plastic deformation results. Find the plastic work per unit volume in terms of σ_1 and $d\epsilon_1$.

Solution With volume constancy, $d\epsilon_1 + d\epsilon_2 + d\epsilon_3 = 0$, so $d\epsilon_3 = -d\epsilon_1 - d\epsilon_2$. From Eq. (8–7), $dw = \sigma_1 \, d\epsilon_1 + (\sigma_1/2)(d\epsilon_2) = \sigma_1[(d\epsilon_1) + (d\epsilon_2/2)]$. From Eq. (8–10), $d\overline{\epsilon} = [2/3(d\epsilon_1{}^2 + d\epsilon_2{}^2 + (-d\epsilon_1 - d\epsilon_2)^2)]^{1/2}$ or

$$d\overline{\epsilon} = [4/3(d\epsilon_1^2 + d\epsilon_1 \, d\epsilon_2 + d\epsilon_2^2)]^{1/2}$$

Since $\sigma_2 = \tfrac{1}{2}\sigma_1$ and $\sigma_3 = 0$, from Eq. (8–12),

$$d\epsilon_2 = (d\overline{\epsilon}/\overline{\sigma})[\sigma_2 - (1/2)\sigma_1]$$

$$\text{so } d\overline{\epsilon} = (2)/\sqrt{3} \, d\epsilon_1$$

and from Eq. (8–8)

$$2\overline{\sigma}^2 = (6\sigma_1^2/4) \text{ so, } \sqrt{3} \, \sigma_1/2$$
$$\text{so } dw = \overline{\sigma} \, d\overline{\epsilon} = [(\sqrt{3} \, \sigma_1)/2](2 \, d\epsilon_1 \sqrt{3}) = \sigma_1 \, d\epsilon_1$$

8.8 BULK FORMING ANALYSES AND OTHER CONSIDERATIONS

8.8.1 Ideal Work Method

The simplest approach to force or pressure measurements in bulk forming operations, *where it is applicable*, involves a force or energy balance assuming 100 percent efficiency; that is why it is called *ideal*. In essence, any effects of friction or *redundant* deformation* are ignored, and what is then necessary is to envision an ideal process that could produce the same *shape change* as the actual process being considered.

As an example, consider the direct axisymmetric extrusion as illustrated in Fig. 8–14. The diameter reduction could be accomplished by pure tensile loading; it is *not necessary* that such a change could *actually* be done by such loading (that is, the induced homogeneous strain could be in excess of the necking strain in pure tension). As discussed in Sec. 8–5, the *plastic work per unit volume* for tensile loading was

* Explained in Sec. 8.8.2.

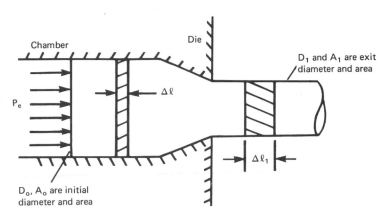

Figure 8–14   Sketch of direct extrusion if ideal deformation is assumed.

$$w_i = \int_{\bar{\epsilon}_1}^{\bar{\epsilon}_2} \bar{\sigma} \, d\bar{\epsilon} \tag{8–13}$$

the subscript i meaning *ideal*. If $\bar{\sigma} = K\bar{\epsilon}^n$ then

$$w_i = [K/(n + 1)[\bar{\epsilon}^{n+1}]] \Big|_{\bar{\epsilon}_1}^{\bar{\epsilon}_2} \tag{8-14}$$

Now in reference to Fig. 8–14, and due to volume constancy, an inlet volume of $A_o \, \Delta\ell_o$ must exit as $A_1 \, \Delta\ell_1$, and the *total* ideal work required is

$$W_i = F_e \, \Delta\ell_0 \tag{8–15}$$

where F_e is the *force* applied to the workpiece and the pressure P_e is simply F_e/A_0. On a per unit volume basis, $w_i = W_i/(A_0 \, \Delta\ell_0)$, so

$$w_i = (F_e \, \Delta\ell_0)/A_0 \, \Delta\ell_0 = P_e \tag{8–16}$$

Thus, the extrusion pressure for ideal conditions is, from Eqs. (8–13) and (8–16),

$$P_e = \int_{\bar{\epsilon}_1}^{\bar{\epsilon}_2} \bar{\sigma} \, d\bar{\epsilon} \tag{8–17}$$

where $\bar{\epsilon}_1$ refers to any strain induced *prior to extrusion* (for example, if the workpiece is initially annealed, then $\bar{\epsilon}_1 = 0$) while $\bar{\epsilon}_2$ is the homogeneous strain based upon the area change (that is, $\bar{\epsilon}_2 = \ell n(A_0/A_1)$).

Rod or wire drawing follows exactly the same arguments, so the outlet *drawing stress* is

$$\sigma_d = \int_{\bar{\epsilon}_1}^{\bar{\epsilon}_2} \bar{\sigma} \, d\bar{\epsilon} \tag{8–18}$$

since the same ideal process is involved.

Example 8–3

A round bar of an annealed metal is cold extruded from a diameter of 12.7 mm to 11.5 mm in one pass. If $\bar{\sigma} = 100,000\,\bar{\epsilon}^{0.2}$ psi. for this metal, what is the extrusion pressure according to the ideal work method?

Solution   From Eq. (8–17),

$$P_e = \int_{\bar{\epsilon}_1}^{\bar{\epsilon}_2} K\epsilon^n\,d\epsilon = [K/(n+1)][\epsilon^{n+1}]]\,\Big|_{\bar{\epsilon}_3}^{\bar{\epsilon}_2}$$

Here, $\bar{\epsilon}_1 = 0$ and $\bar{\epsilon}_2 = 2\ell n(12.7/11.5) = 0.199$

So $P_e = (100,000/1.2)(0.199)^{1.2} = 11,970$ psi

The pressure needed in forging would use frictionless compression as the ideal process, whereas rolling would utilize a plane-strain tension test. In all cases, the ideal work per unit volume is based upon Eq. (8–13), and all that is essential is to employ a process that *could* produce such a shape change and permit evaluation of the induced homogeneous strain. Note that in an operation such as hardness testing, the deformation caused by indenting would not permit a calculation of strain based upon *geometric* changes; thus, the ideal work method could not be used in such a situation. Finally, whenever equations such as Eqs. (8–17) or (8–18) are used, they *always* give a prediction that is *less than* that required for the operation; in this sense they provide a "lower bound."

8.8.2 Efficiency Factors

Besides the work or energy needed to cause homogeneous deformation, two other demands are involved. Friction at the work-tool interface consumes energy, and *redundant deformation* also occurs. The latter is simply due to internal distortion of the work in excess of that needed to produce the external shape change. In Fig. 8–14, it is easy to envisage friction at the interface, while Fig. 8–15 illustrates redundant deformation.

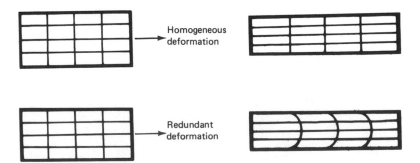

Figure 8–15   A comparison of ideal and actual deformation to show the meaning of redundant deformation.

If all of these factors are considered, then the *actual* work per unit volume is

$$w_a = w_i + w_f + w_r \qquad (8\text{-}19)$$

where the subscripts are self-evident. In practice, w_f and w_r are difficult to estimate and are also not mutually independent, so a simple way to circumvent this is to define an *efficiency factor* η as

$$\eta \equiv w_i / w_a \qquad (8\text{-}20)$$

thereby lumping together the non-ideal work terms. Parameters such as die angle, lubrication, and reduction will influence the value of η, but in practice a value between 0.5 and 0.65 is reasonable. If such a factor is introduced, equations such as (8–17) or (8–18) would be modified by placing a factor of $1/\eta$ in front of the integral, so that the actual values of P_e or σ_d would be greater than those based upon ideal deformation. This use of an efficiency factor, though a bit ill-defined in a fundamental sense, is an attempt to include non-ideal effects that are often difficult to assess by themselves.

Operations such as extrusion, rolling, and rod or wire drawing involve flow through a converging deformation zone. In Fig. 8–16, the three individual work contributions are indicated schematically, and one can see that as the die angle (α in Fig. 8–14) increases, w_f decreases while w_r increases. As a result, there will be an optimum angle leading to the lowest force or pressure in such operations; Fig. 8–17 shows the results found in one study noting that both α and reduction r influence the *optimum* value of α. The reason for the influence of α on w_f and w_r is easily explained: For a given reduction, a decrease in α means a greater contact length between the die and work; since the compressive stress along that interface is essentially constant, the *force* must therefore increase with the greater length. If the coefficient of friction μ does not radically change, then as α decreases, w_f must increase. With redundant deformation, as α increases there is a large degree of internal deformation due to shearing effects; as a consequence, w_r increases with α. Note that as α approaches zero, this approximates a tensile test in which $w_r = 0$.

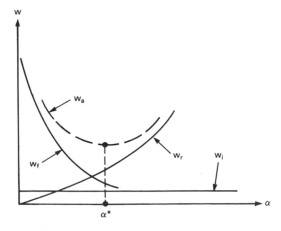

Figure 8–16   Influence of the semi-die angle α on the actual work w_a, during drawing, where the individual contributions of ideal work w_i, frictional work w_f, and redundant work w_r are shown.

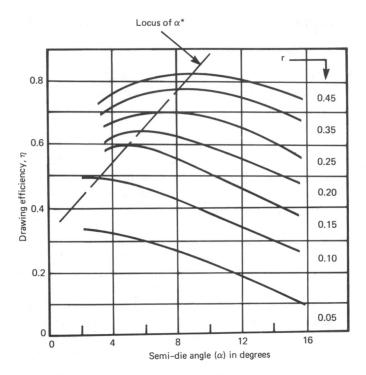

Figure 8–17 Effect of semi-die angle on drawing efficiency for various reductions; note the change in the optimum die angle α^*.

8.8.3 Force Balance or Slab Analysis

As contrasted with the ideal work method, this technique does include the effects of friction at the work-tool interface; in essence, a force balance is made by using a thin slab of metal of differential thickness. This produces a *one-dimensional* differential equation which is then solved using proper boundary conditions. The limitations of this technique arise because of the assumptions involved, these being:

1. The applied load acts in a principal direction and planes perpendicular to that direction define the other two principal directions; thus, the directions of the principal stresses are assumed from the outset.
2. Surface friction effects enter into the force balance, but these have no influence on internal distortion or upon the orientation of principal planes (that is, directions).
3. Homogeneous deformation prevails, so redundant effects are neglected.

Figures 8–18(a) and 8–18(b) illustrate plane-strain strip or sheet drawing where Coulomb friction is assumed at each interface so that the shear stress τ is taken as μP. The

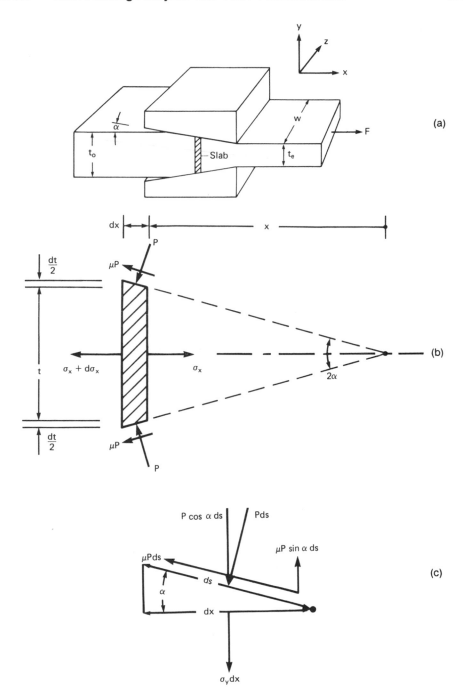

Figure 8–18   (a) Essentials of plane-strain strip or sheet drawing, (b) an enlarged view of the slab used for a force balance analysis and (c) vertical force balance.

enlarged element then is used to give the following force balance in the x-direction.

$$(\sigma_x + d\sigma_x)(t + dt)w + 2P \sin \alpha \,(dx/\cos \alpha)w$$
$$+ 2 \,\mu P \cos \alpha (dx/\cos \alpha)w = \sigma_x wt \qquad (8\text{--}21)$$

which after expanding and neglecting differentials of higher order leads upon integration to

$$\sigma_d /2k = (1 + B/B)[1 - \exp(-B\epsilon_h)] \qquad (8\text{--}22)$$

where σ_d is the drawing (exit) stress, k is the shear yield strength, $B = \mu \cos \alpha$, and $\epsilon_h = \ell n(t_0 /t_e)$. In this derivation, the following were assumed:

1. μ represents a constant coefficient of sliding friction.
2. Either work hardening is neglected or k is taken as the average value of shear yield between the states of the inlet and outlet material.
3. Conical dies are used, so α is constant.

The many details involved with this derivation may be found elsewhere [1]. Equation (8–22) is also used for *wire or rod drawing*, the only difference being that $\epsilon_h = \ell n(A_0/A_e)$ because of the geometry difference.

Next consider direct compression under plane strain ($\epsilon_z = 0$) as shown in Fig. 8–19. A force balance in the x direction is

$$-\sigma_x h + 2\tau \, dx = -(\sigma_x + d\sigma_x)h \qquad (8\text{--}23)$$

With $\sigma_y = -P$ and $\sigma_x - \sigma_y = 2k$ and using Eq. (8–3), we obtain

$$dP = 2\tau \, dx/h \qquad (8\text{--}24)$$

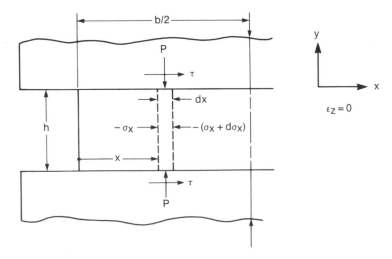

Figure 8–19   Essentials for a slab force analysis in plane-strain compression.

For *sliding* friction, $\tau = \mu P$, and since $P = 2k$ when $x = 0$,

$$P/2k = \exp(2\mu x/h) \tag{8–25}$$

noting that P is maximum at the centerline, $x = b/2$. A plot of Eq. (8–25) is shown in Fig. 8–20, and the increase of P towards the centerline is called the *friction hill*. If we assume that sliding friction does occur over the entire contact region, the *force* to cause deformation must equal the contact area times the *average pressure* P_a, where, as shown in Ref. [1], P_a is found to be

$$P_a/2k = (h/\mu b)\,[\exp(\mu b/h) - 1] \tag{8–26}$$

which for $\mu b/h \ll 1$ can be approximated by

$$P_a/2k \approx 1 + (\mu b/2h) \tag{8–27}$$

Note that since τ cannot *exceed* μP, and the latter cannot be $>k$ (shear strength of the workpiece), then for $P \ge 2k$, μ must be $\le \frac{1}{2}$.

  If lubrication is poor and sliding does not occur, the τ in Eq. (8–23) is replaced with k and *sticking* friction takes place (literally, shear occurs in the metal adjacent to the loading surfaces). Expressions similar to Eq. (8–25) for $P/2k$ and Eq. (8–26) for $P_a/2k$ become

$$P/2k = 1 + (x/h) \tag{8–28}$$

and

$$P_a/2k = 1 + (b/4h) \tag{8–29}$$

Figure 8–21 shows the linear fraction hill associated with Eq. (8–28); note how it differs from Fig. 8–20. It is possible to have mixed conditions where sliding prevails in one region while sticking occurs elsewhere. This will result if predictions based upon Eq. (8–25) give higher values of $P/2k$ than Eq. (8–28), since μP can never exceed k.

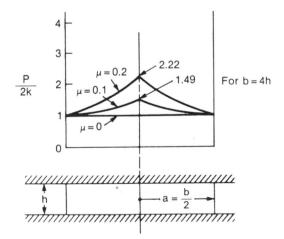

Figure 8–20 Example of the friction hill in plane-strain compression for different values of the coefficient of friction.

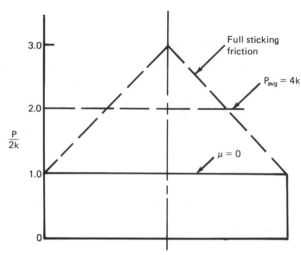

Figure 8–21 The friction hill in plane-strain compression with sticking friction.

Example 8–4

A block of metal whose yield shear strength k is 105 MPa is subjected to plane-strain compression. The block is 250 mm wide by 25 mm high. Find the average pressure at the start of plastic flow
a. If sliding friction occurs and $\mu = 0.10$.
b. If sticking friction occurs.

Solution
a. With Eq. (8–26)

$$P_a = (2kh/\mu b)[\exp(\mu b/h) - 1]$$

where $\mu b/h = 0.1(250)/25 = 1.0$

$$P_a = 2(105)/1\ [\exp(1) - 1] = 361\ \text{MPa}$$

Note, with Eq. (8–27),

$$P_a = 2k[1 + (\mu b/2h)] = 2(105)[1 + (0.1)(250)/50] = 315\ \text{MPa}$$

Here, Eq. (8–27) is not accurate, since $\mu b/h$ is *not* $<<1.0$.
b. With Eq. (8–29),

$$P_a = 2k[1 + (b/4h)] = 2(105)(1 + [250/100]) = 735\ \text{MPa}$$

A similar approach for axisymmetric compression (see Fig. 8–22 for specifics) leads to the following:

$$P/Y = \exp[(2\mu/h)(R - r)] \qquad (8\text{–}30)$$

and

$$P_a/Y = (1/2)(h/\mu R)^2\ [\exp(2\mu R/h) - (2\mu R/h) - 1] \qquad (8\text{–}31)$$

for *sliding friction;* here Y is the *tensile* yield strength which equals $\sqrt{3}k$, according to the Mises criterion.

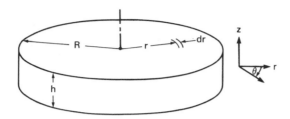

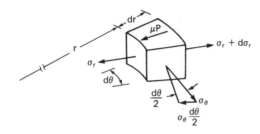

Figure 8–22   Element used in a slab analysis for axisymmetric compression.

With this geometry and *sticking friction*,

$$P = Y + (2k/h)(R - r) \tag{8–32}$$

and

$$P_a = Y + (2kR/3h) \tag{8–33}$$

The concept of a friction hill also enters into flat rolling; Fig. 8–23 is a schematic of this process. In such an operation, it is usual that $w >> h_0$, and as a consequence the inlet and outlet widths are practically equal; see Fig. 8–1. This *plane-strain* behavior results because of the constraint imposed upon the metal in the plastic zone by metal on each side of it (that is, the incoming and outlet metal). Thus, the change in thickness is

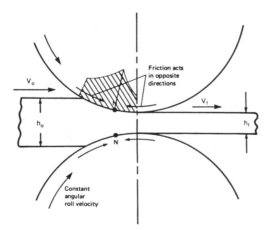

Figure 8–23   Schematic of flat rolling; note the neutral point *N*.

essentially accounted for by an increase in the length of the sheet, so $\epsilon_y \approx 0$ and $\epsilon_z = -\epsilon_x$. Note on Fig. 8–23 that there is one point of contact N in the deformation zone, which effectively separates the directions in which frictional effects act. It represents the point of maximum pressure of the friction hill. In addition, to the left of N (incoming), the *linear* velocity of the rolls is greater than that of the metal; once past N, the metal moves at a higher linear velocity than that of the rolls (which move at a constant *angular* velocity). N is called the *neutral point*. Of course, due to volume constancy, $V_0 h_0 = V_f h_f$.

The *average pressure* in rolling can be well approximated by

$$P_a = (h/\mu L)[\exp(\mu L/h) - 1](\sigma_p) \tag{8–34}$$

where from Fig. 8–24

$$L = (R\,\Delta h)^{0.5}$$

$$h = (1/2)(h_0 + h_f)$$

$$\Delta h = (h_0 - h_f)$$

$$\sigma_p = \text{the plane-strain flow stress (i.e., } 2k)$$

Often, to account for work hardening, σ_p is taken as the average of the flow stresses prior to and after deformation. In practical rolling operations, the strip is effectively pulled

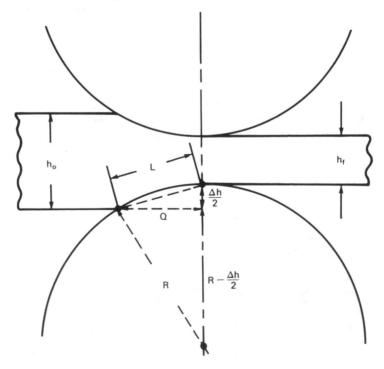

Figure 8-24   Dimensional relations in the roll gap.

through the rolls as its thickness is being decreased. Modifications to the above analysis that account for front (or also back) tension may be found in reference [1].

If P_a is multiplied by the contact area, effectively Lw, this leads to the force that tends to *separate* the rolls by elastic deformation. This force bends the rolls, causing the outlet (rolled) material to be thicker at the center than the edges. Roll *cambering*, shown in Fig. 8–25, is performed to reduce or eliminate such an effect (that is to roll sheet of more uniform cross section). Proper cambering must account for differences in material properties, so a rolling mill whose rolls are cambered correctly for aluminum will not necessarily produce a desired sheet of rolled steel. The major effects of either over- or undercambering are shown in Figs. 8–26 and 8–27. With this number of possible problems, it can be appreciated that rolling is not as simple an operation as it might first appear.

Example 8–5

A metal whose plane-strain flow stress σ_p is 20,000 psi is to be flat rolled from an inlet thickness of 0.125 in. to a thickness of 0.100 in. in a single pass using rolls of 10-in. diameter; the sheet is 18 in. wide, and spreading is negligible. If the average coefficient of friction in the roll gap is 0.08, estimate the roll separating force F_s.

Solution   First determine the average pressure from Eq. (8–34), where

$$h = (0.125 + 0.100)/2 = 0.1125 \text{ in.}$$
$$\Delta h = (0.125 - 0.100) = 0.025 \text{ in.}$$
$$\text{so } L = (5 \times 0.025)^{0.5} = 0.354 \text{ in.}$$
$$P_a = [0.1125/(0.08)(0.354)][\exp [(0.08)(0.354)/0.1125] - 1] \, 20,000$$
$$\text{so } P_a = 22,740 \text{ psi}$$

The roll separating force $F_s = P_a wL$, so

$$F_s = 22,740(18)(0.354) = 145,000 \text{ lbf}$$

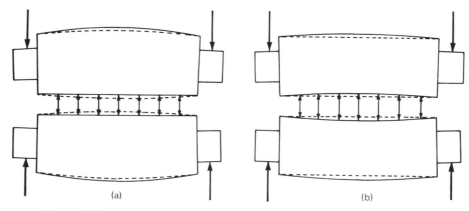

Figure 8–25   (a) Use of cambered rolls to compensate for roll bending and (b) the variation in thickness that occurs if uncambered rolls are used.

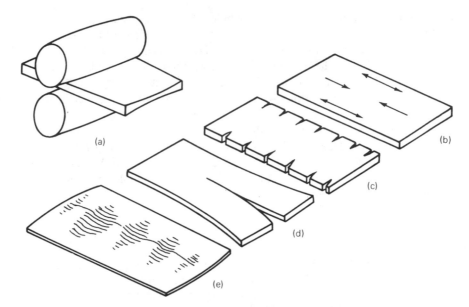

Figure 8–26 Possible effects if rolls are overcambered.

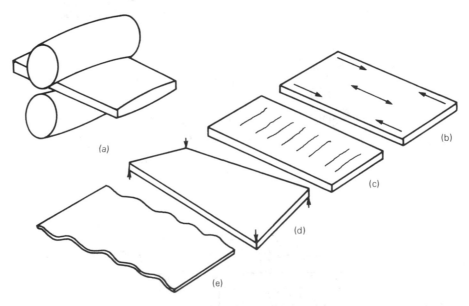

Figure 8–27 Possible effects if rolls are undercambered.

8.8.4 Deformation Zone Geometry

From previous sections, it was shown that the forces in certain bulk operations, as well as frictional and redundant effects, are all influenced by the *shape* of the deformation zone, where in all instances the material was forced to *converge* before exiting in the desired shape. For such operations there is a simple and useful parameter which characterizes certain effects when the zone geometry is altered. It carries the symbol Δ and is defined as the ratio of the *mean* thickness *or* diameter h to the length of contact between work and tool in the zone of deformation; that is

$$\Delta \equiv h/L \qquad (8\text{--}35)$$

For drawing or extrusion, $h = \frac{1}{2}(h_0 + h_f)$ and $L = (h_0 - h_f)/(2 \sin \alpha)$; thus

$$\Delta = (h_0 + h_f) \sin \alpha /(h_0 - h_f) \qquad (8\text{--}36)$$

where α is the *semi-die angle*. If these are plane-strain operations, $r = (h_0 - h_f)/h_0$, so

$$\Delta = (2 - r) \sin \alpha/r \qquad (8\text{--}37)$$

whereas for axisymmetric conditions, where h refers to *diameters*, that is, $r = (d_0{}^2 - d_f{}^2)/d_0{}^2$,

$$\Delta = \sin \alpha \, (1 + \sqrt{1 - r})^2/r \qquad (8\text{--}38)$$

With rolling, and using the chordal length L (discussed in Sec. 8.8.3) instead of the arc of contact (this really causes little error, since the roll radius $R \gg h$ in most cases), then we obtain

$$\Delta = (2 - r)(h_0/rR)^{0.5}/2 \qquad (8\text{--}39)$$

where all parameters have been defined earlier. In a general sense, Δ is proportional to α/r, so small angles and large reductions tend to cause small values of Δ while the reverse holds for large α and small r. Earlier it was mentioned that redundant deformation was also influenced by α and r, so it is not surprising that such effects are tied together with Δ. Note that small values of Δ cause less redundant deformation, where unity is considered small. Perhaps the most direct method to infer redundant deformation in a quantitative way is illustrated by Fig. 8–28. There, the tensile behavior of a fully annealed specimen of stainless steel was first obtained. Then by drawing solid, round, annealed rods by different amounts, the yield strengths of the drawn rods were found; for the two cases shown, the *homogeneous* strains due to drawing were 0.090 and 0.422, respectively. Yet the strengths of the *drawn* rods implied actual strains of 0.185 and 0.500. If we consider the *redundant deformation factor* ϕ to be

$$\phi \equiv (\epsilon_r + \epsilon_h)/\epsilon_h \qquad (8\text{--}40)$$

this would imply that $\phi = 0.185/0.090 = 2.5$ and $\phi = 0.500/0.422 = 1.18$ for these two cases. Using numerous combinations of α and r, and thus a wide range of Δ [via Eq. (8–38)], we obtain the results for three different metals shown on Fig. 8–29. A general expression of the form

$$\phi = C_1 + C_2 \Delta \qquad (8\text{--}41)$$

results, where the constants must be evaluated from experimentation. Note that ϕ is *never* less than unity, since that implies no redundancy, so there is a lower limit on such

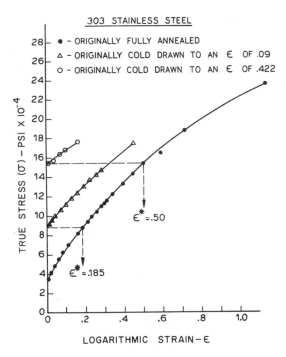

Figure 8–28 Stress-strain curves for 303 stainless steel in the annealed state and after cold-drawing.

equations. Other techniques have also been used to assess the φ-Δ relationship, and any user of such equations must consider how they were determined, since geometric differences and experimental techniques can lead to somewhat different forms of Eq. (8–41). More importantly, values for the empirical constants may differ. In the study related to Fig. 8–28, the empirical equations for the three metals used are

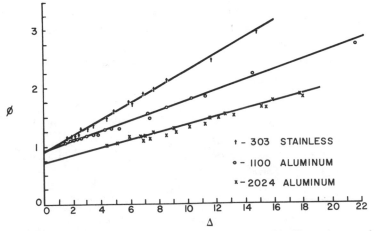

Figure 8–29 Influence of Δ on the redundant strain factor in cold rolling various metals.

303 stainless steel $C_1 = 0.87,$ $C_2 = 0.15$ (8–42)

1100 aluminum $C_1 = 0.89,$ $C_2 = 0.092$ (8–43)

2024 aluminum $C_1 = 0.72,$ $C_2 = 0.067$ (8–44)

Reference [4] gives the details.

In another study [5], using hardness measurements, the influence of Δ on ϕ was determined for both strip and wire drawing. There, Δ was found from Eqs. (8–36) through (8–38), except that α (in radians) replaces sin α. Unless α is large, these forms are quite similar. Those results gave

$$\phi = 1 + C(\Delta - 1) \quad \text{for } \Delta \geq 1 \quad \text{and} \quad \phi = 1 \text{ for } \Delta \leq 1 \quad (8\text{–}45)$$

where $C = 0.21$ for plane-strain and $C = 0.12$ for axisymmetric drawing.

Example 8–6

A one-inch round of commercially pure aluminum is extruded to a diameter of ½ inch through a die of 8 deg semi-angle. Is redundant deformation likely to be significant?

Solution   First find Δ, using Eq. (8–38).

$$\Delta = \sin \alpha \, (1 + \sqrt{1 - r})^2/r$$

where $r = (1^2 - 0.5^2)/1^2 = 0.75$

$$\Delta = \sin \alpha \, (1 + \sqrt{1 - 0.75})^2/0.75 = 0.42$$

With Eqs. (8–41) and (8–43),

$$\phi = 0.89 + 0.092(0.42) = 0.93$$

Also, from Eq. (8–45) since $\Delta < 1$, $\phi = 1$; therefore, redundant deformation is negligible. Note that on physical grounds ϕ can never be less than one, so the value of 0.93 merely implies uniform straining.

The effect of redundant deformation can be seen in hardness variations, Fig. 8–30; internal damage illustrated by density variations, Fig. 8–31; residual stresses, Fig. 8–32; finally, severe internal cracking or fracture, Figs. 8–33 and 8–34. Rolling under high Δ conditions can also lead to *alligatoring,* as shown in Fig. 8–35.

8.8.5 Formability

Although force calculations are often useful, the most important concern in bulk forming is whether the desired deformation can be performed without cracking or fracturing the workpiece. Under equivalent Δ conditions, some of the consequences discussed in the previous section will occur with some metals but not others. In addition, the strains that induce failure in a single metal often depend upon the process employed. Finally, processing variables, such as geometric differences, can be important.

Results such as those shown in Fig. 8–36 indicate that in some cases, fracture strains correlate quite well with the value of ϵ_f based upon the value of A_r from a tensile test.

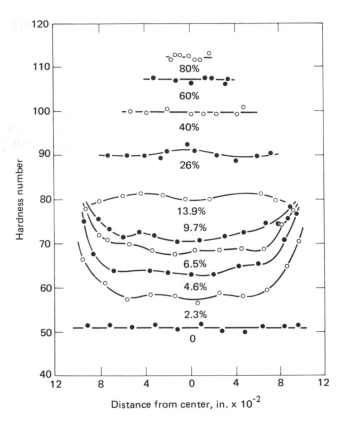

Figure 8–30 Hardness gradients in copper after cold rolling in one pass to the reductions indicated. Initial strip thickness was 0.2 in. and the roll diameter was 10 in.

However, differences in processing variables (for example, square versus round-edged strips) are also important, as that figure shows. Proper treatment of ductile fracture involves both strain and stress, as discussed in Ref. [6].

To discuss in *detail* the influence of metallurgical factors, such as the size, shape, and type of inclusions, is beyond the intent of this text; they are important, and a coverage of this can be found elsewhere [1].

The preferential alignment of grains and inclusions caused by the forming operation (this is called *texturing*) affects both the flow and fracture behavior of the formed workpiece. For example, in flat rolling, extension of the internal structure occurs in the rolling direction, and the *shape* of the grains and inclusions can change quite drastically. It is important to specify whether one is concerned with certain properties of the rolled sheet *or* fracture behavior. It is usual to expect that a cold-rolled sheet will exhibit greater *strains* to fracture if subsequent loading is parallel to the rolling direction as compared to testing *transverse* (that is, perpendicular) to the rolling direction.

The picture is not so clear in terms of mechanical properties. If tensile tests are conducted *after* rolling, one metal may display a higher yield strength in the direction of rolling than in the transverse direction, but the exact *opposite* behavior has been found in other cases.

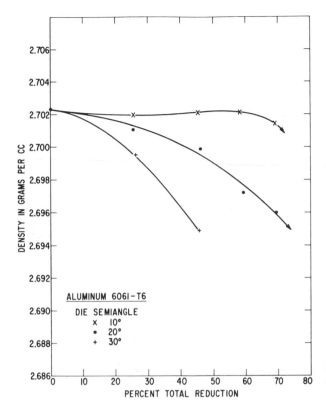

Figure 8-31 Density changes in 6061–T6 aluminum caused by drawing. Note the greater loss of density at high die angles.

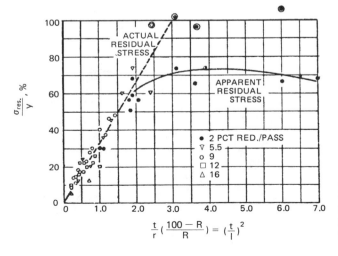

Figure 8–32 Residual stresses at the surface of rolled strip. The residual stresses are normalized by the yield strength, and the abscissa is Δ^2.

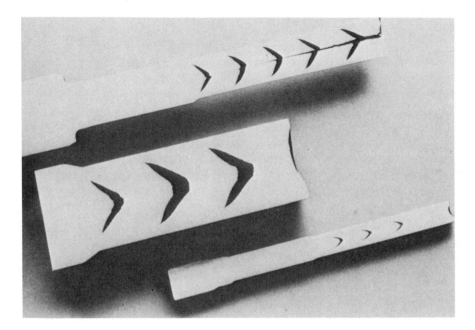

Figure 8-33   Centerline cracks in extruded steel rod.

One definite way to improve formability, though often difficult to do so, is to decrease the magnitude of the mean normal stress, σ_m (that is, to make it more *compressive*). A practical operation where this is done is called *hydrostatic extrusion*. The design of such equipment permits surrounding the workpiece with a high fluid pressure; this tends to *close up* existing voids and to suppress their growth during deformation. Figures 8–37 and 8–38 illustrate two sets of findings. In the first figure, an obvious *increase* in ϵ_f results as σ_m becomes more negative; the second figure illustrates that the workpiece *density* increases with pressure, undoubtedly due to the closing up of initial voids. Hydrostatic extrusion is especially beneficial when brittle materials are involved.

It is unfortunate, but true, that bulk formability cannot be assessed simply in terms

Figure 8–34   Cracks formed in a molybdenum bar cold rolled under high-Δ conditions.

Figure 8–35   Molybdenum rod that "alligatored" from rolling under high-Δ conditions. The "teeth," caused by edge cracks, are not typical.

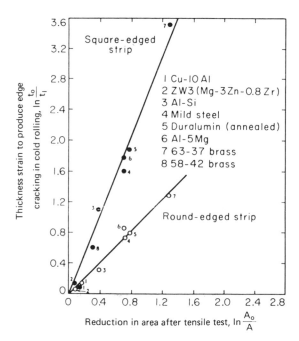

Figure 8–36   Correlation of strain at which edge cracking occurred in flat rolling with reduction of area in a tension test.

of A_r or ϵ_f found from a tension test. As one alternative, compression of solid cylinders has been used. Figures 8–39 and 8–40 illustrate the type of results from such tests (those near $\epsilon_2 = 0$ were found by bending wide specimens to fracture). In the first of those figures, the values of ϵ_1 (circumferential) and ϵ_2 (axial) were found after a barrelled specimen had fractured. Different fracture strain combinations resulted by altering the starting height-to-diameter ratio or the lubrication at the contact surfaces. The second figure illustrates how anisotropy can influence fracture.

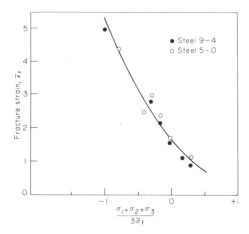

Figure 8–37 Correlation of fracture strain with ratio of the mean normal stress to effective stress.

8.9 SHEET FORMING

Products manufactured from thin metallic sheet usually experience different types of deformation in various regions during processing. That is, bending may predominate in one section, stretching elsewhere, and drawing in other regions. In some cases a combination of deformation modes may occur simultaneously during various stages of the process.

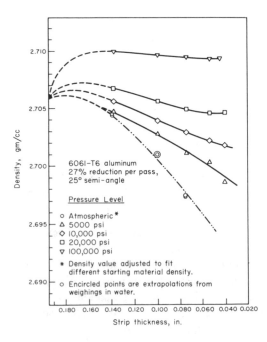

Figure 8–38 Loss of density during strip drawing and the effect of superimposed hydrostatic pressure on diminishing density loss.

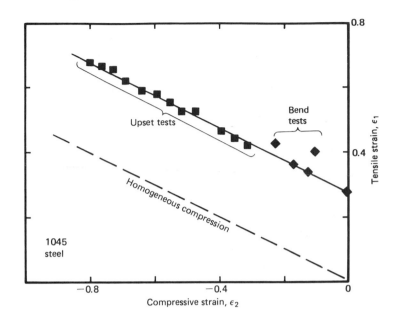

Figure 8–39   Forming limits defined by strains at fracture in upset and formability tests.

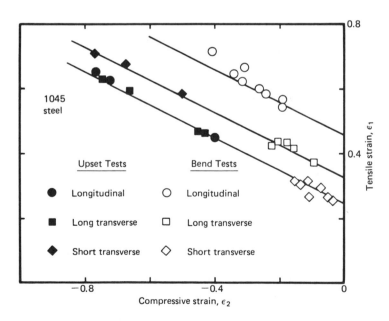

Figure 8–40   Effect of specimen orientation on forming limits.

8.9.1 *Bending*

The simplest process to analyze (yet this is *not* as simple as it might appear) is bending; this is used to produce items such as brackets, cabinets, and furniture components. As a thin strip is bent, tensile strains occur at the outer surface and compressive strains at the inner. If the bend is too severe, the increased tensile strains lead to *fracture;* this is usually the governing constraint. On occasion, as with bending a thin-walled tube, the compressive strains can cause buckling of the inner surface, thereby causing *failure* from that viewpoint. Even when neither of these limiting conditions is reached, two other important results must be considered. The first has to do with *springback,* which is the elastic recovery that follows removal of the applied forces or moments. This obviously leads to a final product whose bend angle differs from that under load. The second consideration relates to the change in the final cross-sectional *shape* that results in the region of the bend. As will be shown later, a section initially rectangular may be *decidedly* altered as the severity of bending is increased.

Consider springback where a thin sheet or strip is subjected to a pure bending moment using the designations in Fig. 8–41. Here any shift of the neutral axis is neglected

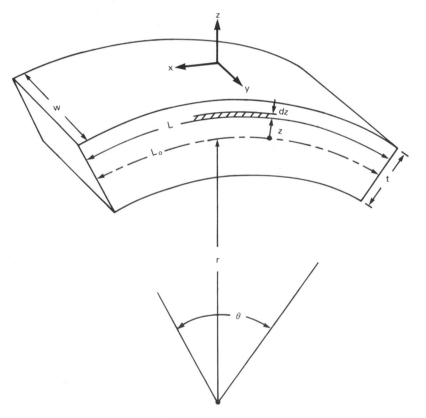

Figure 8–41 Coordinate system for analysis in bending.

and any effects of work hardening are ignored for now; in addition, if $w \gg t$, $\epsilon_y \approx 0$, so a condition of plane strain exists.* If the bend is modest, such that $r \gg t$, then

$$\epsilon_x = t/2r \tag{8–46}$$

at the outer fibers (tensile on outside and compressive on the inside) as shown in Fig. 8–42(a). For a *nonwork hardening* metal, whose stress-strain behavior is illustrated in Fig. 8–42(b), a stress distribution *under* load (plastic deformation is assumed) is given in Fig. 8–42(c). Note that the assumption of plane strain means that the flow stress σ_p equals $(\sqrt{4/3})\,Y$ if Y is the tensile yield stress [check this with Eq. (8–3)]. Thus the full section is at a stress $\sigma_x = \pm\sigma_p$, except for the elastic *core* at the center. This core becomes smaller as the bend becomes more severe.

The relation between the radius under load and after unloading, as taken from reference [1],

$$(1/r) - (1/r') = 3\sigma_p /tE' \tag{8–47}$$

where r is the radius under load, r' is the radius *after* springback, t is the thickness, and E' is the *plane-strain modulus,* which equals $E/(1 - v^2)$. The resulting residual stress distribution, shown in Fig. 8–42(d), comes from

$$\sigma_{x'} = \sigma_P(1 - [3z/t]) \tag{8–48}$$

noting it is compressive $(-\sigma_p/2)$ at the outside and tensile $(+\sigma_p/2)$ at the inside. If work hardening is considered and $\bar{\sigma} = K\bar{\epsilon}^n$ describes such behavior, equations similar to Eqs. (8–47) and (8–48) are given by

$$(1/r) - (1/r') = (6/[2 + n])(K'/E')(t/[2r])^n(1/t) \tag{8–49}$$

and

$$\sigma_{x'} = K'(z/r)^n[1 - (3/(2 + n)(^{2z}/t)^{1-n}] \tag{8–50}$$

where $K' = K(4/3)^{(n+1)/2}$.

---

* At the free surfaces, plane stress prevails.

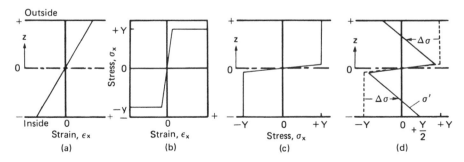

Figure 8–42 Strain and stress distribution across the sheet thickness. Bending strain varies linearly across the section (a). For a nonwork-hardening stress-strain behavior (b), the bending moment causes the stress distribution in (c). The residual stress distribution (d) results from elastic springback.

Figure 8–43 illustrates the residual stress distribution for a work hardening metal. Note that the degree of springback predicted by Eqs. (8–47) and (8–49) can be very large.

A useful method to reduce springback is to apply tensile forces simultaneously with bending; this is typical of *stretch forming*. Owing to the tensile stresses, the neutral axis moves towards the inner surface and, with sufficient tension, the neutral axis moves out of the sheet and the entire cross section yields in tension. Figures 8–44(a) through 8–44(d) illustrate various conditions when one is bending with superimposed tension using a work hardening metal. An analysis of springback can be simplified by approximating the stress-strain curve in the region of loading by a straight line of slope $d\sigma/d\epsilon$. This results in

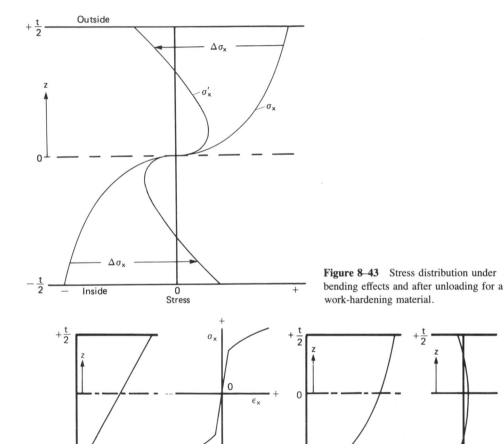

Figure 8–43 Stress distribution under bending effects and after unloading for a work-hardening material.

Figure 8–44 Bending with superimposed tension. With sufficient tension, the neutral axis moves out of the sheet so the strain is tensile across the full section (a). With the σ-ϵ curve in (b), the stress distribution in (c) results. Elastic unloading leaves minor residual stresses, as seen in (d).

$$r'/r = [1 - (d\bar{\sigma}/d\bar{\epsilon})/E']^{-1} \tag{8–51}$$

Because $d\bar{\sigma}/d\bar{\epsilon} << E'$, springback is greatly reduced.

Example 8–7

A piece of steel whose flow stress is 45 ksi is to be bent to a final radius r' of 10 in.; the piece is initially 0.040 in. thick. What tool radius is needed:
a. If pure bending is used?
b. If tension is applied and induces a net tensile strain of 0.025 at the centerline? Assume that $\bar{\sigma} = 100,000\ \bar{\epsilon}^{0.2}$ pertains.

Solution   a. Using Eq. (8–47), with $r' = 10$ in., $\sigma_p = 45$ ksi, $t = 0.040$ in., and $E' = E/(1 - \nu^2) = 30,000$ ksi/$(1 - 0.3^2) = 33,000$ ksi, we obtain

$$\frac{1}{r} - \frac{1}{10} = \frac{3(45)}{0.04(33,000)} = 0.1023$$

so $1/r = 0.2023$ or $r = 4.9$ in.
b. Using Eq. (8–51), with $d\bar{\sigma}/d\bar{\epsilon} = nK\epsilon^{n-1} = 0.2(100)(0.025)^{-0.8}$ (note that K has units of ksi here), we find that $d\bar{\sigma}/d\bar{\epsilon} = 383$ so

$$\frac{10}{r} = (1 - \frac{383}{33,000})^{-1} = 1.013$$

therefore, $r = 9.87$ in.

This requires a tool radius that causes much less bending than in (a) and would ease the problem accordingly.

One study concerned with the minimum possible bend radius short of fracture on the outer fibers showed a reasonable correlation between sheet thickness t, percent reduction of area at fracture in tension A_r, and the radius at the inside of the bend R; this was determined by using an externally applied moment only. Results are shown in Fig. 8–45, the solid line being expressed by

$$R/t = (50/A_r) - 1 \tag{8–52}$$

Note that if $A_r > 50$ percent this analysis implies the sheet coud be bent to a zero radius without fracturing (that is, r cannot be physically less than zero). Figure 8–46 shows a ½ in. by ¼ in. bar of annealed aluminum bent to fracture; notice the final cross section, which was rectangular prior to bending.

8.9.2 The R-Value

With sheet-forming operations more complicated than bending, a material parameter that finds particular use is called the R-value or *strain ratio*. R relates to anisotropy of mechanical properties and is determined from the three orthogonal strains in a flat tensile specimen taken from the plane of the sheet, that is, from

$$\epsilon_\ell = \ell n(\ell/\ell_0), \qquad \epsilon_w = \ell n(w/w_0), \quad \text{and} \quad \epsilon_t = \ell n(t/t_0) \tag{8–53}$$

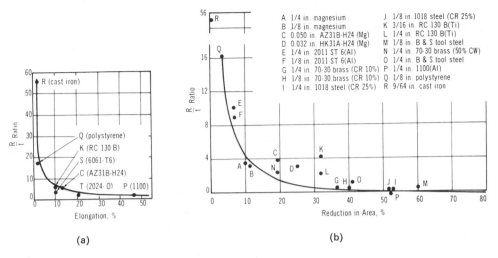

Figure 8-45   Correlation of limiting bend severity R/t, with tensile ductility.

where ℓ, w, and t refer to longitudinal, width, and thickness directions, respectively; Fig. 8–47 shows such a specimen. Now,

$$R = \epsilon_w/\epsilon_t = \epsilon_w/-(\epsilon_\ell + \epsilon_w) \qquad (8\text{–}54)^*$$

and since $\epsilon_\ell + \epsilon_w + \epsilon_t = 0$, R is a positive number. The physical significance is that if $R > 1$ the material is more resistant to thinning than to deformation in the plane of the sheet; the reverse is true if $R < 1$. If the sheet is truly isotropic, $R = 1$. For most ductile metals it is found that R varies with direction of testing, and the most common practice is to define an average value designated $\bar{R}$. Here,

$$\bar{R} = (\bar{R}_0 + R_{90} + 2R_{45})/4 \qquad (8\text{–}55)$$

where the subscript $_0$ refers to the rolling direction, $_{90}$ to the transverse, and $_{45}$ to the angle midway between the others. Observe the following:

1. If $R = 1$, the metal is isotropic
2. If $R \neq 1$, but it is constant regardless of the angle at which the tensile specimen is made, the sheet is said to have *planar isotropy* but *normal anisotropy*.
3. If $R \neq 1$, and it varies with orientation in the plane, then the sheet exhibits both planar *and* normal anisotropy.
4. Measurements of the yield strength Y may often show little variation with orientation, and one might conclude that the sheet displays isotropy. This can be erroneous, since in these circumstances R might not be unity.

---

* Strictly, $R = d\epsilon_w/d\epsilon_t$ and may change with continual deformation. We assume that this ratio remains constant, so total strains are used.

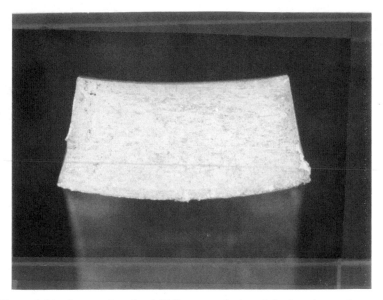

Figure 8–46   Cross section of an initially rectangular bar of aluminum bent to fracture. Photo courtesy of W. H. Durrant.

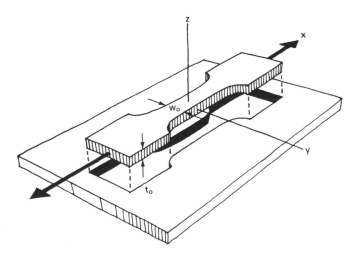

Figure 8–47   Strip tensile specimen cut from a sheet. The R-value is defined as the ratio of width to thickness strains; $\epsilon_w/\epsilon_t = \epsilon_y/\epsilon_z$.

Example 8–8

From a 0.100-in. thick sheet of metal, tensile specimens are cut at 0, 45, and 90 degrees to the rolling direction. Each specimen is then subjected to the same tensile load and plastic deformation results. Prior to loading, a rectangle of 2-in. length and 0.750-in. width was scribed on the specimen surface, where the larger dimension was parallel to the gage section. After loading, these two dimensions changed as indicated on the following page.

At 0 deg $\ell = 2.5$ in., $w = 0.65$ in.
At 45 deg, $\ell = 2.55$ in., $w = 0.63$ in.
At 90 deg, $\ell = 2.45$ in., $w = 0.67$ in

Find the average strain ratio $\bar{R}$ for this sheet.

Solution   Using Eq. (8–53) for the 0-deg direction, we have

$$\epsilon_\ell = \ell n(2.5/2) = 0.223$$
$$\epsilon_w = \ell n(0.65/0.75) = -0.143$$

and from Eq. (8–54),

$$R_0 = -(0.143)/-(0.223 - 0.143) = 1.79$$

For the 45-deg specimen, $\epsilon_\ell = 0.243$, and $\epsilon_w = -0.1744$, so $R_{45} = 2.54$. R_{90} is found to be 1.26.
With Eq. (8–55)

$$\bar{R} = (1.79 + 1.26 + 2[2.54])/4 = 2.03$$

8.9.3 Anisotropic Yield Criterion

In sheet-forming analyses it is usually essential to invoke some form of a yield criterion; otherwise, only descriptive or qualitative comments are possible.

The most widely used anisotropic yield criterion was proposed by Hill [7] in 1948; other criteria have also been proposed [8,9]. In terms of principal stresses, Hill's criterion is

$$F(\sigma_2 - \sigma_3)^2 + G(\sigma_3 - \sigma_1)^2 + H(\sigma_1 - \sigma_2)^2 = 1 \qquad (8\text{–}56)$$

where F, G, and H are parameters that characterize the anisotropy. In terms of R-values,

$$R_0 = R = H/G \quad \text{and} \quad R_{90} = P = H/F \qquad (8\text{–}57)$$

and if *planar isotropy* occurs (that is, $R = P$), then for the special case of plane stress (say $\sigma_3 = 0$), Eq. (8–56) becomes

$$\sigma_1^2 + \sigma_2^2 - (2R/[R+1])\,\sigma_1\sigma_2 = X^2 \qquad (8\text{–}58)$$

where X is the uniaxial tensile yield strength in the 1 direction.

8.9.4 Cupping or Deep-Drawing

A reasonable classification for many sheet-forming operations is to distinguish them as *deep-drawing* or *stamping*. The deep drawing of flat-bottom, cylindrical cups (also called *cupping*) is a process used to produce beverage cans, flashlight cases, and the like. There, the plane of the sheet experiences one principal strain that is tensile while the second is compressive; since any thickness change is small, that strain is usually assumed to be zero.

Stamping operations are generally complex in terms of deformation. In regions where biaxial stretching occurs, planar strains are both tensile, whereas the thinning of the sheet causes a compressive strain through the thickness. Bending and unbending can also occur and cause redundant deformation. It should be noted that operations can include all of these factors (in addition to frictional effects), so analyses become involved.

Starting with cupping, Fig. 8–48 illustrates the coordinate system of concern while Fig. 8–49 gives details of a partially drawn cup. The crucial regions to assess are the *flange,* where major deformation occurs (the flange is forced to move in radially as the cup is drawn), and the *wall,* which must support the force needed to draw in the flange. Two of the major possibilities that must be prevented for most cupping operations are *wrinkling,* Fig. 8–11, and *fracture,* Fig. 8–50. In some applications, wrinkling is not a detriment, as noted by the flanges on aluminum foil dishes (pies, TV dinners). However, if this is to be avoided, an adequate level of hold-down force must be applied. A rule of thumb is that the hold-down-*pressure* should be about *1 percent of the yield strength* of the metal; however, parameters such as sheet thickness, cup diameter, die radius, and condition of lubrication can all influence proper hold-down pressure, so this is only a ballpark suggestion. If this pressure is excessive, greater drawing forces are needed to cause flange deformation and may cause failures, as shown in Fig. 8–50.

The parameter most commonly used to assess formability in cupping is called the

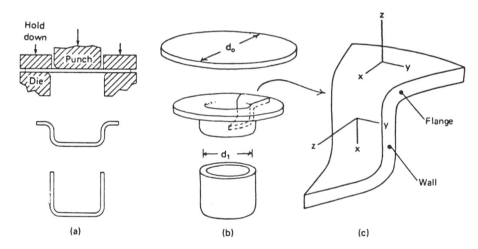

Figure 8–48   Schematic of a cupping operation showing the coordinate axes.

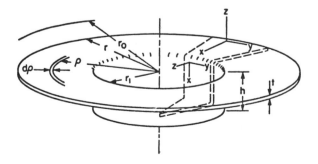

Figure 8–49   Schematic of a partially drawn cup and the coordinate system.

Figure 8–50 Drawing failures by necking at the bottom of cup wall. From D. J. Meuleman, Ph.D. thesis, the University of Michigan (1980).

limiting drawing ratio (LDR); this is the largest ratio of blank-to-cup diameters that can be drawn successfully. Too large an LDR leads to fracture. If the wall and blank thickness are assumed equal as a first approximation, the total volume of the blank can be related to surface areas by

$$\pi r_0^2 t = \pi r_1^2 t + \pi r_1 th \tag{8–59}$$

where r_0 and r_1 are the blank and cup radii, h is the cup height, and t is the thickness as in Figs. 8–48 and 8–49. Then, in terms of diameters,

$$(h/d_1) = [(d_0/d_1)^2 - 1]/4 \tag{8–60}$$

so for a draw of $d_0/d_1 = 2$, the height-to-diameter ratio of the cup will be about 0.75. Often, deeper cups may be needed, but this could demand diameter ratios that would exceed the LDR. Other operations, to be discussed, are then necessary.

An analysis due to Whiteley [10] gives insight into the LDR. Although it involves certain simplifying assumptions (see, for example, reference [1], it provides a basis of understanding. It considers the LDR to be governed by the ratio of the flow stress *in-the-wall* σ_w to that *in-the-flange* σ_f and results in

$$\ell n(\text{LDR}) = \sigma_w/\sigma_f = \beta = \ell n(d_0/d_1) \tag{8–61}$$

where β is taken as the stress ratio. For an isotropic metal, $\sigma_w = \sigma_f$, so the LDR = 2.718. In practice it is closer to 2.1 to 2.2, because the effects of friction plus the bending and unbending over the die lip were neglected. If a *deformation efficiency* η is used, then

$$\ell n(\text{LDR}) = \eta\beta \tag{8–62}$$

and a reasonable value of η based upon practice is of the order of 0.75. To handle anisotropy, Eqs. (8–55) and (8–58) can be employed to give

$$\ell n(\text{LDR}) = \eta\sqrt{(\bar{R} + 1)/2} \tag{8–63}$$

Although many experiments show that LDR does increase with larger $\bar{R}$, the effect is not as great as indicated by Eq. (8–63). For further discussion on this point see reference [1].

Example 8–9

You wish to produce a cup having a height-to-diameter ratio of one. What limiting drawing ratio (minimum value) is needed? Assume that constant thickness prevails.

Solution   Using Eq. (8–60), with $d_0/d_1 = \text{LDR}$ (see Eq. 8–61), then we obtain

$$1.0 = [(d_0/d_1)^2 - 1]/4$$

so

$$(d_0/d_1)_2 = 5 \quad \text{or} \quad d_0/d_1 = 2.24$$

Often, a successful cup can be drawn without fracture, but *earing* occurs. Figure 8–51 shows this, the ears being high regions separated by valleys. Ear formation is undesirable, since the ends must be trimmed; this is an expense, and metal is wasted. It has been shown that earing correlates with the angular variation of R in the sheet and, although four ears are most common, other even multiples have been observed. The empirical parameter that seems to correlate best with the position of ears (with respect to orientation in the sheet) is ΔR, where

$$\Delta R = (R_0 + R_{90} - 2R_{45})/2 \qquad (8\text{–}64)$$

Figures 8–52 and 8–53 relate to this discussion.

8.9.5 Redrawing and Ironing

To produce cups of depth greater than is attainable with a single drawing pass, *redrawing* and, perhaps, *ironing* are often employed. Redrawing (direct) was shown in Fig. 8–12, while a full draw, redraw, and ironing sequence using concentric punches was illustrated in Fig. 8–13. Note that ironing tends to produce a uniform wall thickness, since in typical drawing operations the wall is thicker near the top. Analyses for limiting redrawing and ironing ratios have been proposed [11] but are beyond the scope of this text. However, a few words are added here on the effects of work hardening. It has been shown [11] that in simple cupping, the value of the strain hardening exponent n has only a very minor effect on LDR over the common range of n-values observed. With regard to redrawing and ironing, however, heavily cold-worked sheets will *outperform* annealed ones. Since in this text, n is treated as a fixed value for a given metal, it is not correct to talk about one n-value for an annealed metal and a different one for that metal in some condition of cold work. Rather, it could be stated that a metal which possesses a greater degree of *uniform* tensile deformation (that is, annealed) will not be as successful in redrawing or

Figure 8–51   Earing behavior of cups made from three different copper sheets. Arrow indicates rolling direction of sheets.

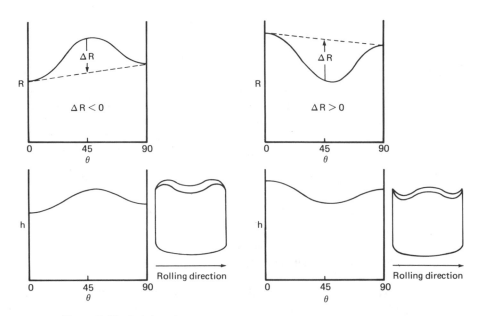

Figure 8–52   Relation of earing with angular variation of R; h is the wall height.

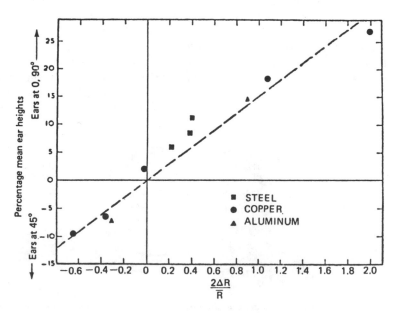

Figure 8–53   Correlation of extent of earing with ΔR.

ironing as the same sheet possessing little or no ability to strain uniformly in tension (that is, work-hardened). To support this observation, it is noted that *very heavily cold-rolled* sheet is used in the deep drawing and ironing of steel and aluminum beverage cans.

8.9.6 Forming Limit Diagrams

Many shapes formed from sheet metal are far more complicated than simple cupping, and analytical description, such as that given above, is lacking. Instead, resort is made to what is called a *forming limit diagram* (FLD). The basis behind such an approach is to print a grid of circles on the sheet. Whether the sheet is to be stamped into a prototype of an actual production part or a shape of simpler contour (usually used in research work), the procedure is identical. As the shape is formed, the circles distort into ellipses whose major and minor axes indicate the directions and magnitudes of the principal strains in the *plane* of the sheet. In such operations, the onset of any *local necking* is generally considered *failure,* although continued deformation must be induced to cause actual *fracture.* Regardless of where the process is terminated, the principal nominal strains are calculated from

$$e_1 = (d_1 - d_0)/d_0 \quad \text{and} \quad e_2 = (d_2 - d_0)/d_0 \qquad (8\text{--}65)$$

where the subscripts $_1$ and $_2$ relate to the major and minor axes of the ellipse and d_0 is the original circle diameter. If true strains were used,

$$\epsilon_1 = \ell n(d_1/d_0) \quad \text{and} \quad \epsilon_2 = \ell n(d_2/d_0) \qquad (8\text{--}66)$$

noting that length and *not area* changes are of concern here. Although forming limit strains are usually large, and true strains are more appropriate, many workers employ nominal strains for no apparent reason.

Figure 8–54 illustrates an example where the sheet was fractured; the plotted circles indicate strain combinations that were *safe,* those that caused *fracture,* and those very near the fracture zone that were considered borderline. In a practical sense, such tests can be used as a diagnostic tool to determine if failures are likely in production runs or if the severity of stamping is approaching a probable failure. Such tests can often save a great deal of expense, since it is of greater advantage to spot potential problems *before* a stamping line is put into actual production. Thus, the results of such tests can be used to make modifications in die design or in part requirements early in the process.

Figures 8–55 and 8–56 illustrate experimental FLDs for several commonly used metals. Factors such as sheet thickness, strain hardening exponent $n,$ and strain-rate exponent $m,$ all appear important. Also, the more *uniformly* strain is distributed in the plane of the sheet, the deeper may parts be stamped. In connection with these two figures, note that *lowest* values of e_1 occur when $e_2 = 0$ (that is, plane strain).

8.9.7 Concerns in Sheet Forming

As we discussed in Sec. 8.8.5, the primary concern in bulk forming operations is to avoid cracking of the workpiece. Other concerns related to residual stresses and preferred orientation of grains and inclusions are also important, *but* the avoidance of fracture must

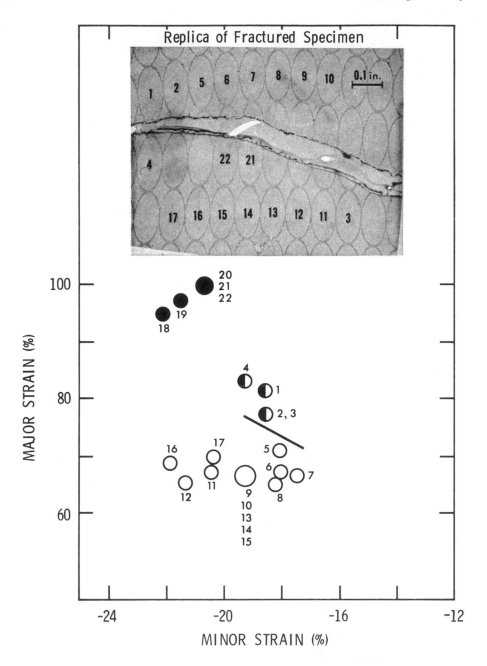

Figure 8–54 Distortion of printed circles near a localized neck (top) and a plot of the strains in the distorted circles (bottom). Solid points indicate failure; open points are away from the failure, and partially filled points are near the failure zone.

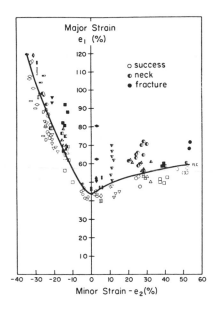

Figure 8–55 Forming limit diagrams for low-carbon steel obtained from data similar to Fig. 8–54, Strains below the curve are acceptable, while those above indicate regions affected by localized necking.

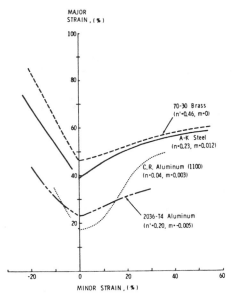

Figure 8–56 Forming limit diagrams for several metals.

come first. With sheet-forming operations the avoidance of local necking is a major concern, since that often dictates the extent of useful deformation, and the influence of $\overline{R}$, m, and n (average strain ratio, strain-rate exponent, and strain-hardening exponent) have been discussed in that regard.

Two important possibilities that influence the *appearance* of the finished surface are called the *orange-peel* effect and *stretcher strains*. The first is a surface roughness on the scale of the grain size. Due to differences in the orientation of adjacent grains on the surface, a tendency to undergo differing degrees of deformation leads to a roughened surface, somewhat like that on an orange. This result can occur only if there is a *free* surface that is not in intimate contact with tool surfaces. Using a metal with a finer grain size will reduce the overall roughening.

Stretcher strains are incomplete Lüders bands* that occur during forming. Metals such as low-carbon annealed steels display a pronounced yield point, and Lüders bands form during early straining up to strain levels of about 0.02. Because strains induced during many sheet-forming operations can vary greatly from one region to another, those regions that are strained to quite low levels are apt to display stretcher strains. One widely used method to avoid this phenomenon is to first subject the sheet to a *roller-leveling* operation. Basically, this amount of light cold working removes the pronounced yield-point affect (that is, the sheet is strained just a little). Now when the subsequent forming operation is conducted, stretcher strains do not occur. Figures 8–57 and 8–58 show these two surface defects.

* See Chapter 6.

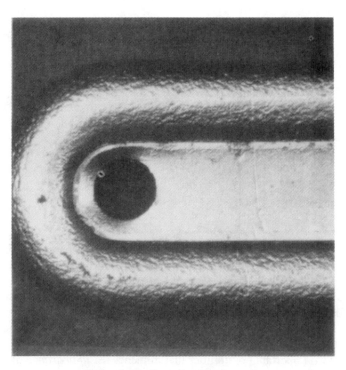

Figure 8–57 Example of orange peel.

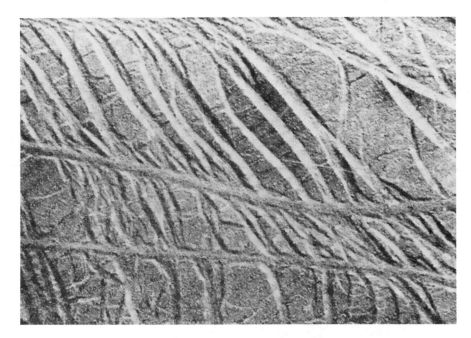

Figure 8–58   Stretcher strains on a 1008 steel sheet stretched slightly beyond the yield point.

8.10 STRAIN RATE

In many analyses, the effect of strain rate, $\dot{\epsilon} = d\epsilon/dt$, is ignored, since its affect on the responding yield or flow stress is small. However, in hot working operations and, in studies connected with forming limit diagrams, the effect of $\dot{\epsilon}$ can be important. Often, the effect of $\dot{\epsilon}$ on flow stress is presented as

$$\sigma = C\dot{\epsilon}^m \tag{8–67}$$

where a series of tensile tests is conducted at different strain rates and the resulting stresses are chosen at some *constant* strain, as illustrated in Fig. 8–59. Implicit in Eq. (8–67) is a straight line plot on logarithmic coordinates. C is called the *strength constant*, while m is the *strain-rate* exponent. High values of m are important in *superplasticity,* where extremely large elongations result without local necking. Figure 8–60 shows a tensile specimen of a superplastic metal elongated 1950 percent. In studies devoted to forming limit diagrams, the influence of m is finding increased meaning, since it appears that metals with relatively large m values do promote large total elongation and thereby superplastic behavior.

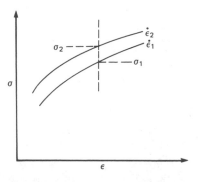

Figure 8-59 Stress-strain behavior as a function of two different strain rates, each under continuous loading. At a common strain, the strain rate exponent, $m = \ell n(\sigma_2/\sigma_1)/\ell n(\dot{\epsilon}_2/\dot{\epsilon}_1)$

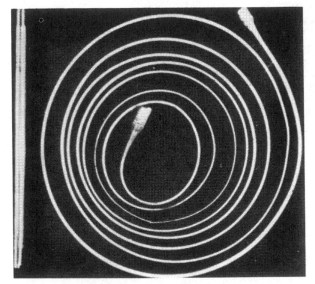

Figure 8–60 Tensile bar of Bi-Sn eutectic showing superplastic extension.

8.11 BAUSCHINGER EFFECT [12]

If a metal having an initial yield strength of Y_0 is plastically deformed, say by tensile loading, and then is unloaded, its new yield strength under subsequent tensile loading will be higher than Y_0 due to work hardening. The use of $\overline{\sigma} = K\overline{\epsilon}^n$ would permit a calculation of the *new* yield strength, which is shown as Y_1 in Fig. 8–61. Suppose, however, that the plastically strained metal is loaded in compression following the unloading from Y_1. It is often found that the compressive yield strength Y_2 will be *less than* Y_1 but greater than Y_0 (assuming initial isotropy). This is the Bauschinger effect. An explanation at the microscopic level, based upon dislocation theory, can be found elsewhere [2], while an explanation using macroscopic arguments is discussed in Ref. [13].

Much confusion exists regarding the Bauschinger effect. In these authors' experience, cold working a ductile metal by any method will usually *tend* to increase the yield strength in all directions as compared with their prestrained values, but the increase will not be necessarily uniform. It is possible that misconceptions have arisen in the translation

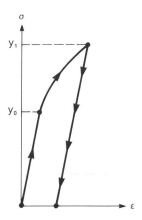

Figure 8-61   Stress-strain behavior of a bar subjected to tensile of deformation; Y_1 is the tensile yield strength of the deformed bar.

of Bauschinger's original paper (in German), and Fig. 8–62 is used to illustrate this. Because sketches in this regard are usually not drawn to scale, the unloading from Y_1 is often exaggerated. It is true that a small nonlinearity is sometimes observed, and if it is carried into the compressive region seems to imply that the *new* compressive yield strength is *less than* the original value Y_0 after cold working in tension. If this argument were carried to the extreme, where a very large degree of tensile deformation was first induced, it would be concluded that the new compressive yield strength is zero or even tensile. This is, of course, absurd. Perhaps the confusion arises from the manner in which yield strength (*not* proportional limit or elastic limit) is defined. The most sensible way to check this concept would be to take an annealed metal and determine the original Y_0 values in tension and compression by using the offset method discussed in Chapter 6. Next take another sample of the same annealed metal and subject it to plastic deformation well beyond the initial Y_0. This could be done by either tensile or compressive deformation (tensile is perhaps easier here). Now, from the *cold-worked specimen* produce a new

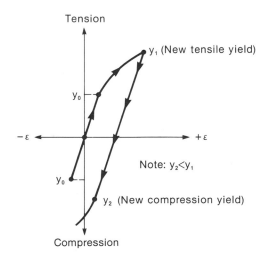

Figure 8–62   An illustration of the Bauschinger effect.

tensile specimen *and* a new compression specimen; then subject one to a tensile test and one to a direct compression test. If these results are plotted as a stress-strain curve (either nominal or true) and the new yield strengths are defined by an offset, it would be found that Y_1 (tension) is usually $>Y_2$ (compression) but *both* are $>Y_0$. Note that if prestraining is done by compression rather than tension, and the same subsequent testing is conducted, then Y_2 (compression) $> Y_1$ (tension). This is as far as the concept of the Bauschinger effect need be taken; if it is limited to this extent, confusion will be avoided.

8.12 *HOT AND COLD WORKING*

When metal is deformed at temperatures above that at which *recrystallization, R_x* occurs, it is called hot working; if deformation proceeds below the R_x region, it is called cold working. These are general terms; more complete coverage of R_x was presented in Chapter 7. Typical metal-forming operations induce deformation so quickly that even at elevated temperatures there is inadequate time to cause R_x *during deformation*. Instead, recrystallization occurs as the deformed metal cools down from these high temperatures. In any event, the net result is that work hardening does not occur. Much greater deformations, short of fracture, are possible with hot working, because the stresses necessary to cause flow are much lower compared with cold working; this results because the yield strength of metals is drastically lowered at typical hot working temperatures and, as mentioned above, work hardening does not occur. There are occasions, however, where failure at low strains can result during hot working; this is often called *hot shortness*. This comes about if a liquid phase is present at these elevated temperatures and results because the metal is heated into a two-phase, liquid-solid region. Understanding equilibrium diagrams makes this clear and is one reason why they were discussed in Chapter 7. One practical example of this phenomenon relates to hot working many steels. If the sulfur combines with iron to form FeS, this tends to form liquid films at grain boundaries and, at typical temperatures used, leads to hot shortness during deformation. It is common practice to add manganese to most steels, since sulfur reacts more readily with this element to form MnS *rather than* FeS. At elevated temperatures used in hot working operations, the MnS remains solid so hot shortness is avoided.

A general summary, including the advantages and disadvantages of these two broad categories of deformation, follows:

1. *Hot working*
   a. The flow stress or yield strength of the metal is lowered whether R_x occurs during or after deformation. As a consequence, much lower inputs of energy are needed to cause the desired deformation, so the size of necessary equipment and power sources is reduced.
   b. The end product is in an annealed or softened state, so any subsequent operations are begun on a structure of relatively low yield strength. Additionally, residual stresses are nonexistent for all practical purposes.
   c. Oxidation of the surface occurs, being more severe with some metals than others. This causes a rough surface finish and poor dimensional control and is detri-

mental in any subsequent machining operations, since it accelerates wear of the cutting tool.

    d. The use of lubricants is practically precluded, an exception being special glasses that have been developed for this purpose (they are relatively expensive).

2. *Cold Working*

    a. The initial yield strength and its increase due to deformation are greater during cold working than during hot working; thus, for similar deformations, cold working demands larger power inputs.

    b. Both the strength and ductility of the end product are different from the initial material.

    c. Surface finish and dimensional control are vastly superior to their counterparts of hot-worked products.

    d. Due to work hardening, the yield strength of the end product is greater than the initial material.

A few other comments related to this section are

1. Certain reactive metals, such as titanium and zirconium, are sometimes hot worked under a protective atmosphere to eliminate oxidation; this does introduce an added cost that would be prohibitive in most cases.

2. Large ingots of steel are initially hot worked to the nearly desired size, *pickled* in acid to remove the oxide scale, and then cold worked to final size.

8.13 FORMING OF POLYMERS AND CERAMICS

As mentioned at the start of this chapter, and reflected in the coverage so far, the forming of metals, certainly on a tonnage basis, predominates this field of processing. However, the same types of processes are employed in the production of parts made from polymers and ceramics; a brief coverage involving applications to these classes of solids will conclude this chapter.

8.13.1 Polymers

Many thermoplastics (see Chapter 7) are produced in various cross-sectional shapes by *extrusion*. There it is usual to feed small pellets of the polymer into a heating chamber after which a force is applied that causes the pliable polymer to exit from a die whose shape governs that of the extruded workpiece. Although different *methods* are used to apply the necessary force, the fundamentals of extrusion are really identical to those covered in earlier sections.

    In *compression molding,* a quantity of the polymer is placed into a cavity, heated, and then subjected to compressive forces; in essence, this is a forging process, as shown in Fig. 8–63.

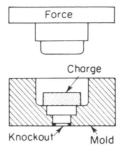

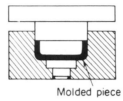

Figure 8–63 Schematic of compression molding.

Blow molding, shown in Fig. 8–64, is used to make items such as plastic bottles; it involves the use of a tubelike shape which is first heated and then placed into a mold. The application of internal pressure causes the tube to bulge until it touches the walls of the mold. Since *recovery** after bulging is to be minimal, the polymer must be strained to a level that induces permanent deformation. This may require stable neck propagation, as discussed in Chapter 6.

Thermoforming involves the use of sheet material which is first heated. With the application of air pressure, the sheet is forced to fit over a shaped mold; essentially this is a stretch-forming operation. Often, instead of air pressure, the opening between the sheet and mold surface is evacuated, and since many polymers used with this process have low strength, atmospheric pressure on the top of the sheet is sufficient to force the workpiece to make contact with the mold surface. Figure 8–65 illustrates the details.

Polymeric sheets are often made by *calendering*. There, a heated mass of bulk polymer is fed into and through a series of rolls to produce some final thickness; this is a rolling process.

8.13.2 Ceramics

The basics behind powder processing, usually involving *sintering,* are covered in Chapter 10. Here we will discuss only those aspects involving forming procedures. Prior to the final operations that cause ceramic products to become extremely hard and brittle (at that point they *cannot* be subjected to further plastic deformation), the ceramic powder is mixed with various additives to bind the particles into a coherent but somewhat porous mass. This mass is usually placed into a mold cavity and then subjected to compressive

* See Chapter 7.

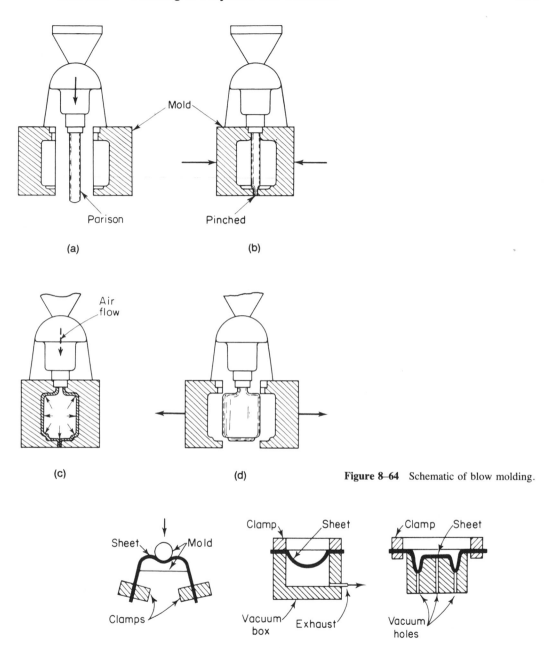

(a)

(b)

Air flow

(c)

(d)

Figure 8–64   Schematic of blow molding.

(a) Mechanical forming

(b) Vacuum free forming

(c) Drape vacuum forming

Figure 8–65   Examples of thermoforming.

loads which compact the material. A reasonable increase in density results that can approach the bulk density of the raw material; in addition, this *preforms* the material to some desired shape prior to producing the compact into its final hardened form. This is really a closed-die forging operation. Extrusion of a ceramic powder-additive mixture is also done. Regardless of the process used, forming of ceramics is *always* accomplished prior to the latter operations that lead to the typical brittleness of ceramic products.

REFERENCES

1. W. F. Hosford and R. M. Caddell, *Metal Forming: Mechanics and Metallurgy*. Englewood Cliffs, N.J.: Prentice-Hall, Inc., 1983, Chapters 9 and 14.

2. R. M. Caddell, *Deformation and Fracture of Solids*. Englewood Cliffs, N.J.: Prentice-Hall, Inc., 1980, Chapter 4.

3. R. von Mises, Gott. Nach; math.—phys., Kasse, 1913, p. 582.

4. R. M. Caddell and A. G. Atkins, *Journal of Engineering for Industry, Transactions of ASME,* Series B, **90** (1968), p. 411.

5. W. A. Backofen, *Deformation Processing*. Reading, Mass.: Addison Wesley, 1972, p. 137. Also, J. Burke, ScD. thesis, Dept. of Metallurgy and Materials Science, Massachusetts Institute of Technology, 1968.

6. A. G. Atkins and Y. W. Mai, *Elastic and Plastic Fracture*. Chichester, U.K.: Ellis Horwood Ltd.: New York: John Wiley and Sons, 1985.

7. R. Hill, *Proceedings of the Royal Society*, London, **193 A** (1948), p. 281.

8. W. F. Hosford, *7th North American Metal Working Res. Conference Proceedings*, SME. Dearborn, Mich., (1979), p. 191; also R. Logan and W. F. Hosford, *International Journal of Mechanical Science*, **22** (1980), p. 419.

9. R. Hill, Math. Proc. Cambr. Phil. Soc., **85** (1979), p. 179.

10. R. Whiteley, *Trans. ASM*, **52** (1960), p. 154.

11. W. F. Hosford, ''Formability; Analysis, Modeling and Experimentation,'' Metall. Soc. of AIME, N.Y. (1978), p. 78.

12. J. Bauschinger, *Civilingenieur*, 1881, p. 289.

13. N. H. Polakowski and E. J. Ripling, *Strength and Structure of Engineering Materials*. Englewood Cliffs, N.J.: Prentice-Hall, Inc., (1966) pp. 144–46.

PROBLEMS

8–1. Each face of a metal cube has an area of 3 in.^{2}, and the tensile yield strength of this metal is 40 ksi. For normal loading, frictionless conditions may be assumed and an x-y-z coordinate system indicates directions normal to the faces. If *compressive* loads $F_x = -20,000$ and $F_y = -25,000$ pounds force are applied to appropriate faces, what *tensile* force F_z is required to cause yielding according to the Mises criterion?

8–2. A thin-walled tube having closed ends is made from a metal whose yield strength in pure shear is $k = 20$ ksi. The diameter of the tube is 4 in. and the wall thickness t is 0.050 in. For such a geometry under internal pressure P the stress state can be approximated quite accurately by

$$\sigma_\theta = (Pr/t) \text{ (hoop stress) where } r = \text{radius}$$
$$\sigma_\ell = (Pr/2t) \text{ (axial or longitudinal stress)}$$
$$\sigma_r \approx 0 \text{ (radial direction)}$$

Determine the value of P to cause yielding.

8–3. The tensile yield strength Y of a metal is 140 MPa. In service it will experience known stresses of $\sigma_x = 50$ MPa, $\sigma_y = -80$ MPa, $\tau_{xy} = 30$ MPa and $\tau_{yz} = \tau_{zx} = 0$. What magnitude of a tensile stress σ_z will cause yielding?

8–4. Prior to the loading of a testpiece, a square grid having dimensions of 6 mm per side is marked on one region at the surface. Under loading, the grid changes into a rectangle of 7 mm by 5 mm. Determine the effective strain in the region of the grid under this plastic deformation.

8–5. Determine the plastic work per unit volume for the situation where $\sigma_1 = 2\sigma_2$ and $\sigma_3 = 0$; express the answer in terms of σ_1 and $d\epsilon_1$.

8–6. For the stress state $\sigma_x = 10$, $\sigma_y = 5$, $\tau_{xy} = 3$ (all in MPa), $\sigma_z = \tau_{yz} = \tau_{zx} = 0$, determine the magnitudes of the principal stresses from a plot of Mohr's circle.

8–7. A solid shaft of 2 in. diameter is subjected to the simultaneous loading of an axial tensile force of 80 kN and a torque that causes a shear stress at the surface of 20 MPa. With a plot of Mohr's circle find the magnitudes of the principal stresses and the largest shear stress.

8–8. Aluminum is extruded from a 4 in. to a 1 in. diameter in a single stroke. If the average yield stress is 10 ksi and the efficiency factor η is 0.50, what extrusion pressure is required according to the ideal work method?

8–9. A sheet of metal 24 in. wide and 0.150 in. thick is cold rolled to a thickness of 0.100 in. in a single pass; the 24-in. dimension remains essentially constant. The inlet speed of the sheet is 300 ft/min and $\bar\sigma = 25{,}000\ \bar\epsilon^{0.2}$ is the strain hardening relation. Here, $\eta = 0.65$ and the von Mises criterion is applicable. Ignoring any effects of strain rate, find the horsepower needed to perform this operation, using the ideal work method.

8–10. You are to estimate the force needed to coin an American quarter; the process is done cold. The action involved may be considered as axial compression, and the entire workpiece flows plastically. The outer diameter is about 0.95 in., the mean thickness after forming is 0.060 in., the average flow stress is 25 ksi, and sticking friction is reasonable.

8–11. A metal has a constant flow stress of 35 MPa. It is to be drawn from a diameter of 100 mm to 50 mm in a single pass using a die of 30 deg semi-angle. Using the slab method, compute the drawing stress if the frictional condition at the interface is
(a) frictionless
(b) $\mu = 0.20$

8–12. Hot forging is to be done on a slab of metal whose initial cross section is one inch by one inch, the length being 10 in.; after forging, the length remains at 10 in. while the cross section is 0.500 in. high by 2 in. wide. A flat-faced *drop hammer* supplies the necessary force, and sticking friction at the interface may be assumed. The flow stress of this hot billet is 2000 psi.
(a) Determine the *force* needed to perform this operation.
(b) Determine the *work* needed.
Use the slab method of analysis.

8–13. Wire drawing is to be performed by using a die of semi-angle (α) of 6 deg and a reduction $r = 0.20$.

(a) Calculate the ratio of the contact area between the tool and workpiece to the average cross-sectional area of the deformation zone.

(b) Determine Δ.

8–14. The parameter Δ has been defined as the ratio of the length of an arc across the midsection of the deformation zone to the contact length, the arc being centered on the apex of a cone or wedge formed by extrapolating the die walls.

(a) For wire drawing, show that this definition leads to

$$\Delta = \alpha(1 + \sqrt{1 - r})^2/r$$

(b) Using this definition, compare the values of Δ with that given by Eq. (8–38), where

(1) $\alpha = 10$ deg, $r = 0.25$
(2) $\alpha = 45$ deg, $r = 0.50$

8–15. As sheet is cold-rolled it is wrapped around a coiler (physically much like a roll of scotch tape). The coil diameter must be large enough so that coiling involves only elastic bending. If the sheet is 4 feet wide, 0.035 in. thick, and has a yield strength of 30 ksi, what should be the minimum diameter of the coiler?

8–16. A strip tensile specimen has a surface grid imprinted on its face where $\ell_o = 2$ in. and $w_o = 0.75$ in. When it is loaded into the plastic region at a given instant, it is observed that $\ell = 2.35$ in. and $w = 0.710$ in. What is the R-value based upon these observations?

8–17. Calculate the height-to-diameter ratios for drawn cups with drawing ratios of 1.8 and 2.5. Assume constant thickness.

8–18. A typical beer can (aluminum) is 5.25 in. high by 2.438 in. in diameter; the wall thickness is about 0.005 in. while the thickness on the bottom is 0.016 in. What starting blank diameter would be needed to make this can?

___ *9*
machining
or
cutting _____

9.1 INTRODUCTION

Machining can be envisaged by considering an example that should be known to the reader. Starting with a block of wood and a sharp knife, the process of whittling involves the cutting or carving of small pieces (or chips) of wood from the initial block until some desired shape is produced. In the nomenclature of machining, the initial block is the *workpiece,* while the knife is the *cutting tool.* In this process, one hand holds the block and is considered as the *workholder,* and the second hand holds the knife and provides the necessary force *and* relative motion between the work and tool; this hand is viewed as the *toolholder.* All of the essential components involved in any machining setup regarding the workpiece, cutting tool, their relative positions, and motion to produce some desired surface shape are embodied in this simple illustration.

Using this example, we may define *machining* as that process which produces a part of a desired shape and size by *removing* material (called chips) from a parent workpiece through the use of sharp-edged cutting tools. In effect, an oversized workpiece is carved to its desired shape. Since the vast number of industrial materials that are machined are metals, and most experimental or empirical information is available for these solids, the expressions ''metal cutting'' and ''machining'' are often used interchangeably. Near the end of this chapter some comments will be made about the machining of plastics, since interest in these materials has been on the increase.

The three basic components of any common machining operation are the machine tool, the cutting tool, and the workpiece. The *machine tool* consists of a group of subassemblies whose purpose is to maintain the relative positions of the workpiece and the cutting tool, and to provide the necessary power input and relative motions between them in order to produce the shape and size desired. The *cutting tool* may have one or more edges, the geometry of which may be accurately and deliberately controlled or which may have a random configuration.

Finally, it is crucial to realize that whatever shape is produced from the initial workpiece, it is *always* the result of the relative motions of the workpiece and cutting tool. From certain *basic* machine tools, modifications have been made to produce differently named tools, yet if one considers the relative motions involved, it will be apparent that these more advanced machine tools are really off-shoots of the basic ones we shall emphasize in this text.

9.2 BASIC MACHINE TOOLS AND INVOLVED MOTIONS

9.2.1 Engine Lathe

Figure 9–1 is a sketch of a typical engine lathe where the major components are indicated. There is no need to be too concerned about the various component names (there are several texts—for example, Refs. [1 through 4]—which do this in great detail, and these can be consulted if the reader wishes to pursue this in depth); what is more important is

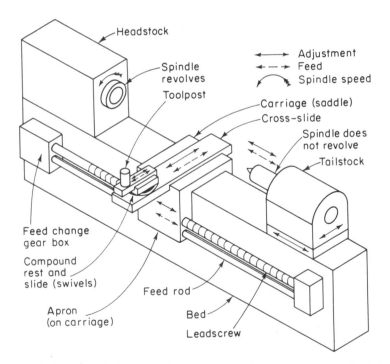

Figure 9–1 Schematic showing major components and movements of an engine lathe.

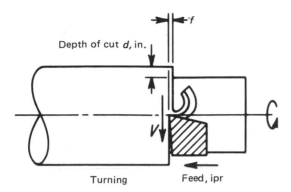

Figure 9–2 Illustration of turning, showing tool and workpiece motions.

to note that machine tools which are categorized as a lathe, or an offshoot of a lathe, possess one thing in common. The workpiece is *rotated* and the cutting tool is *almost* always fed or moved in a straight line.* Any surface that results from this combination of motions can, at least theoretically, be produced on a lathe. Note that this may not always be the most efficient way to machine such a surface, since other machine tools, by using their relative motions, may provide a more economical way to do this. Of the more common *operations* performed with lathes are

1. Turning.   Here the work is rotated and the cutting tool moves in a straight line parallel to the axis of rotation of the work. The net result, regardless of what cross-sectional shape the initial workpiece had, is a cylindrical surface.† In most cases, the workpiece would be initially cast, rolled, extruded, and so on to be approximately circular prior to machining. Figure 9–2 is a schematic of a turning operation.

2. Tapering.   Here the work rotates and the tool moves in a straight line at some angle with the axis of rotation of the workpiece. The manner by which the tool is made to move at such an angle can be done by offsetting the workpiece through the tailstock or

---

* One obvious exception is where curved surfaces are produced by guiding the tool to move in two directions in a plane simultaneously so as to produce a curve-like surface.

† Note that if the workpiece had a square cross section, the cutting tool must project far enough into the section if the resulting shape is to be cylindrical throughout. See Fig. 9–3.

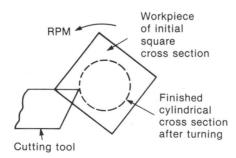

Figure 9–3 Turning an initially square workpiece to produce a cylindrical surface.

by using a taper attachment. Again we will not stress the hardware, but Fig. 9–4 illustrates this process schematically.

    3. Facing.   Often, the end face of the workpiece, rather than the length, must be machined. This is done by rotating the work and moving the cutting tool in a direction perpendicular to the axis of rotation; see Fig. 9–5.

    4. Drilling.   By adapting a drill bit to the tailstock and then advancing the drill in a straight line into the center of the rotating workpiece, a hole is produced; Fig. 9–6

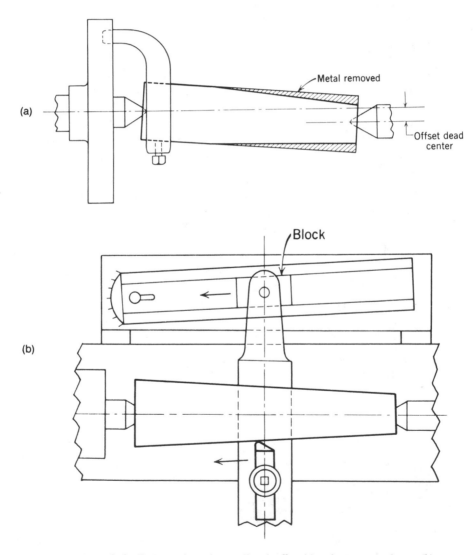

Figure 9–4   Taper turning using a tailstock offset (a) and a taper attachment (b).

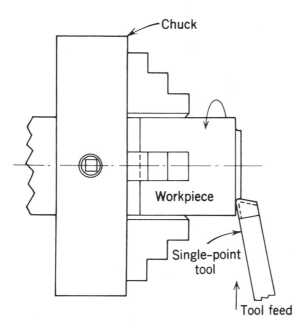

Figure 9–5 An illustration of facing.

illustrates drilling. Drill presses, to be discussed shortly, are used to machine most *initial* holes, but in cases where a hole is to be made in reference to other surfaces machined on a lathe, it is most sensible to drill the hole with lathe motions. In many instances, a drilled hole possesses adequate dimensional accuracy for a desired function; for example, it may serve simply as clearance for a bolt to fit through. Other requirements may demand a hole of closer size variation and roundness that exceeds the capability of drilling, so *finishing* operations follow drilling. To employ such finishing operations *always* requires a hole to start with; these subsequent operations simply enlarge the initial hole to some final desired size and surface finish. Two such operations are discussed next.

5. Boring.   This is really *internal* turning. A workpiece containing a drilled hole is rotated while the cutting tool moves in a straight line parallel to the axis of rotation. Rather than a cylindrical surface on the outer diameter (turning), an internal cylindrical hole results. Figure 9–7 shows a boring operation.

Figure 9–6 The essentials of drilling, showing tool and workpiece motions.

Workpiece

Tool

Straight
boring

Figure 9–7   An illustration of boring.

6. Reaming.   On occasion, functional needs may require a hole whose desired size variation and surface finish exceed those produced by boring.* In single point operations such as boring, elastic deflection of the cutting tool can cause dimensional variations on the cut surface. In reaming, a tool containing a number of cutting edges (see Fig. 9–8) is adapted to the tailstock and moved in a straight line into the previously bored hole. Each cutting edge removes a chip of small thickness and since the cutter is balanced in a radial sense regarding the forces involved in cutting the chips, a final hole of quite uniform diameter and *smoothness*† results.

7. Threading.   In this operation, the cutting tool is ground as indicated in Fig. 9–9 and, via the lead screw and gearing in the headstock (see Fig. 9–1), is moved at a predetermined rate so as to provide a helical type of cutting action, thereby producing a thread of a constant pitch. The full thread is not cut in one pass of the cutting tool; instead, after each pass, it is moved back to the beginning of the cut and advanced radially inward, so that each successive pass for the full length of thread increases its height until the final depth is achieved. Note that there are other methods for producing threads.‡

8. Other comments.   It is important to highlight certain aspects involved in the discussion up to this point. First, the motions involved have two general characteristics. The relative *linear* velocity, called the *cutting speed,* between the cutting tool and workpiece is *high.* Although this word is ill-defined at this point, we shall return to it later in this chapter. In addition, the velocity at which the cutting tool moves into the workpiece (called the *feeding* motion or velocity) is always *much lower* than the cutting velocity (see Fig. 9–11). These motions have tremendous significance in machining operations regardless of the type of machine tool employed.

Additionally, two broad types of cuts can be defined. In one, except for the surface finish that results, the major shape of the final surface is caused by a *generating* type of cut. There, the shape of the cutting tool has a negligible effect on the resultant surface

---

* In certain current practices, boring at very high speeds using diamond cutting tools produces a finished surface of excellent finish and dimensional consistency.

  † Surface finish was discussed in Chapter 3.

  ‡ *Taps* for internal threads and *dies* for external threads are also used (see Fig. 9–10). With taps or dies, the full thread depth is cut in one pass. Rolling of threads can also be accomplished, but this is not pursued further.

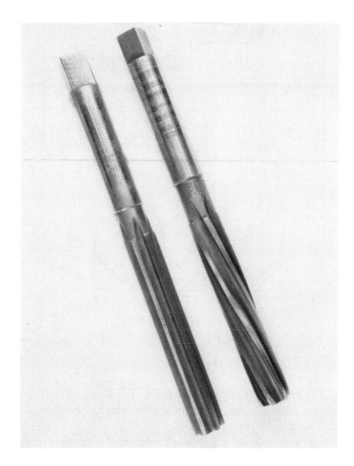

Figure 9–8   Several types of reamers. Photo courtesy of W. H. Durrant.

Figure 9–9 A schematic of threading using a single tool.

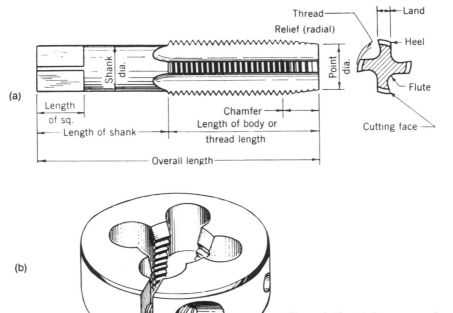

(a)

(b)

Figure 9–10   (a) Components of a tap for cutting internal threads and (b) a die for cutting external threads.

shape. Figure 9–12 illustrates this point. For some operations, the ground shape of the cutting tool produces the final shape of the workpiece. These are called *forming* cuts, and Fig. 9–13 illustrates this point.

What now follows is a similar coverage of other basic machine tools and operations; again the emphasis is not upon detailed descriptions of machine tool components; rather the relative workpiece-tool motions are stressed.

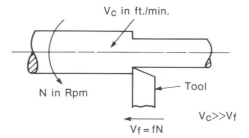

Figure 9–11   Schematic to illustrate the cutting and feeding velocities in turning.

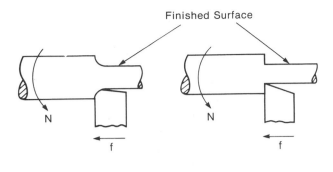

Finished Surface

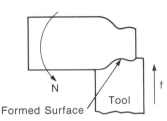

Figure 9–12   Illustrations of generating cuts.

Formed Surface   Tool

Figure 9–13   Illustration of a forming cut.

9.2.2 Milling Machines

By far the most common motions involved here utilize a cutting tool that *rotates* and a workpiece that moves in a *straight line*. Figure 9–14 illustrates a number of the more common combinations. In most instances, flat surfaces are generated, although formed surfaces can be produced. The cutting tool may rotate about a vertical axis (vertical milling) or a horizontal axis (horizontal milling). In fact, if a drill bit is adapted to the rotating head, holes can also be produced; this is done on occasion, but is usually not the primary manner for cutting holes. One of the principal differences between lathe and milling operations is that the latter utilizes a cutting tool composed of a *number* of cutting edges rather than a single cutting edge. Later, when rates of metal removal are discussed, the impact of this difference will be more meaningful.

9.2.3 Drill Presses

The most common method used to produce a hole employs a type of drill press. Here, the cutting tool called a *drill* (see Fig. 9–15) is fitted into the spindle of a drill press shown in Fig. 9–16, most often by the mating of two tapered surfaces (that is, the drill and the spindle of the press). In the majority of cases the workpiece is clamped to a supporting surface called the *table,* and the drill itself is then rotated about a fixed axis and moved in a straight line motion so as to sever chips and produce a hole. Although numerous names are given to many types of drill presses, the fundamental motions just discussed are similar to all.

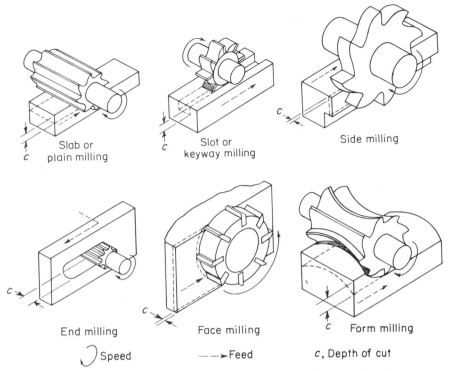

Figure 9–14 Various types of milling operations to produce different surface shapes.

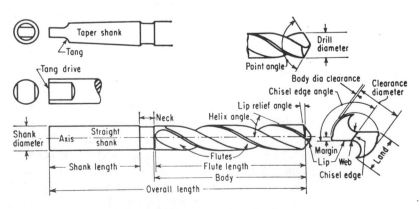

Figure 9–15 Twist drill nomenclature.

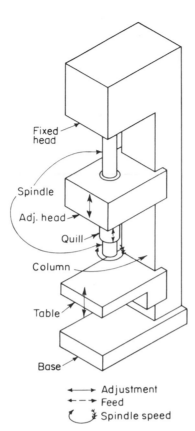

Fixed head

Spindle

Adj. head

Quill

Column

Table

Base

←——→ Adjustment
←— —→ Feed
↻ ↕ Spindle speed

Figure 9–16 Schematic of a drill press showing major components and movements.

9.2.4 Shapers, Planers, and Broaches

Shaping is illustrated in Fig. 9–17. There, a single point cutting tool (that is, one cutting edge) reciprocates back and forth on each *stroke* of the tool. During the return portion of the stroke, when *no* cutting takes place, the workpiece, which is securely clamped to the table, moves or *feeds* at right angles to the tool motion. Flat, vertical, or angled surfaces can be produced by shaping, as indicated in Fig. 9–17. Because the cutting action is not continuous, as in turning, and since only one cutting edge removes chips during a *portion* of each back and forth stroke, the rate at which metal is removed by shaping is rather low. A *planer,* although usually larger in size, is very similar to a shaper; the major difference is that while the work reciprocates back and forth, the tool *moves* during each return stroke in a direction perpendicular to the cutting direction. Note that while the tool is moving prior to each cut, the workpiece is clamped to a table that simply moves back and forth along a fixed path (see Fig. 9–18). Of all the machine tools discussed to this point, the *rate* at which metal is removed is lowest for a shaping operation. By making one major alteration, this same operation produces a great increase in metal removal. This operation, called *broaching,* is illustrated in Fig. 9–19. In essence, a number of cutting edges are

SURFACES MACHINED

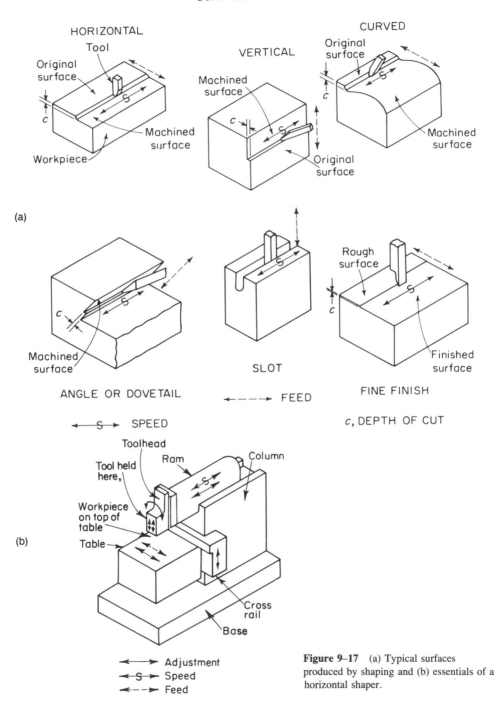

Figure 9–17 (a) Typical surfaces produced by shaping and (b) essentials of a horizontal shaper.

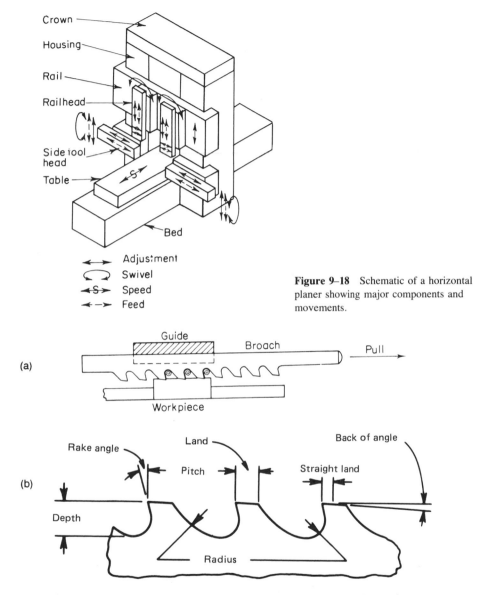

Figure 9–18 Schematic of a horizontal planer showing major components and movements.

Figure 9–19   (a) Schematic of a broaching operation and (b) nomenclature of one type of cutter.

adapted to a single toolholder and moved into and through the workpiece much like a shaping operation. Each cutting edge removes a chip successively lower than the previous edge. Since a number of edges are employed, the rate of metal removal increases dramatically per stroke as compared with shaping, yet the fundamental work-tool motions are identical.

9.2.5 Grinders

In one respect, *grinding machines* are similar to milling machines since the cutting tool, here called the *grinding wheel,* rotates while the work usually moves in a straight-line motion. However, there are at least two fundamental differences between grinding and all the basic operations discussed to this point. The first has to do with the cutting tool itself. In all previous operations, the edges of the cutting tool (whether one, as in turning, or multiple, as in milling) are ground carefully to some desired shape; that is, they are geometrically controlled edges. Grinding wheels are usually composed of small particles of aluminum oxide, silicon carbide, or diamond which are *bonded* in a matrix to form the grinding wheel. Here the small bits are random in terms of orientation; thus a geometrically controlled series of cutting edges does not exist. Besides this randomness of cutting-edge orientation, grinding, as with other abrasive operations called *honing* and *lapping,* always involves the removal of *very small* chips compared with the other basic operations. In terms of rate of metal removal, these abrasive operations involve exceedingly low levels, but this is the fundamental reason that they produce exceptional size control and surface finish as compared with other machining operations. This will become more evident when cutting forces are covered later. Also, grinding is usually essential where extremely hard metals or ceramics are to be machined. Figure 9–20 illustrates several grinding operations. Any engineer involved with the manufacture of products should be aware that machining, and *especially* abrasive operations, should be used only when functional necessity demands the type of dimensional control and surface finish that can be produced by such methods. They are expensive and should not be used indiscriminately.

9.2.6 A Short Wrapup

To this point, machining has been defined and certain *basic* machine tools have been discussed in terms of the relative motions between the workpiece and the tool.

As an illustration of how the basic machine tools have been altered to increase production rates, consider two modifications of the basic lathe. Figure 9–21 shows what is called a *turret* or capstan lathe. There, depending upon the cutting operations required, a number of different tools are positioned and set on various faces of a movable tool holder—a turret. Suppose an initially round workpiece is to be finished as shown in Fig. 9–22(a). As with a basic lathe the work is held in position and is driven by the spindle. Now, however, there is no tailstock to support the opposite end of the workpiece. Note that this limits the free length that can be machined in this manner, since the workpiece is analogous to a cantilever beam; deflections during cutting can be excessive if the "beam" is too long. Instead, the turret replaces the tailstock but provides no end support. Figure 9–22(b) illustrates the various tools mounted in position on the turret. Beginning with the starting drill and proceeding in sequence through the tapping of the thread, all internal operations are accomplished as shown in Figure 9–22(c). Turning of the outer diameter and then cutting off the part to the desired length are accomplished with tools adapted to the *square* turret. Note from Figure 9–21 that this square toolholder, which is

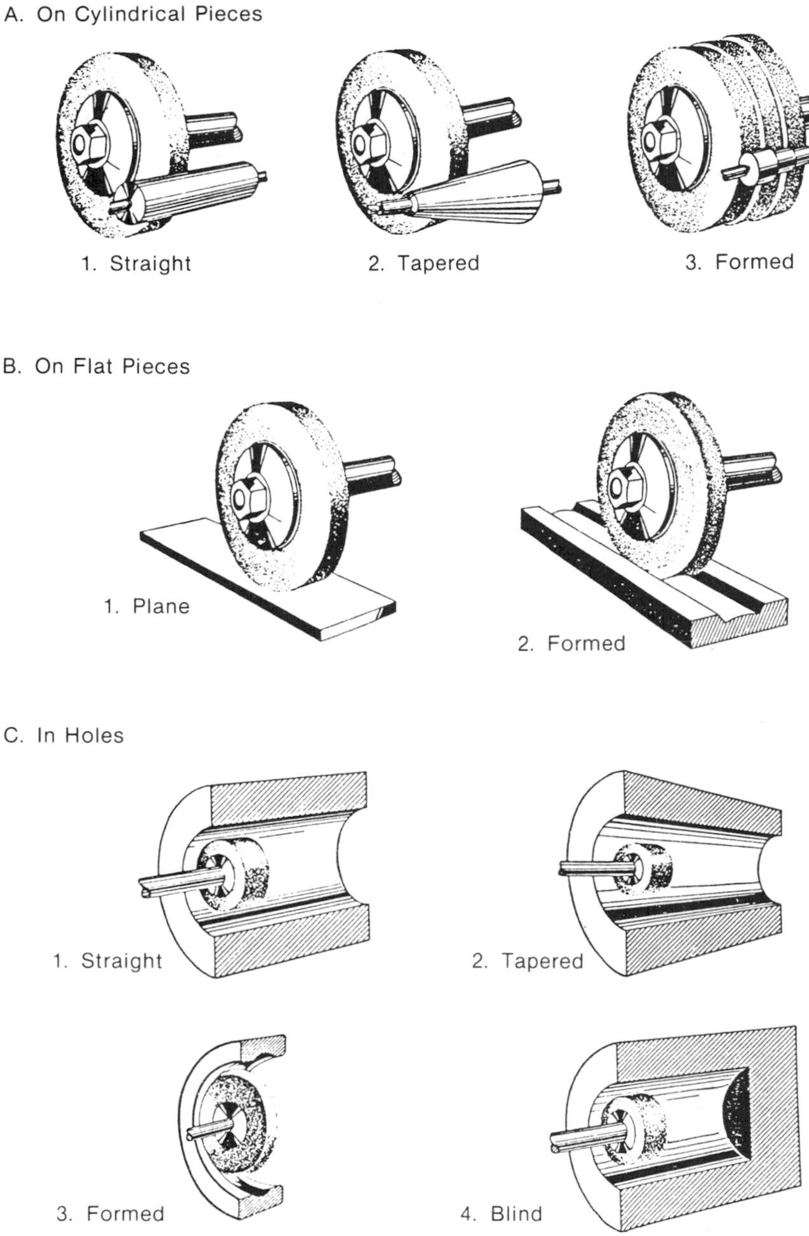

A. On Cylindrical Pieces

1. Straight   2. Tapered   3. Formed

B. On Flat Pieces

1. Plane

2. Formed

C. In Holes

1. Straight   2. Tapered

3. Formed   4. Blind

Figure 9–20   Various types of grinding wheels and operations.

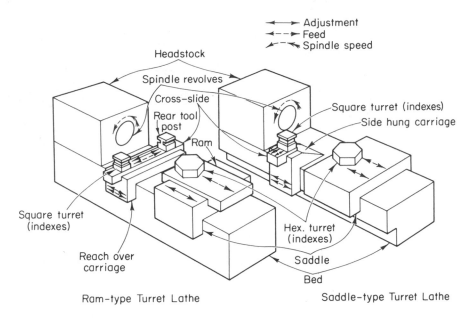

Figure 9–21 Schematic of a horizontal turret lathe showing major components and movements.

connected to the cross-slide, can be indexed in 90-degree increments, thereby providing the use of up to four other tools in addition to those mounted on the hexagonal turret.

Several comments are necessary to complete this discussion. First, each individual tool, whether adapted to the turret or the square toolholder, must be preset to govern the desired length and size of cut. As an example, the turning tool must project beneath the work surface a specified amount to produce the desired diameter; in addition, it must travel a particular distance parallel to the axis of the workpiece to produce the desired length of the turned surface. Various types of *stops* are used for the latter purpose; in essence, these stops cause a disengagement from the drive source at the desired point of travel. Second, after any cutting action is completed by a tool on the hexagonal turret, the turret is returned to its original position (by the hand of the operator or often by automatic means) at which point it rotates to bring the next preset tool into position. During this return, no *cutting* takes place as far as the hexagonal turret tools are concerned. Finally, if this part were to be produced with a basic lathe, it might be necessary to set one tool in position, complete that operation, then break down that setup, and so forth. With the turret lathe, this constant revision of setup for each tool is avoided, since the initial setup, although sometimes lengthy, positions all tools at one time. It is for this reason that a turret lathe will outproduce a basic lathe in terms of pieces per hour; however, this advantage of a turret lathe diminishes if only a few pieces are to be made. As an extreme example, if only one part were needed, a skilled machinist could produce this on a basic lathe probably before the tooling could be completely set on the turret lathe. Therefore, the *volume* of parts may often dictate which machine tool is most efficient.

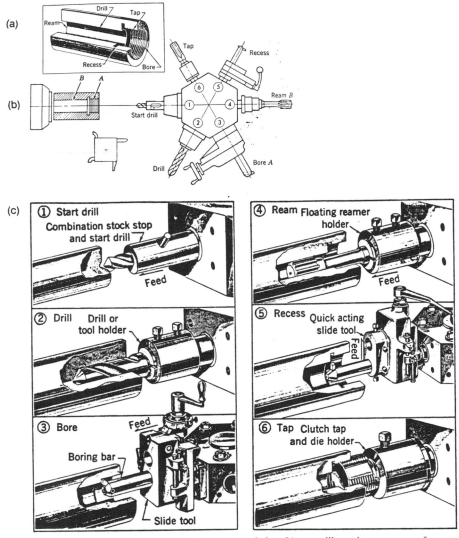

Figure 9–22 (a) Part to be produced on a turrent lathe, (b) setup illustrating sequence of operations related to turrent tolls, and (c) details related to internal cutting operations.

As a final step in evolving from a basic lathe to a similar machine tool capable of extremely high production rates (that is, pieces per hour), the *automatic turning (or screw) machine* was developed. Rather than employing a turret, all tools are prepositioned on various independent toolholders whose feed rate and length of travel are governed by various devices such as cams and screws. Unlike the turret lathe, where the turret must be retracted to position the next tool, all tools on an automatic screw machine are already positioned. Since there is less *noncutting* time involved, production rates are higher.

The principal purpose in this comparison is to illustrate how a basic machine tool may be modified to give increased production, *yet* the relative work-tool motions are identical. Other basic machines have been modified in an analogous manner, and it is hoped that this one illustration will prove adequate in presenting the concepts behind such changes. Note also that with modern control devices, NC or CNC may be added to the name of the machine tool, yet it is best to think of the machine in terms of relative motions, thereby considering it as an offshoot of the more basic machine tools.

9.3 CUTTING TOOLS

9.3.1 Materials

In all machining operations, the removal of chips involves an input of work from some power source. If we restrict the emphasis to metals, there is usually a large amount of plastic deformation involved in producing and severing the chip from the workpiece; the work or energy input is converted to heat in the cutting zone. Added heat is developed as the chip rubs over the tool. (Chip formation is discussed later.) This somewhat complex combination of high temperatures and frictional effects at the tool-work interface requires an acceptable cutting tool produced from a material possessing certain basic characteristics. These include the following:

1. The tool *must* be *harder* and *more wear resistant* than the workpiece. If we try to cut steel with a tool made from commercially pure aluminum, we will not get very far.

2. The tool should *not* be susceptible to undesirable changes in microstructure at the temperature prevailing during cutting. As an example, a tool of plain-carbon steel may possess a martensitic structure initially, but at elevated temperatures this structure may change due to tempering,* and the initial hardness and wear resistance will decrease considerably.

3. The tool should possess adequate *toughness* to avoid fracture, especially with intermittent cutting (such as in milling), since impact loading can induce severe stresses in the tool.

4. If, in addition to the above properties, the tool possesses high thermal conductivity, heat may be more rapidly transferred from the cutting zone into other members of the machining setup. This would aid in prolonging the life of the tool.

Based upon these requirements, the various tool materials are as follows:†

9.3.1.1 Plain High-Carbon Steels.   These contain about one percent carbon and are initially heat-treated to produce a microstructure of martensite and cementite (iron

---

* Discussed in Chapter 7.

† For typical industrial use, some of these tool materials find much greater application than do others.

carbide). Their initial hardness is of the order of 60 to 64 Rockwell C. However, at temperatures in the range of 200°C to 300°C and for modest time at such temperatures, these materials become softer and less wear resistant. This change is due to tempering of the martensite and since higher cutting speeds at the work-tool interface lead to higher temperatures, this tool material must be used at low speeds when cutting metals (thus they find limited use in most industrial applications). With low-strength materials such as wood and plastics, they can be used successfully at higher speeds.

9.3.1.2 High-Speed Steels (HSS).   In addition to containing carbon of the order of 0.7 to 0.8 percent, these materials utilize various alloying elements that may total as much as 25 percent of the mass. Although their proper heat treatment is a bit more involved than that used with plain carbon steels, the basic maximum hardness is similar. Perhaps the key advantage in such a comparison is the ability of high-speed steels to maintain their initial hardness and wear resistance at temperatures up to 500°C. Thus, they can perform satisfactorily at higher cutting speeds than can the plain-carbon type; that is why they are called high-speed steels. There are literally hundreds of commercial grades of these tool materials, the major types being designated T, where tungsten is the major alloying element, or M, where molybdenum predominates. More specific and detailed information about compositions and recommended uses can be best obtained from the producers of these tool steels.

9.3.1.3 Cast Nonferrous Alloys.   These contain the elements cobalt, chromium, and tungsten, which comprise about 95 percent of the total structure. A hardness of about 60 Rockwell C is typical and is obtained without heat treatment; in essence they are cast and then ground to the desired shape. They can be satisfactorily used at temperatures of 600°C but are more brittle than high-speed steel tools. Their introduction somewhat overlapped that of the next tool material, sintered carbides, and since these carbides were, in general, superior cutting tools, the cast nonferrous alloys never caught on to the extent anticipated.

9.3.1.4 Sintered (Cemented)* Carbides.   Elements such as tungsten, titanium, tantalum, and chromium are combined with carbon to form hard, wear-resistant carbides. They are *sintered*† to shape and, in the presence of a binder (usually cobalt), display a hardness and wear resistance in excess of all previous materials. (They also have an elastic modulus about three times that of steel, which leads to smaller deflections under a given load.) Perhaps their major drawback is their relatively low fracture toughness, which makes them susceptible to fracture or chipping under impact loading. As with high-speed steel tools, there are so many types of carbides that recommendations as to specific use should be sought from the producers.

9.3.1.5 Ceramics.   Aluminum oxide is the most widely used material of this

---

* *Coated* carbides usually consist of a base of tungsten carbide that is coated (using chemical vapor deposition) with a thin layer of titanium carbide, titanium nitride, or aluminum oxide.

† See Chapter 10.

type. Not only is it used in grinding wheels (as are silicon carbide and diamonds) but is sometimes used as single point cutting tools. Ceramics are extremely hard and wear resistant and can operate best in continuous cutting (for example, turning) rather than interrupted cutting.w

9.3.1.6 Cubic Boron Nitride. A more recent development in cutting tool materials involves the bonding of a thin layer of cubic boron nitride (CBN) to a carbide base material. Shock resistance is provided by the carbide substrate, while the CBN (which next to diamonds is the *hardest* material currently available) provides extremely high wear resistance. It is also chemically inert to elements such as iron and nickel and displays excellent oxidation resistance at elevated temperatures.

9.3.1.7 Diamonds. Man-made diamonds as well as low-grade natural diamonds are the hardest and most wear resistant of all tool materials; however, they are also very brittle. This usually restricts their usage to grinding wheels or in final finishing operations, such as precision boring, where they operate at high speeds but at a small size of cut. In such situations they can produce parts of close tolerances and excellent surface finish.

In closing this section it is important to note that no single tool material is best in all situations. Depending upon the operation, conditions of cutting speed and size of chip, intermittent or continuous cutting, and so forth, any one of the materials mentioned may prove most economical in a particular situation. For example, diamonds find little success in machining ferrous metals such as steels, while carbides often chip if the cutting speed and feed rate are too low. No attempt is made here to provide a detailed list of the pros and cons of each type of tool material regarding the most appropriate application.

Figure 9–23 illustrates the influence of temperature on the hardness of some of the common tool materials. Since temperature increases with cutting speed, it is obvious why, for example, sintered carbide tools can operate satisfactorily at much higher speeds than can plain-carbon tool steels. Ceramics, CBN, and diamonds are used at even higher speeds.

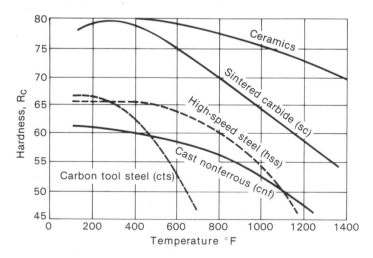

Figure 9–23 Effect of temperature on hardness of various cutting tool materials.

9.3.2 Tool Shape (Signature)

On several past occasions, mention was made of the shape of the cutting tool and whether it was geometrically controlled or of a random nature. Let's dispense with the *random* type first. In abrasive operations* such as grinding, particles such as aluminum oxide are, in fact, the cutting edges. (*Note:* they do remove *chips* just as in turning, but the chip size is so small it is hard to see by eye.) These particles are bonded together, perhaps in the form of a circular disc or *grinding wheel.* As the wheel rotates and is fed into the workpiece, all edges at the wheel periphery enter and then exit from the cutting zone, each removing a small chip. The key point to emphasize is that the *orientation* of each particle as it proceeds to cut is not controlled; that is, angular relationships between the cutting particle and the workpiece do vary.

With controlled cutting edges, we can more logically discuss tool geometry. Included here in some detail are single point tools, drills, and milling cutters. Again, there is no attempt made to cover *all* types of such cutting tools. Certain handbooks and texts of a descriptive nature can be consulted to fill in missing gaps. Here, we are attempting to show that although different names and nomenclature are involved, there is still a degree in similarity of *purpose* for producing controlled angles on cutting tools.

9.3.2.1 Single Point Tools   Although there are different systems† used to describe the *nomenclature* of single point tools, our preference is the one discussed below since, for an introduction to this subject, it is more direct and simpler to envisage than other systems.

Figure 9–24(a) shows three views of a single point tool, while Fig. 9–24(b) provides an overall perspective; also shown are the tool angles given in a sequence called the *tool signature.* Each of the three faces constitutes a plane whose relation to the initial tool blank is developed by a pair of angles. (The faces are usually produced by grinding away material from the initial blank.) For example, the back and side rake angles produce a surface where both angles are measured from the top of the blank, whereas the end cutting edge and end relief angles are referred to the original front of the tool blank. If a nose radius other than zero is specified, this refers to the radius of an arc that is ground to blend in smoothly with the end and side cutting edges. The combined influence of the *rake* angles is important in controlling the direction of chip flow; in addition, these angles also influence the tool life and magnitude of the forces needed for chip removal.‡ The basic function of the *relief* angles

---

* *Honing,* most often used to finish holes, and *lapping,* used to finish flat surfaces, are, in essence, fine-scale grinding.

† See for example, G. Boothroyd, *Fundamentals of Metal Machining and Machine Tools.* New York: McGraw-Hill Book Co. (1976), pp. 167–84.

‡ It is more correct to consider an *effective rake angle,* which results from the values of the back and side rake angles. For our introductory purposes this refinement need not be considered. We note too that as tool hardness increases but toughness decreases, smaller rake angles are used. This is why carbides, ceramics, and diamonds often possess *negative* rake angles to avoid early chipping of the cutting edge.

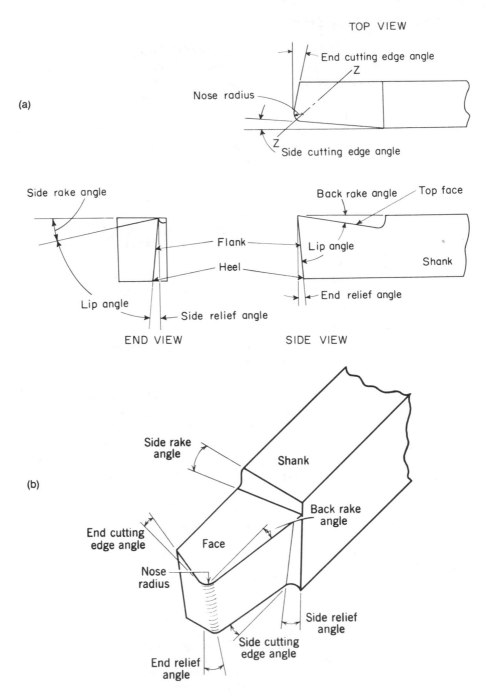

Figure 9–24   (a) Three views of a single-point cutting tool showing the tool signature and (b) overall view of a finish-ground tool.

is to avoid rubbing between the tool and workpiece; the *end* cutting edge angle is ground for the same reason. For cuts such as turning, the *side* cutting edge angle provides the major cutting edge, and this angle also influences the direction of chip flow and contact area between the chip and tool. In addition, as discussed later, tool life and surface finish are both affected by the side cutting edge angle. Finally, the primary effects of the nose radius are connected with tool life and surface finish. Although most of these effects will be covered later in more detail, the reason for introducing them at this time is to indicate that there are important consequences connected with the manner in which a single point tool is ground to a certain signature.

9.3.2.2 Drill Geometry.

Figure 9–15 shows the geometry and nomenclature related to a two-fluted twist drill. Although it appears very different from a single point tool, there are definite analogies involved. Each cutting edge angle (two here) provides the same function as the side cutting edge angle of the single point tool, while the helix angle is a ''combination'' of the end and side rake angles. The flutes, which provide the means for the chip to be removed from the cutting zone, have no direct counterpart with single point tools unless one considers them analogous to the top face of such tools. There is, however, no correspodence between the nose radius and the chisel edge. Note that drills do not have a conical ''point''; we shall discuss why later.

9.3.2.3 Milling Cutters.

Since there are a variety of milling operations used in machining (see Fig. 9–14), various cutting tool descriptions are encountered. For illustrative purposes, consider a face milling cutter, as in Fig. 9–25, where the pertinent angles are indicated. Note that for the cutter shown, a number of single point tools would be inserted and clamped into a solid body. Although this is a multiedged tool, just as is the two-fluted twist drill, the entire unit is called a *milling cutter*. Table 9–1 provides a useful comparison of the approximate equivalence of the angles mentioned to this point.

With operations such as shaping and planing, the tool nomenclature is really identical to the single-point signature discussed earlier, whereas broaching and reaming include certain modifications. Yet with all of these operations, the basic angles provide similar functions.

9.3.2.4 Grinding Wheels.

We have noted earlier that a grinding wheel consists of many particles which are held in position by some type of bond. Abrasive particles most commonly used are aluminum oxide, silicon carbide, and diamond; cubic boron nitride is also used on occasion. Of the important characteristics of any grinding wheel (besides the *type* of abrasive particles) are the following:

1. The particle size called the *grit number*. In a relative sense this ranges from coarse, number 10, to very fine, number 500. Going to finer grit numbers within that scale improves the capability of getting better surface finish which, in turn, requires smaller sizes of cut (that is, feed rates and depth of cut).
2. The type of *bond*. There are a number of materials used to bond the abrasive grains

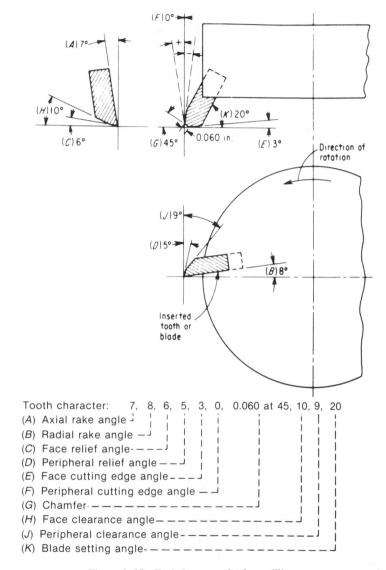

Tooth character: 7,   8,   6,   5,   3,   0,   0.060 at 45, 10, 9, 20

(A) Axial rake angle
(B) Radial rake angle
(C) Face relief angle
(D) Peripheral relief angle
(E) Face cutting edge angle
(F) Peripheral cutting edge angle
(G) Chamfer
(H) Face clearance angle
(J) Peripheral clearance angle
(K) Blade setting angle

Figure 9–25 Tool signature of a face milling cutter.

together so as to form a solid mass that becomes a grinding wheel. Among the more common types of bonds are the *vitrified, resinoid,* and *rubber;* other materials are used, so the ones stated serve as illustrations. Vitrified wheels are quite strong but also brittle, since the bond is basically a glass. The resinoid type employs a thermosetting resin (see Chapter 7) as the primary bond material. Because the elastic modulus is lower than the vitrified type, these wheels have greater flexibility

TABLE 9–1* COMPARISON OF APPROXIMATELY EQUIVALENT TOOL ANGLES FOR VARIOUS TYPES OF CUTTING TOOLS

Single Point Tool	Drill	Milling Cutter
Back rake	⎰ Helix	Axial rake
Side rake	⎱ Helix	Radial rake
End relief	⎰ Lip relief	End relief
Side relief	⎱ Lip relief	Peripheral relief
End cutting edge	Semi-point	End cutting edge
Side cutting edge	Semi-point	Corner angle

*For greater detail, see *Machining Data Handbook.* Metcut Research Associates Inc., Cincinnati, Ohio, 1966, Section 2, p. 411.

and somewhat lower strength. With the rubber-bonded type, even lower stiffness and greater flexibility results; however, they can be made quite thin and these are often used as *cutoff* wheels (in essence, they perform as a type of circular saw).

3. The *grade* of a wheel. This is a measure of the strength of the bond or tenacity with which the bond resists a tearing out of particles from the wheel. Grade is also related to the so-called *hardness* of a wheel, where harder wheels imply higher bond strength.

4. The *structure* of a wheel. This designates the relative porosity, which has several implications. A relatively *dense* structure indicates a greater number of abrasive particles per unit volume, but a lesser degree of surface area into which particles of the work material (called *swarf*) can fit as they are removed. The reverse comments apply as porosity increases; this implies a more *open* structure.

We again make no attempt to document the numerous combinations of such designations used; many types of wheels are commercially available, and proper advice from producers of grinding wheels must often be sought. However, there is one overall concept to keep in mind. Ideally, one would like to use a grinding wheel until the major number of abrasive particles that contact the workpiece become worn (or are dulled) to the point that effective grinding diminishes. At that time, the worn particles would literally be pulled away from the bond so as to reveal new, sharp particles for subsequent grinding. A proper combination of grade and structure thus becomes important in such a self-sharpening wheel. Often the wheel has an overall combination of characteristics such that a self-sharpening action does not occur. Then it is necessary to *dress* the wheel by some additional act that severs the worn grains in order to expose fresh ones. Dressing can be accomplished by moving a supported and shaped diamond (for example, pointed) across the periphery of the wheel so as to remove a thin layer from the outer diameter. This also *trues* the wheel; that is, it tends to return the wheel to a circular configuration. If wheels are not *self-sharpening* or *dressed,* small pieces of the material being ground will *load* the regions between the abrasive particles; at the same time, the particles themselves become worn. Several consequences can result. First, the temperature of the workpiece in the area or region being ground can increase to undesirable levels. With steel, for example, the temperature can reach a level that will transform the original structure to austenite; then as the wheel passes that region, the rate of heat transfer into the body of the workpiece can be fast enough to convert the austenite to martensite. This can induce undesired residual surface stresses or even cracking of the workpiece. In general, by maintaining sharp cutting edges, flooding the cutting region with a copious flow of coolant, and reducing the size of cut will preclude such undesirable results.

9.4 CUTTING FLUIDS

A truly large number of cutting fluids is available commercially; to document all would be a prohibitive and, for our purposes, an unnecessary task. In many industrial situations, users will most often consult with manufacturers of fluids to seek advice as to which is most suitable for a particular machining operation.

In a most general sense, cutting fluids are used for one of two purposes. For operations where low friction is of greatest concern, thereby leading to excellent dimensional control and surface finish, so-called *lubricants* find major use. Many such operations are carried out at relatively low speeds. At the other extreme are those fluids whose primary purpose is to conduct heat away from the cutting zone, thereby cooling the workpiece; they are often called *coolants*.*

Lubricants are often composed of mineral oil, sulfur, chlorine, and fatty oils and, in comparison with coolants, have lower thermal conductivity. Coolants often are composed of a mixture of water and soluble oils; in many cases the composition involves water, emulsifiers, fatty oils, and rust inhibitors. Note that water itself is an excellent coolant but, because it would produce rust on the machine tool, cannot in any practical sense, be used by itself. Perhaps the most important point to remember is that coolants are used where tool life is of primary concern, whereas lubricants find major use where surface finish and dimensional control are most important. One can certainly take exception to such a division, since instances can be cited wherein the use of coolants at cutting conditions of high speeds and a small size of chip can produce excellent surface finish. Yet to provide the most *general* insight regarding cutting fluids, the distinction proposed is certainly reasonable.

9.5 CHIP FORMATION AND TOOL WEAR

From previous discussions, one can now envisage just what is used to conduct a machining operation. Depending upon the desired shape of the part to be produced, a machine tool, in conjunction with a cutting tool made from a particular tool material and having a certain signature, plus the use of a pertinent cutting fluid, are selected. Now we are ready to start cutting and a good question to ask is ''just what happens when material is cut?'' First, two (but not all) answers are now given.

---

* Cutting fluids also flush chips away from the cutting zone.

9.5.1 Chip Formation

To cause a chip, enough power must be supplied by the machine tool, since a certain amount of energy is needed to sever or shear the chip from the workpiece; in addition, the cutting tool must possess certain desirable properties, as we discussed earlier. Depending to a large extent on the properties of the workpiece, two broad classifications of chips can result. In machining *brittle* metals such as gray cast iron (which has limited ductility), the chips tend to come off in small pieces and are called *discontinuous* chips; this is illustrated in Fig. 9–26(a). In contrast, the machining of ductile metals often causes a continuous ribbonlike chip to form*; thus, the word *continuous* is used in this case.

It is sensible to further subdivide continuous chips into those which do *or* do not form with a *built-up edge* (BUE). Figures 9–26(b) and 9–26(c) illustrate this point. BUE formation is a phenomenon of decided significance, since its presence can have a pronounced effect on both tool life and the surface finish of the machined part. Most probably, a built-up edge results when metal from the underside of the chip adheres to the tool due to the effects of friction at the tool-chip interface and the extremely high pressure that can result at that interface. Although this problem has been studied for many years, it is still found as a subject in learned publications. It does appear, however, that the BUE reaches some critical size; then some of it breaks loose and is deposited on the finished work surface as well as the underside of the chip; see Fig. 9–27. This overall process repeats itself as cutting continues. The results of much research indicate that to diminish or to preclude BUE formation, one should:

1. Increase the cutting speed.
2. Decrease the thickness of the chip (that is, use smaller feed rates).
3. Use a lubricant.

It is perhaps ironic that in view of certain detrimental aspects of the BUE, it actually covers the cutting edge during machining and forces the moving chip to make contact on the top face of the tool; this may actually prolong tool life. The manner by which *tool wear* occurs, as discussed next, is also affected by the presence of a BUE.

9.5.2 Tool Wear

Due to the combined effects of elevated temperature, forces of rather large magnitude, and frictional effects between a chip and cutting tool, it should not be surprising that cutting tools wear as the operation proceeds. Two general types are usually specified, these being *flank* and *crater* wear. Figure 9–28 illustrates this. Flank wear (for example, with a single point turning tool) results on the side face in a direction downward from the cutting edge. Literally, tool material is worn away by the rubbing (frictional) action at the flank-tool

---

* Such a result is detrimental in practice, since chips must be removed from the machine tool periodically. Continuous chips, being like tangled wire, are an expensive nuisance. To avoid this, *chip breakers* are often used. Regardless of how a chip breaker is introduced, the major effect is to force the chip to bend so severely that it periodically fractures into a number of discontinuities, thereby avoiding a long, continuous chip.

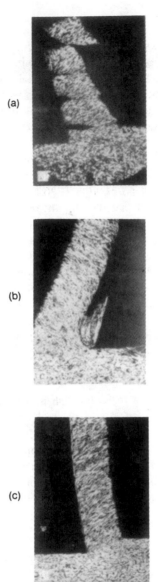

Figure 9-26 Examples of a discontinuous chip (a), a continous chip with a built-up edge (b), and a continuous chip without a built-up edge (c).

interface. Crater wear, also illustrated in Fig. 9–28, most often results when a BUE forms, thereby causing the major frictional effect between the chip and tool to occur on the top face of the tool (that is, away from the cutting edge). Of course, either type of wear is undesirable, and either may dictate the useful time a tool operates before it must be removed and reground, or replaced, to produce a new or fresh signature. Perhaps crater wear can be considered more serious, since if it becomes excessive, the remaining section of the tool can no longer support the forces involved and catastrophic fracture of a section

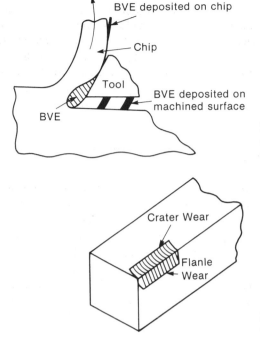

Figure 9–27 Sketch showing built-up edge deposits on underside of chip and on the finished work surface.

Figure 9–28 Sketch showing the difference between flank and crater wear.

of the tool results; this requires excessive regrinding, which leads to greater expense. Although machining economics is covered in Chapter 13, it is prudent at this point to state a sensible general philosophy. One should *never* use tools to catastrophic failure* in *industrial* practice, since machining costs will increase.

9.6 BASIC MACHINING PARAMETERS

The word *basic* as used here may seem confusing, since one could argue that inputs such as the tool material and signature, type of cutting fluid, and work material are all important in analyzing any machining process. This is certainly true. However, *once* the machine tool and workpiece are chosen (this comes about usually from design specifications including the *shape* of the desired part and the mechanical properties needed), and once a sensible tool material and shape are specified, along with an acceptable cutting fluid, there then remain only *three* parameters that can be adjusted to influence such things as tool life, power requirements, and surface finish among other concerns. These three parameters are the *feed rate, depth of cut,* and *cutting velocity* (or cutting speed), and they are the only factors that can be manipulated or controlled by alteration, once all other parameters have been defined (that is, work material, tool material, and the like). In fact,

* In tool life testing, this is often done (see Sec. 9.9).

the combined effects of these basic quantities are the final *inputs* that govern just what goes on at the cutting edge. Much has been written about numerically controlled (NC) machines or computer-aided manufacturing (CAM),* and one may get the impression that somewhat exotic control devices are all that is needed to understand machining. (Many experts in controls or computers also seem to believe this.) Yet if one does not understand the importance of the *basic machining parameters,* which are the factors to be properly controlled, all else can become secondary. After all, a computer does what it is told to do, so if an improper combination of feed, depth, and velocity are fed into a computer, it will be controlling poor inputs.

9.6.1 Feed or Feed Rate

In its most general sense, the *feed* is analogous to the thickness of a chip and is most often defined as the distance that the cutting tool moves into or along the workpiece surface per some relative unit time. Several examples should help to clarify this. Turning cuts involve a feed rate that is set on the machine tool in terms of *inches per revolution* (ipr).† During each revolution, the tool advances parallel to the axis of rotation a certain distance at a constant rate; this means that the thickness of the chip is uniform for whatever feed rate is set. With the spindle speed set for a particular rpm (call this N), then in one minute, the tool would have moved fN inches; Fig. 9–29 shows this. With drilling operations the concept is identical; if we set a feed rate of so many ipr and a spindle speed of N, the drill advances fN inches into the workpiece in one minute. Figure 9–30 illustrates this.

With milling, the situation is not so simple, for several reasons. Here, as the cutter rotates (see Fig. 9–31), and the workpiece, which is clamped to the table, is fed into and past the cutter, the chip thickness (that is, feed) is not constant; rather, it varies during the cutting action. It is most common to select a desired *average* thickness of chip (usually specified as *inches per tooth*), then to multiply this by the number of teeth in the cutter to give inches/rev., and finally to multiply by N (rpm of the cutter as set by the operator) leading to a feed in *inches/min.* It is this feed that is *set* on the machine which then, by this backward calculation, leads to the desired average chip thickness, so

$$F = fnN \qquad (9\text{-}1)$$

where F = feed of table *set* in in./min
$\quad\ \ N$ = spindle speed in rpm as *set*
$\quad\ \ n$ = number of teeth in the cutter
$\quad\ \ f$ = resultant *average* chip thickness in inches per tooth

With these three widely used operations as a guide, the feed rates for other operations should be understood if it is kept in mind that the feed always comprises one of the dimensions related to the cross-sectional *area* of the chip as it is removed from the

---

* See Chapter 5.

† Many machine tools used in the United States are still calibrated in English (not SI) units; thus we employ such units here. Whether inches or mm are used does not alter these basic concepts. Conversion to SI units is a simple matter.

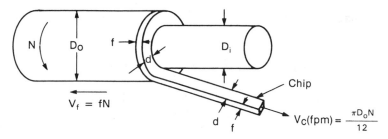

Figure 9–29   Machining parameters related to a turning cut.

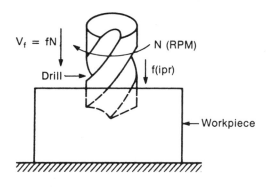

Figure 9–30   Machining parameters related to a drilling cut.

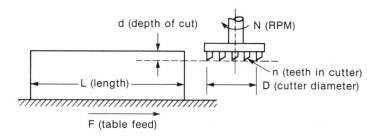

SIDE VIEW

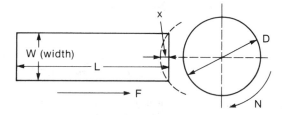

TOP VIEW

Figure 9–31   Machining parameters related to a face-milling cut.

workpiece. In nearly every case, the feed is the smaller of the two dimensions and carries the symbol *f*.

9.6.2 Depth (or Width) of Cut

In conjunction with the feed, the depth of cut is the other dimension that defines the area of the chip. For turning, as in Fig. 9–29, the depth of cut is the *radial* distance that the tool point is set from the outer diameter; that is it is one-half $(D_0 - D_i)$; it will be designated by *d*.

For drilling, the depth of cut is literally the radius of the drill, whereas in milling, as in Fig. 9–31, the depth of cut is the distance the tools project beneath the original work surface. A *note of caution* about milling. From geometric considerations, the *area* of chip removed by each *tooth* per revolution is the product of *fw*, where *w* is the *width* of the workpiece. This illustrates one instance where definitions must be used with care; this should become clearer when *rates of metal removal* are discussed shortly.

9.6.3 Cutting Velocity (Speed)

The cutting velocity *V* will, in this text, *always* be considered as the *maximum linear* speed existing between the tool and workpiece.

Continuing with turning, drilling, and milling as above, the cutting velocity *V* is given by

$$V = \pi dDN/12 \qquad\qquad (9\text{--}2)*$$

where

V = cutting velocity in *feet* per minute
N = rpm of either the workpiece (turning) *or* the tool (drilling and milling)
D = diameter in *inches* of the component being rotated (outer diameter of work for turning, but the diameter of the drill or milling cutter for those operations)

For any other machining operation, if one carefully notes the relative geometries involved, it should not be difficult to use some common sense in order to determine the values of *f*, *d*, and *V*.

9.7 CUTTING ZONE GEOMETRY AND FORCES

Now that topics of tool angles, wear, and basic machining parameters have been covered, we continue with the question of just what goes on at the cutting edge. In Fig. 9–32(a), the basic manner by which chips are removed is illustrated.† The cutting tool, having a

---

* If SI units are desired, *V* in feet/min can be converted to m/s by multiplying by 0.0051.

† Strictly, this illustrates what is called orthogonal or *two-dimensional* cutting. Turning the full wall thickness of a tube or shaping, where the width of the cutting edge is greater than the width of the workpiece, both constitute orthogonal cutting. Although most operations are three-dimensional, the concepts presented here still portray a realistic picture of the basic mechanism of chip removal.

rake angle α and *relief angle* θ, projects below the original work surface at some distance t (strictly, this is the feed or chip thickness as set), and moves through the workpiece at a velocity V. Chips form by causing material ahead of the tool to shear along the *shear plane* which lies at a *shear angle* Φ that is oriented as shown. The actual thickness of the chip t_c depends upon α and Φ. Often, the ratio of t to t_c is called the *cutting ratio r* and in actual practice r is always less than one; that is, $t < t_c$. Reference to Fig. 9–32(b) where the length of the shear plane is taken as unity, leads to

$$r = t/t_c = \sin \Phi / \cos (\Phi - \alpha), \qquad (9\text{–}3a)$$

so

$$\tan \Phi = (r \cos \alpha)/(1 - r \sin \alpha) \qquad (9\text{–}3b)$$

Since α is ground to a desired angle and t can be set accurately, a reasonable measure of t_c allows a prediction of the shear angle Φ with Eq. (9–3b). Also, only when $\alpha = \Phi$ is t_c equal to the length of the shear plane.

Figure 9–32(c) is a velocity diagram, or *hodograph,* which is used to determine velocities of concern. Relative to the tool, the workpiece moves at a velocity V taken in the horizontal direction. As the chip forms, the metal is suddenly forced to move at a velocity V_c parallel to the tool face; this requires a jump discontinuity along the shear plane V_s at an angle Φ to the horizontal. By setting V to some scaled value and using the angles Φ and α as shown, the magnitudes of V_c and V_s can be measured from the holograph as functions of the value of V. An alternative approach is to note that continuity of mass, if we assume no change in the width of the chip, requires that

$$Vt = V_c t_c \quad \text{or} \quad V_c = rV \qquad (9\text{–}3c)$$

Noting that $V_s = A + B = V \cos \Phi + V_c \sin (\Phi - \alpha)$
and

$$V_c = V \sin \Phi \, / \cos (\Phi - \alpha)$$

leads to

$$V_s = V \cos \alpha \, / \cos (\Phi - \alpha) \qquad (9\text{–}3d)$$

Although not derived here,* an expression for the *shear strain* involved in machining can be found in terms of Φ and α. It is

$$\gamma = \cot \Phi + \tan (\Phi - \alpha) \qquad (9\text{–}3e)$$

For example, if Φ is 30 deg and α is 10 deg, then γ is 2.1. What this implies physically is that the shear strain in machining is *extremely high* relative to other deformation processes; values as high as five are not uncommon. This would be roughly equivalent to a reduction of area exceeding 90 percent in a tensile test, so the amount of strain hardening induced in a chip is very large. Hardness measurements of chips and the region surrounding the shear plane support this observation.

Figure 9–32(d) illustrates force relationships pertinent in orthogonal cutting. The

---

* See S. Kalpakjian, *Mechanical Processing of Materials* Van Nostrand Co., Inc., 1967), pp. 245–50.

total force F_t acting between the chip and tool can be resolved into different sets of components. First consider the forces that are horizontal and vertical; these are the cutting force F_c and vertical or thrust force F_v. Next, the force F_t could be resolved into components parallel and normal to the tool face, these being F_f and F_n, respectively. Finally, F_t can be resolved into components normal and parallel to the shear plane, these being N and F_s. Depending upon which pair of components is of greatest interest, various relationships as functions of F_t, Φ, α, and the *friction angle* β can be developed where the coefficient of friction between the chip and tool is taken as tan β. One illustration of such relationships has been the extensive research work, conducted for many years, concerned with predicting the relationship or interconnection of Φ, α, and β. Using the model shown

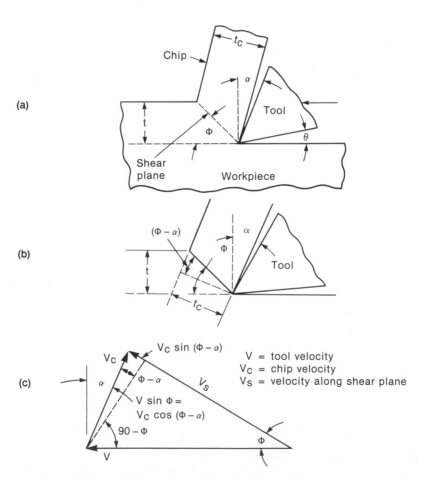

Figure 9–32   (a) Orthogonal machining indicating machining parameters and angles of concern, (b) geometry relating chip thickness t_c, to thickness as set t, (c) velocity vector diagram or hodograph, and (d) force relationships in plane and reduced to a force circle.

in Fig. 9–32(d), assuming that β is independent of Φ, and considering that the shear angle results so that the maximum shear stress occurs on the shear plane (or alternately that the cutting force F_c is a minimum) leads to

$$\Phi = 45 + \alpha/2 - \beta/2 \qquad\qquad (9\text{-}3f)^*$$

Many other expressions have also been developed where other models and assumptions were employed. Unfortunately, none is in acceptable agreement with experimental results using a *wide range* of cutting conditions. Yet, in a qualitative sense, Eq. (9–3f) provides some correlation with reality. If α is fixed for a given tool, then decreasing β, perhaps by using an effective lubricant, should cause Φ to increase, with a corresponding decrease in the chip thickness t_c (t being constant here). Thinner chips, leading to less deformation and lower forces, should, if anything, improve surface finish. Such a result is observed in practice.

* This is often called Merchant's analysis after M. E. Merchant, *Journal of Applied Physics* 16.5 (1945) p. 267.

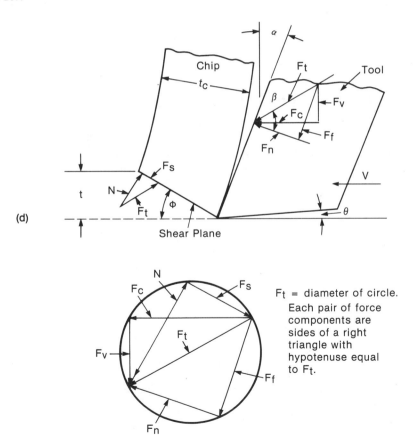

Figure 9–32 (Continued)

In closing this brief section, it is appropriate to discuss why analyses and models such as those used above do not always provide the degree of agreement with experimental results one would desire.

1. Assuming that shear occurs on a single plane is not verified in practice; rather, a shear *zone* is observed. Thus considering that there is a distinct shear angle Φ is incorrect.
2. Plastic deformation ahead of the shear zone as well as in a small substrate beneath the finished work surface is not taken into account.
3. Built-up edge formation is not considered.
4. Tools are not perfectly sharp, regardless of the care exercised in grinding the various angles. In addition, tool wear begins quickly as cutting commences.

In spite of these shortcomings, such efforts are still beneficial in terms of improving our understanding of some of the basic mechanisms of machining, and for that reason alone they have been worthwhile. Perhaps more realistic models and assumptions will one day lead to predictions that agree more closely with experimental data. We note here that considerations of cracking and the inclusion of fracture toughness in such studies is one such attempt.*

9.8 RATE OF METAL REMOVAL (RMR)

The importance of RMR is that it governs the rate of production. Higher rates of metal removal lead to producing a larger number of pieces per unit of time, say parts per hour. Usual practice is to employ RMR with units of cubic inches per minute.

First consider turning, as in Fig. 9–29, where values for feed and depth have been set. Imagine, as the work begins to rotate at N rev/min and the tool feeds into the work, that you grabbed the end of the chip and led it away from the work for one minute. (*Don't* ever try to do this because chips can be razor sharp and are also quite hot.) What results, for all practical purposes, is a chip whose area is *fd* square inches and whose length is V feet. Thus an excellent approximation for the volume removed per minute is

$$RMR = 12Vfd \qquad (9\text{-}4)\dagger$$

where

* A. G. Atkins, "Fracture Toughness and Cutting," *Internnational Journal of Production Research* **12**, 2 (1974), p. 263. See also A. G. Atkins and Y. W. Mai, *Elastic and Plastic Fracture* (Chichester: Ellis Horwood, New York: John Wiley and Sons, 1985), Chap. 10.

† The data in Tables 9–2 to 9–5 were originally obtained by using English units. To avoid unduly small or large numbers, velocities, for example, have traditionally been specified in feet per minute, whereas feed and depth are given in inches. This requires the use of the coefficient 12 to give a velocity in inches per minute. Such a conversion is needed in a number of places in sections that follow. One could argue that the exclusive use of SI units would avoid such conversion factors, but to convert all such original data is unnecessary. If an answer is desired in SI units, then one can find the answer in English units and simply make the proper conversion at that time.

V = cutting speed in *feet*/min
f = chip thickness (feed) in inches
d = depth of cut in inches
RMR = in.^{3}/min

To preempt a question that almost *always* arises at this point, let's compare the approximate value for RMR above with an exact calculation, using reasonable values for the various parameters.

Suppose a four-inch-diameter bar of a solid is to be turned with a feed rate of 0.015 ipr, depth of cut of 0.075 in. and an rpm of 150. From Eq. (9–2) the value of V based upon the outer diameter is 157 fpm. Then, using Eq. (9–4), we obtain

$$\text{RMR} = 12(157)(0.015)(0.075) = 2.12 \text{ in.}^3/\text{min}$$

Now consider a more exact calculation. Since the annular area of metal removed is $(\pi/4)(4^2 - 3.85^2)$ and the tool moves fN inches in one minute, then the exact value is

$$\text{RMR} = (\pi/4)(4^2 - 3.85^2)(0.015)(150) = 2.08 \text{ in.}^3/\text{min}$$

This leads to an error of less than 2 percent and is one of the reasons that the simplicity of Eq. (9–4) finds wide use. Of course, as the relative ratio of the depth d and diameter D increases, Eq. (9–4) becomes less accurate.* For most practical situations and certainly for our purposes, we will use Eq. (9–4) when computing values of RMR in *turning*.

With drilling, and starting at the time when the full drill diameter is cutting, the value of RMR is simply the area of the hole times the linear velocity of drill movement into the work piece. Thus,

$$\text{RMR} = (\pi D^2/4)(fN) \qquad (9\text{-}5)$$

where D is the drill diameter in inches and f and N carry their usual meaning.

At this point, we stress that memorizing a number of such equations is not essential and can lead to errors if *used in situations where they do not apply*. It is best to look at the given operation and sensibly determine which dimensions and speeds are directly tied to the rate of metal removal. With this suggestion, let's return to the milling operation shown in Fig. 9–31. As the workpiece enters the cutting zone, it is moving at some speed of F in./min as set on the machine. Once the full width of workpiece is being cut (see dotted line on Fig. 9–31), in each succeeding minute a volume of metal equal to the cross section of dw times the value of F will be removed; thus,

$$\text{RMR} = Fdw \qquad (9\text{-}6)$$

where d and w are the linear dimensions shown. Two points are instructive here. First, no single equation satisfies RMR for all operations, so a little logic must be used for operations other than those described. The second point is that regardless of the operation, RMR is a direct function of feed, depth of cut, and cutting speed (that is, rpm here). It

---

* This is because the velocity based upon D deviates more from the *mean* velocity based upon the mean diameter.

is hoped that these few illustrations will provide adequate background to this important concept.

We close this section with a few comments related to *machining time,* that is, how long does it take to actually make the desired cut? In all cases it is simply a matter of determining the length of workpiece to be machined and then dividing by the velocity *parallel* to or *in the direction* of that length. For example, if a part that is L inches long is to be turned, then the machining or cutting time t_m is simply

$$t_m = L/fn \qquad\qquad (9\text{-}7)$$

Similarly, if a hole of depth L is to be drilled, $t_m = L/fN$ also. This does neglect the small distance the drill point must move before the full drill diameter comes into play. This can be determined with a little trigonometry and added to L if greater accuracy is desired. Milling as in Fig. 9–31 requires a bit more explanation. When the cutter first touches the work, it must travel a distance x before the full width is being machined. (This is similar to the distance the drill point must move as just mentioned, although x is usually much larger.) Again, x can be found because pertinent dimensions are known. Now if the cutter travels the distance L, then $t_m = (L + x)/F$, F being the table feed. In many cases, however, the cutter is moved the entire length of the work to produce a somewhat more uniform surface. In that case, $t_m = (L + D)/F$, where D is cutter diameter.

It is of interest here to note that if certain other parameters are altered, that alone will *not* change the rate of metal removal *or* the machining time. For example, using carbide tools will, in general, permit the use of higher cutting speeds as compared with high-speed steel tools. But *unless* the cutting velocity itself is increased, the use of the same values of feed, depth of cut, and cutting speed will still produce the same RMR and machining time. This is one reason that we have considered these parameters as being basic.

Example 9–1

A block of metal, 4 in. wide by 11 in. long by 3 in. high, is to be reduced to a height of 2.8 in. Either a shaper or milling machine could be used and the sketches indicate the manner by which this might be done. In shaping, the single point tool must have a small overtravel at each end of the stroke (use 0.5 in. here); thus the length of either the cutting or return stroke is 12 in. in this case. In milling, consider that the approach distance a is also 0.5 in. and that

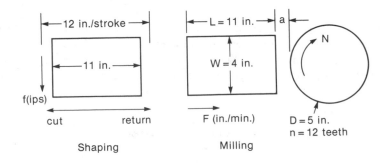

the milling cutter of diameter D will travel the full 11-in. length *plus* the cutter diameter. If the feed rate is to be 0.006 (inch per *stroke* on shaper and inch per *tooth* on the mill) and the cutting velocity is to be 60 feet per minute, find:

1. The time to machine the piece by each method.
2. The rate of metal removal when full cutting applies.

Solution   1. First consider the shaper. The tool is set to reciprocate back and forth in terms of strokes per minute (spm), where the linear velocity goes from zero to a maximum and then back to zero at each end of the stoke. If we base strokes per minute on an *average* cutting velocity of 60 fpm (that is, only the cutting portion of the stroke), then

$$V \text{ (fpm)} = S \text{ (spm)} \, L \text{ (length of stroke in feet)}$$
$$V = 60 \text{ fpm} = S(11 + 1)/(12) \quad \text{so} \quad S = 60 \text{ spm}$$

The time to complete the 4-in. width (if it is assumed that the first stroke makes a full cut) is

$$t = W/fS = 4/(0.006)(60) = 11.1 \text{ min}$$

With the milling operation, the rpm comes from

$$V = \pi DN/12 \quad \text{or} \quad N = 12(60)/5\pi = 46 \text{ rpm}$$
$$F \text{ (ipm)} = fnN = 0.006 \, (12)46 = 3.3 \text{ ipm}$$

The total distance to travel is $0.5 + 11 + 5 = 16.5$ in.

$$t = 16.5/3.3 = 5 \text{ min}$$

2. RMR for the shaper is the volume per stroke times the strokes per minute, or

$$\text{RMR}_s = 11(0.006)(3 - 2.8)(60) = 0.79 \text{ in.}^3/\text{min}$$

For the milling machine,

$$\text{RMR}_m = (\text{width})(\text{depth})F = 4(0.2)(3.3) = 2.64 \text{ in.}^3/\text{min}$$

This example illustrates that milling is faster than shaping, because the multitoothed cutter provides a higher RMR and thus lower machining times.

9.9 TOOL LIFE

It can be argued that the single most critical aspect of machining is the time* that the cutting tool operates in an acceptable manner; this is called the *tool life*. Various criteria have been used to define tool life, some of which are

1. Complete destruction of the cutting edge.
2. A specified amount of flank wear.
3. A loss of dimensional accuracy.
4. Degradation of surface finish to an unacceptable level.
5. An increase in the forces associated with cutting to an undesirable level.

---

* Number of pieces produced or the production rate, say pieces per hour, are basically equivalent.

In this text, the first criterion will be used. All that is involved is to measure the accumulated time that the tool actually cuts under a combined set of cutting parameters until failure occurs. Not only is this simple to do and easy to understand, but it negates the need for special sensing or measuring devices which are inevitably associated with most of the other definitions. Finally, more quantitative information, based upon *experiment,* is available in the literature where this definition has been used.

9.9.1 Tool Life (Taylor) Lines

In 1906, F. W. Taylor* showed that for a fixed set of cutting conditions (that is, workpiece, feed, depth of cut, tool material and signature, and cutting fluid), tool life and cutting velocity were directly related. The plot or line describing such a relation is called a *tool life line*.

Since far more empirical information has been published on this subject, where turning served as the cutting operation, we shall restrict our coverage to that one process. However, the ideas and concepts developed are applicable for other operations such as drilling, milling, and the like. The word *machinability* has been around for many years; we note this here because in assessing the ease with which various metals are machined, a relative ranking has most often been made in terms of some measure of tool life.

Now suppose that we intend to turn a material, say an annealed low carbon steel, using a coolant as a cutting fluid and high-speed steel tools ground to a recommended signature. For discussion purposes, a particular feed rate and depth of cut have been selected. With all of the above conditions kept *constant,* we are ready to obtain a tool life line. Using a freshly ground tool, a cut is started at some initially arbitrary cutting velocity, and the total cutting time is measured until tool failure.† Once the failure zone is cleaned up,‡ a new tool is inserted, the cutting velocity is altered (either higher or lower than the first), and a second test is conducted. Repeating this procedure leads to a number of combinations of cutting velocity-tool life values. With few exceptions, these data points, when plotted on *logarithmic* coordinates, can be reasonably described by a straight line whose equation has the form

$$VT^n = C \tag{9-8}$$

In Fig. 9–33, for example, the data give $VT^{0.1} = 100$. Here, V is the velocity in ft/min, T is the tool life in minutes, n is the tool life exponent, and the value of C accounts for the affects of all of the other parameters which were *fixed* for *this series of tests.*

The numerical value of C is simply the magnitude of V for a tool life of one minute (that is, the intercept of the line where $T = 1$) while n is the *absolute* value of the slope

---

* F. W. Taylor, "On the Art of Cutting Metals," *Transactions of American Society of Mechanical Engineers* (ASME), **28** (1906), p. 31.

† Tool life is sometimes expressed in cubic inches of metal removed to failure. For reasons that will be shown, we prefer to keep the machining parameters as discrete quantities rather than lump them in a volume rate.

‡ Just prior to tool failure, the temperature in the cutting zone is high enough to form a structure of austenite (for steels), and stopping the cutting action can be followed by rapid cooling of this zone. It is not only possible but probable that a thin layer of martensite results. This must be removed prior to the next test, since this is not the structure being studied in this case.

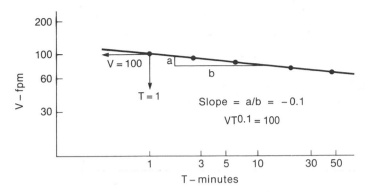

Figure 9–33 A typical tool life line.

of the line. (Strictly, $V = CT^m$, where m is negative as shown. Recasting to $VT^n = C$ means that n is taken as positive.) Thus, for the set of cutting conditions used, once the line or equation is obtained, the cutting velocity needed to produce a desired tool life can be readily found. Certainly, C must have units to make this equation dimensionally correct, but they are rather messy to include. The reader need not be concerned about this if it is realized that the product of V (in feet per minute) and the tool life T (in minutes) raised to the power n will give a numerical value equal to C. *This same physical concept will pertain for using many other empirical expressions that follow in this chapter.* At this point, some practical comments are offered. As with any empirical information, regardless of the field of engineering, the equation of the tool life line describes an observed relationship between two parameters. It is not *exact;* that is, there will always be modest variations in predicted versus observed values, yet such predictions are both useful and reasonable. This often bothers students, since much of their earlier background involved the derivation of equations based upon first principles. Yet one must realize that much of engineering entails the use of empirical work leading to results such as Eq. (9–8). Note also that since the magnitude of n is always less than unity, modest changes in V cause rather *severe* changes in T.

9.9.2 Generalized Tool Life Equations

The development of Eq. (9–8), being based upon a number of constant parameters (that is, f, d, and the like), poses restrictions regarding the use of that equation. If, for example, a constant feed rate of 0.010 ipr had been involved during those tests, and it was desired to use a feed rate of 0.020 ipr, then Eq. (9–8) could not be expected to provide a reliable prediction of T for a certain value of V, if it is assumed that *all* other parameters were identical to the test conditions. Intuition should indicate that cutting a thicker chip would require greater input energy and lead to higher cutting temperatures, thereby giving a shorter tool life for the same cutting speed. This is exactly what does occur. The tool life line developed with the higher feed rate would be displaced below. However, it is usually parallel to the first. In essence, n would be the same, but the magnitude of C would differ and reflect the difference in the combination of cutting parameters. If a constant but different feed rate of, say, 0.005 ipr is used, still a third line results, being displaced *above* the other two (since thinner chips would be cut and less heat would be generated) but again parallel to the others. This *family* of tool life lines is shown on Fig. 9–34(a). Again, it is important to remember that except for the difference in the feed rates, *all* other parameters were held fixed when the three individual lines were determined. Now, as shown in Fig. 9–34(b), if an *arbitrary* tool life of 60 minutes is chosen, and if the three discrete velocities involved are plotted versus the three feed rates, another straight line usually results; for the values used in this discussion this has the equation of

$$V_{60} = K_{60} f^a = 2.6 f^{-0.7} \qquad (9\text{–}9)$$

where V_{60} means the velocity for a 60-minute tool life. If instead we had chosen an arbitrary tool life of 20 minutes and plotted V_{20} versus feed, the resulting line would be parallel to the one for V_{60} but displaced above it (that is, the constant would be larger than 2.6, but the exponent would remain at -0.7). What has been accomplished is that the feed rate is no longer fixed, instead, for each equation such as Eq. (9–9), the *tool life* is fixed. One could rightly ask, just what have we accomplished? To answer this question, let's pursue a similar family of tool life lines where a different depth of cut had been used throughout, but *all* other parameters are still fixed. Arbitrarily we assume that the depth of cut during the first set of experiments was 0.125 inch while the second set used a depth

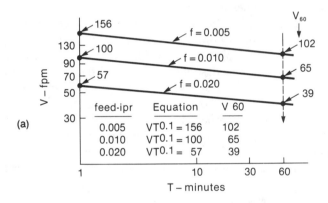

(a)

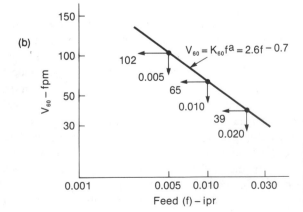

(b)

Figure 9–34 (a) Family of tool life lines where the feed rate is the only variable, and (b) a plot of the velocity for a 60-minute tool life versus feed rate for the data shown in (a).

of 0.250 inch. Figure 9–35(a) includes these findings plus those from Fig. 9–34(a). Following the procedures that led to Eq. (9–9), we see in Fig. 9–35(b) that

$$V_{60} = 2.6f^{-0.7}(0.125 \text{ in. depth}) \tag{9–10a}$$
$$V_{60} = 2.1f^{-0.7}(0.250 \text{ in. depth}) \tag{9–10b}$$

The coefficient K_{60} (that is, 2.6 and 2.1) varies with the depth of cut and plots of these points are shown in Fig. 9–35(c). *Note:* If other values of K_{60} were determined for other values of d, they would also fall on the line shown; only two values are used here to avoid repetitive discussion. The line on Fig. 9–35(c) fits the equation

$$K_{60} = C_{60}d^b = 1.4d^{-0.3} \tag{9–11}$$

Combining Eqs. (9–9) and (9–11) in both symbolic form and using the numerical values deduced from the plots, we obtain

$$V_{60} = K_{60}f^a = C_{60}f^a d^b = 1.4f^{-0.7}d^{-0.3} \tag{9–12}$$

Recalling that V_{60} denotes a 60-minute tool life and with $VT^{0.1} = C$ for any single tool life line, we obtain

$$V_{60}(60)^{0.1} = C \quad \text{or} \quad V_{60} = C/60^{0.1} \tag{9–13}$$

Combining Eqs. (9–12) and (9–13) gives

$$V_{60} = C/60^{0.1} = 1.4f^{-0.7}d^{-0.3} \tag{9–14}$$

or

$$C = (60)^{0.1}(1.4)f^{-0.7}d^{-0.3} \tag{9–15}$$

so

$$C = 2.11f^{-0.7}d^{-0.3} \tag{9–16}$$

But, since $VT^{0.1} = C$, substitution into Eq. (9–16) and rearranging give

$$VT^{0.1}f^{0.7}d^{0.3} = 2.11 \tag{9–17}$$

Thus, we have extended the more restricted, individual tool life lines into a more *generalized* form. However, such an empirical expression is accurate only if the other fixed parameters are maintained (that is, tool signature, work material, and the like). Note that the individual parameters whose numerical values would be introduced into Eq. (9–17) carry the units mentioned in the comments leading up to this equation. The constant of 2.11 does have units which are indeed cumbersome and serve no useful purpose; this need not be carried further.

Example 9–2

The generalized tool life equation for a particular turning operation is $VT^{0.1}f^{0.75}d^{0.3} = K$. For the same work material, tool material and signature, and cutting fluid combination, it is known that $VT^{0.1} = 120$ when the feed rate is 0.010 ipr and the depth of cut is 0.150 in. Using this information, determine the cutting velocity that should be used to produce a 30-minute tool life, using a feed of 0.005 ipr and depth of cut of 0.200 in.

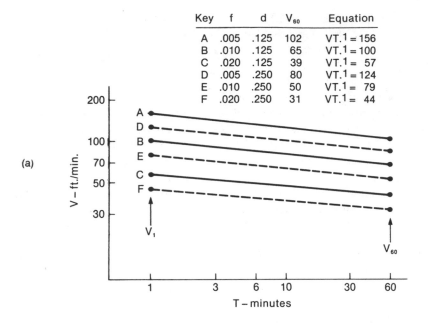

Key	f	d	V_{60}	Equation
A	.005	.125	102	$VT^{.1} = 156$
B	.010	.125	65	$VT^{.1} = 100$
C	.020	.125	39	$VT^{.1} = 57$
D	.005	.250	80	$VT^{.1} = 124$
E	.010	.250	50	$VT^{.1} = 79$
F	.020	.250	31	$VT^{.1} = 44$

(a)

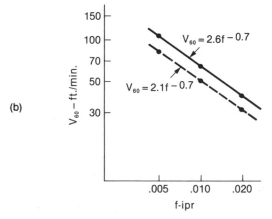

(b)

$$V_{60} = 2.6f^{-0.7}$$

$$V_{60} = 2.1f^{-0.7}$$

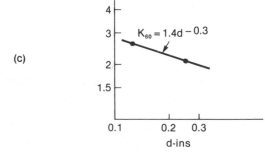

(c)

$$K_{60} = 1.4d^{-0.3}$$

Figure 9–35 (a) Various tool life lines for different combinations of feed rate and depth of cut, (b) plots of V_{60} versus feed rate using two different depths of cut, and (c) a plot of the constant for a 60-minute tool life versus depth of cut.

Solution   Although this is not the quickest way to solve this problem, let's find K first. Since the results from an individual tool life line are known,

$$VT^{0.1}f^{0.75}d^{0.3} = K = 120(0.010)^{0.75}(0.150)^{0.3}$$

so $K = 2.15$.

Then

$$V(30)^{0.1}(0.005)^{0.75}(0.200)^{0.3} = 2.15$$

so

$$V = 2.15/(1.405)(0.019)(0.617) = 131 \text{ fpm}$$

Note also,

$$120(0.010)^{0.75}(0.150)^{0.3} = V(30)^{0.1}(0.005)^{0.75}(0.200)^{0.3}$$

so,

$$V = 120/(30^{0.1})(2)^{0.75}(0.75)^{0.3} = 131 \text{ fpm}$$

Since vast amounts of information can be found in the literature, we include only typical tabular or graphical data here to give some idea of how an equation such as Eq. (9–17) might be modified to reflect the influence of parameters such as work material, signature, and cutting fluids. These are shown in Table 9–2. It is unfortunate that such parameters cannot be introduced directly into Eq. (9–17), but since their effects cannot be adequately described by a power-law expression, such action is precluded.

Even with such limitations there are important insights to be gained from Eq. (9–17). Recall that for turning, the rate of metal removal was given by

$$RMR = 12Vfd \qquad\qquad (9\text{–}18)$$

and, from above,

$$VT^{0.1}f^{0.7}d^{0.3} = 2.11 \qquad\qquad (9\text{–}19)$$

If any of the three parameters V, f, or d is altered, RMR changes in a direct ratio. For example, doubling f would double RMR if V and d are fixed. Note that in Eq. (9–18) all *exponents* are unity. However, because the *exponents* in Eq. (9–19) are different, their influence is not equivalent. To illustrate this point, suppose that a certain tool life is desired and that the turning cut involves a particular depth of cut to produce a desired diameter. Then, since T and d are fixed, Eq. (9–19) reduces to

$$Vf^{0.7} = 2.11/(T)^{0.1}(d)^{0.3} = \text{constant} \qquad\qquad (9\text{–}20)$$

Once we select a feed *or* a cutting velocity, the other parameter is automatically defined. As an illustration, suppose that a feed rate of 0.010 ipr and velocity of 100 fpm provide the desired tool life T. What results if we double f and still want the same T, d being constant? Using Eq. (9–20), we obtain

$$100(0.01)^{0.7} = \text{constant} = V(0.02)^{0.7} \qquad\qquad (9\text{–}21)$$

so the *new* velocity is found to be 62 fpm.

TABLE 9–2*   SOME TYPICAL VALUES FOR C AND n IN $VT^n = C$ FOR VARIOUS METALS, TOOL SIGNATURES, AND SIZES OF CUT

Tool		Work Material	Size of Cut		C	n
Material	Signature		f-ipr	d-in.		
High C steel	8,14,6,6,6,15,3/64	Yellow brass	0.013	0.100	300	0.10
HSS	"	Gray cast iron	0.026	0.050	172	0.10
HSS	8,14,6,6,6,0,0	SAE 1035 steel	0.013	0.050	130	0.11
HSS	8,14,6,6,6,15,3/64	SAE 1045 steel	0.013	0.100	192	0.11
		SAE 3140 steel	0.013	0.100	178	0.16
		SAE 4350 steel	0.013	0.100	78	0.11
		SAE 4350 steel	0.026	0.100	46	0.11
HSS	8,22,6,6,6,15,3/64	Monel metal	0.013	0.100	170	0.08
		Monel metal	0.026	0.150	127	0.07
T64 carbide	6,12,5,5,10,45,0	SAE 1040 steel	0.025	0.062	800	0.16
		SAE 1060 steel	0.025	0.125	660	0.17
		SAE 1060 steel	0.025	0.250	560	0.17
		SAE 1060 steel	0.042	0.062	510	0.16
		SAE 1060 steel	0.062	0.062	400	0.16
High C steel	8,14,6,6,6,15,3/64	Bronze	0.013	0.100	232	0.11
HSS	8,14,6,6,6,0,0	SAE B1112 steel	0.013	0.050	225	0.11
Stellite (Cast nonferrous)	0,0,6,6,6,0,3/32	SAE 3240 steel	0.031	0.187	215	0.19

*Adapted from O. W. Boston, *Metal Processing* (New York: John Wiley and Sons, Inc., 1951), p. 150. The high-speed steel tools (HSS) were all of the 18-4-1 type, and some of the quoted values for f, d, C, and n have been rounded off.

Now compare the rates of metal removal for these two cases, using Eq. (9–18). For the first case,

$$(RMR)_1 = 12(100)(0.01)d = 12d \qquad (9\text{–}22)$$

whereas

$$(RMR)_2 = 12(62)(0.02)d = 14.9d \qquad (9\text{–}23)$$

Thus, the increase in RMR is

$$\% \text{ increase} = 100(14.9 - 12)/12 = 24\% \qquad (9\text{–}24)$$

while the tool life remains the same.

Consider the alternative of doubling the velocity. Then

$$(100)(0.1)^{0.7} = (200)f^{0.7} \qquad (9\text{–}25\text{a})$$

so the computed feed rate is about 0.0037 ipr. Then

$$\text{RMR}_2 = 12(200)(0.0037)d = 8.9d \qquad (9\text{–}25\text{b})$$

Thus, there is a *decrease* in RMR of

$$\% \text{ decrease} = 100(12 - 8.9)/12 = 25.8\% \qquad (9\text{–}26)$$

The lesson to be drawn relates to the most logical *setting* of the machining parameters to produce high rates of metal removal for a desired tool life. It is the *magnitudes of the exponents* in Eq. (9–19) that are significant. Since, in the usual case, the depth of cut has the smallest exponent, feed the intermediate exponent, and velocity the largest exponent, it is *always most sensible* to use the largest depth of cut permissible, and then to use as large a feed rate as acceptable, and, finally, to adjust the velocity to give the desired tool life. The above example shows why *roughing* cuts, where dimensional control and surface finish are not of primary concern, are performed with low speeds, producing chips of relatively large cross-sectional area (that, *fd*). For a desired tool life, the rate of metal removal is of prime importance. So-called *finishing* cuts, designed to satisfy both dimensional control and surface finish specifications, are carried out at high speeds, producing small chips. Operations such as grinding epitomize such a set of conditions; this indicates why abrasive-type operations are often the last employed, since they do produce parts having small dimensional variations and excellent finish.

9.10 FORCE AND POWER REQUIREMENTS

There are a number of practical reasons why it is important to make reliable estimates of the magnitude of forces necessary to cut materials. Some of the most important reasons are

1. To estimate the motor size for new machine tools.
2. To determine the limiting rate of metal removal with existing machine tools.
3. To assist in the design of clamping and workholding devices.

Variations in cutting forces may also be used in determining the effectiveness of cutting fluids, in machinability evaluations of different materials, and as one sensing source in the area of adaptive control of machine tools. Other uses could be stated; however, in this presentation the major emphasis will be placed upon those listed above as 1 and 2.

9.10.1 Single Point Turning

Because it is the simplest operation to understand and because more numerous studies have been reported in the literature in comparison with other cutting operations, the basic understanding of forces is presented first for a single point turning operation. Figure 9–36(a) shows a schematic of the tool-workpiece setup with the feed and rotational motions indicated; the round bar is being reduced from diameter D_1 to D_2 in one pass. The total force acting on the tool is resolved into three components *for convenience*. These are designated as the radial force F_r, which acts at right angles to the axis of rotation, the longitudinal force F_ℓ, which acts parallel to the axis of rotation and in the same horizontal plane as F_r, and, finally, the tangential force F_t, which acts tangent to the rotating workpiece and is perpendicular to the plane containing the other two components.* Figure 9–36(b) illustrates these three components and the total cutting force as viewed head on with the single point tool.

There is no known method to *accurately* predict the magnitude of these force components without resorting to experimentation. This should not be too surprising when one realizes that, to varying degrees, the following list of common variables enters into force requirements:

---

* F_t is often referred to as F_c, the cutting force, while F_ℓ is sometimes noted as F_f, the feeding force; see Fig. 9–36(b).

(a)

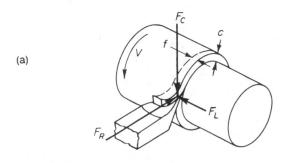

(b)

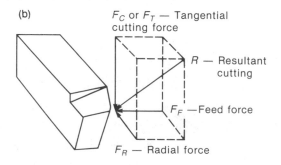

F_C or F_T — Tangential cutting force

R — Resultant cutting

F_F —Feed force

F_R — Radial force

Figure 9–36   (a) Schematic of tool and workpiece to display the three force components, and (b) resolution of the three components into a total force acting on a single-point tool.

1. Feed rate or chip thickness
2. Depth or width of cut
3. Cutting velocity
4. Combination of tool angles (that is, tool signature)
5. Condition of lubrication
6. Work material and/or type of microstructure
7. Sharpness (or dullness) of tool edge

A common practice used in force studies is to employ a *force dynamometer* designed to measure the three components simultaneously for a particular combination of cutting conditions. With a number of such tests on the same workpiece, a range of each important cutting parameter may be investigated to indicate the manner in which the force components behave. Here we shall concentrate on the effects of items 1, 2, and 3 in the above list and, eventually, present some comments on the other parameters.

For a thorough explanation of the manner in which the force components are studied and used, it is necessary to concentrate only on any one of them, since an identical approach would follow for the other two. Because it is the predominant component in power requirements, we shall temporarily concentrate on F_t and then bring in F_r and F_ℓ when the discussion about F_t is completed. Here, the procedure is similar to that used in the discussion on tool life, where a number of experiments are to be conducted using a particular work material, tool material and signature, and condition of lubrication; these parameters are to be fixed for the discussion that follows. Let us further assume that tool dullness, as caused by wear, is not considered for now. What remains as potential variables are the three parameters that constitute the machine tool variables, that is, the feed rate, depth of cut, and cutting velocity.

As an opening investigation, consider that the depth of cut and cutting velocity are fixed, since it is desired to determine the manner in which F_t varies with the feed rate. Once all parameters are fixed, the tool is fed into the workpiece at a selected feed rate, say 0.005 ipr, and when the force output from the dynamometer indicates a quasi-steady-state condition, the tool is disengaged. A number of such tests are conducted over a range of feed rates of interest and what is usually observed, within a very small scatter band, is that the influence of feed rate on F_t can be described quite adequately by a power law expression of the form

$$F_t = K_1 f^{a_1} \tag{9-27}$$

Note the similarity with Eq. (9–8) for a tool-life line, where again the plot is a straight line on logarithmic coordinates.

Before proceeding further, there is a nagging question that sometimes arises here. Does this imply that in order to predict an accurate value of cutting force we must first conduct tests to obtain actual measures of what is to be eventually predicted? The answer is yes, and this should not be too disturbing if one pauses to consider the attainment and use of other engineering parameters. Consider the elastic modulus of *steel*, for example.

For decades, the numerical value of this property has been employed in design calculations, and its average value is usually taken as 30×10^6 psi (207 GPa). Long before calculations based upon atomic bonding forces were made and before experiments using single crystals of iron showed a variation in modulus from about 18 to 42 million psi, experiments with polycrystalline steels led to the average modulus value so commonly used by engineers. It would obviously be pure folly to continually repeat experiments involving this elastic modulus whenever it is used for predictive purposes; in a parallel sense, the use of cutting force expressions developed earlier *from experimentation* also find future use.

Now consider the plot on Fig. 9–37 and let us assume that as the feed rate was varied to produce the individual test values for F_t, the depth of cut and cutting velocity were maintained at 0.075 inch and 75 feet per minute, respectively. The straight line assumed to describe the behavior of F_t versus feed for all of the *fixed* values of the other seven parameters can be expressed as

$$F_t = K_1 f^{a_1} = 3800 f^{0.8} \tag{9–28}$$

where the value of a_1 is the slope as shown and the value for K_1 is equivalent to the value of F_t when f equals unity. Obviously, a *physical* value of f equal to unity is ridiculous; remember it is only the *equation* of this line that is sought.

A good question at this point relates to the usefulness of Eq. (9–28). Just as we discussed the restrictions placed upon an individual tool life line, that is, Eq. (9–8), these also apply to Eq. (9–28). By again determining *families* of curves of cutting force as a function of feed rate, then depth of cut, and finally cutting velocity, plots such as those shown in Figs. 9–38 and 9–39 result. Using the same detailed procedures that lead to Eq. (9–17) (that is, the *generalized* tool life equation), the combined effects of feed, depth,

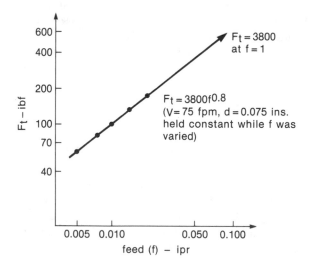

Figure 9–37 Typical variation of tangential or cutting force with feed rate using a constant cutting velocity and depth of cut.

and velocity on the cutting force, for individual materials, can also be expressed in a general form, in terms of the parameters controlled on the machine tool, as

$$F_t = Kf^a d^b V^c \qquad (9\text{--}29)$$

Many published results have shown that the exponent c is approximately zero, that is, the value of F_t is little influenced by V over a broad range of cutting speeds.* For our purposes we shall take this exponent as zero *in all further discussions.*

The magnitudes of K, a, and b are dependent upon the work material, tool signature, and, to a much lesser degree, the cutting fluid and tool *material.* To date, no real success has been found in introducing these parameters into an equation such as (9–29). Again, as with tool life, recourse must be made to available graphical or tabular information; note that such information is not necessarily available for all such parametric variations. A little experimentation may then be necessary.

As to Eq. (9–29), we note the following:

F_t = tangential cutting force (in lbf, N, and so forth)
f = feed rate in ipr
d = depth of cut in inches
V = cutting velocity in fpm

* With many metals, for speeds less than 50 fpm (this is low for most operations), F_t may show an increase with speed, whereas at high speeds (say several hundred fpm), F_t may decrease as speed is increased.

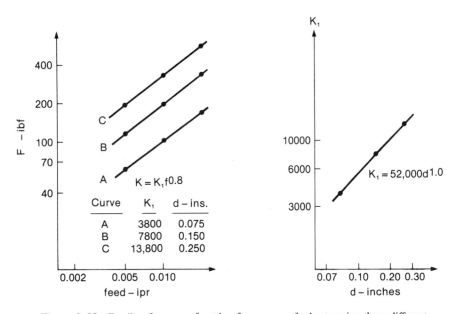

Figure 9–38 Family of curves of cutting force versus feed rate using three different depths of cut but a common velocity.

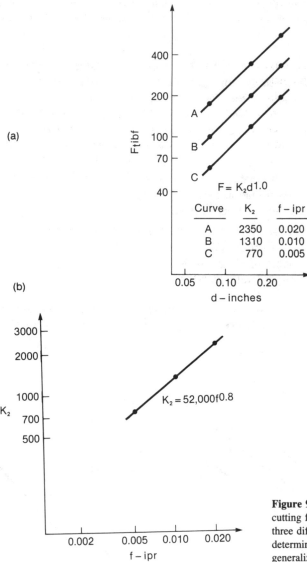

(a)

(b)

Figure 9–39 (a) Family of curves of cutting force versus depth of cut using three different feed rates and (b) determination of the parameter for generalizing the cutting force equation.

a = slope of plot of F_t versus f (typically 0.7 to 0.9)
b = slope of plot of F_t versus d (typically 0.9 to 1.1)
c = slope of plot of F_t versus V (typically zero)
K = coefficient whose magnitude depends primarily upon the work material and the tool signature.

Several qualitative comments are also provided, where for each parameter being discussed, all others are fixed:

1. *Increasing* the side rake angle causes a *decrease* in F_t.
2. *Increasing* the nose radius will usually *increase* F_t.
3. *Stronger* materials increase the value of K and, therefore, F_t (for example, steel requires larger values of F_t than does brass).
4. As tool wear proceeds, F_t increases, and near tool failure this increase can be drastic. We know of no *quantitative* correction that can be made to reflect the influence of tool wear on F_t. An increase in this force can be used as a diagnostic means to forewarn of impending tool failure.
5. Cutting fluids have relatively little influence on F_t.

All that was said about the method for determining F_t as a function of the many variables involved could be applied to the analysis of F_ℓ and F_r. To indicate typical differences or similarities that occur among these three components of the total cutting force it is most direct to include a number of plots. These are shown in Fig. 9–40. Table 9–3 includes a number of empirical expressions for force components for various materials and tool signatures.

9.10.2 Power Consideration in Turning

If for the time being we ignore that fact that no machine tool possesses a mechanical efficiency of 100 percent, it is then possible to concentrate on the horsepower that is demanded at the cutting tool for a particular set of cutting conditions. Since the previous section described how the total cutting force can be resolved into three useful components, the total horsepower at the cutter can be analyzed in a similar manner. It is common practice to consider the force in lbf, the velocity in feet per minute and the horsepower to be expressed as hP = $FV/33{,}000$. In regard to the three components,

$$hp_c = hp_t + hp_\ell + hp_r \qquad (9\text{–}30)$$

with the subscripts having the same meaning as earlier. In terms of specific components,

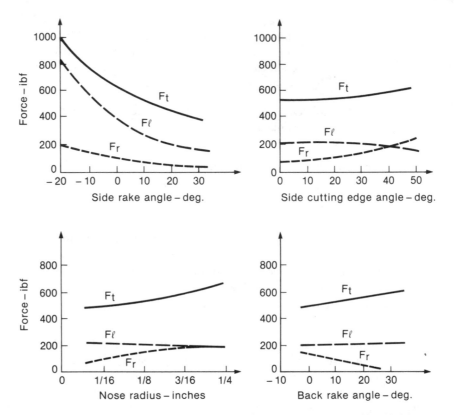

Figure 9–40   Empirical results illustrating how force components vary with tool signature.

$$hp_c = F_t V/33{,}000 + F_\ell fN/(12)(33{,}000) + F_r f_r N/(12)(33{,}000) \qquad (9\text{–}31)$$

Of course, for turning cuts, the feed in the radial direction (that is, f_r) is zero, so no power is consumed for that component. Note too that the feed f times the rpm N divided by twelve gives, in English units, the velocity in the feeding direction in feet per minute. A similar physical approach may be taken with other cutting operations and is, in fact, recommended in preference to memorizing numerous equations.

Before discussing Eq. (9–31) further, consider that the cutting velocity is related to the rpm by Eq. (9–2) as

$$V = \pi DN/12 \qquad (9\text{–}32)$$

If Eq. (9–32) is introduced into Eq. (9–31), there results

$$hp_c = (F_t \pi DN)/12(33{,}000) + F_\ell fN)/12(33{,}000) \qquad (9\text{–}33)$$

In most practical operations the product of πD is of the order of 200 to 1000 times larger

TABLE 9–3* TYPICAL VALUES FOR THE COEFFICIENT C AND THE EXPONENTS a AND b IN THE GENERAL EXPRESSION $F = Cf^a d^b$ FOR VARIOUS METALS. NOTE THAT HIGH-SPEED STEEL TOOLS OF DIFFERENT SIGNATURES WERE USED.

Work Material	Tool Signature†	F_t—Cutting			F_e—Feeding			F_r—Radial		
		C	a	b	C	a	b	C	a	b
Stainless steel (18-8)	1	151,000	0.85	0.96	23,000	0.48	1.26	46,500	1.0	0.7
1020 steel (hot rolled)	1	133,000	0.85	0.98	113,000	0.8	1.46	3,565	0.67	0.5
Brass (yellow)	1	83,400	0.81	0.96	95,100	0.9	1.43	9,860	0.97	0.4
Cast iron (126 BHN)	1	29,400	0.78	0.89	8,000	0.5	1.3	—		
Cast iron (241 BHN)	1	64,000	0.78	0.89	13,500	0.5	1.3	—		
Low C steel (100 BHN)	2	133,000	0.83	1.0	33,700	0.48	1.45	923	0.56	0
1020 steel (annealed)	3	156,000	0.88	1.0	51,000	0.42	1.58	2,020	0.69	0

*Adapted from O. W. Boston, *Metal Processing* p. 166 and *Manual on Cutting Metals* 2nd ed. (ASME, 1952), p. 274. Original data revised to above format.

†Tool Signatures were
1. 8,14,6,6,6,15,3/64.
2. 8,14,6,6,6,0,3/64.
3. 0,14,6,6,6,0,3/64.

than the feed f (for example, with a 3-inch-diameter bar and a feed rate of 0.020 ipr, the ratio is about 450 to 1) so as a sensible approximation, the horsepower at the cutter can be taken to be the power demanded by the tangential component only. *Note,* however, that this is because the *velocity* in the longitudinal direction is so small in comparison with the velocity in the tangential direction and *not* because of the relative magnitudes of these force components which may be fairly *similar.*

With the above observation, a reasonable approximation is

$$hp_c = F_t V_c/33,000 = F_t \pi DN/396,000 \qquad (9\text{–}34)$$

Now if the effects of machine parameters on F_t [that is, Eq. (9–29)] are used in Eq. (9–34), there results

$$hp_c = \frac{Kf^a d^b V^{c+1}}{33,000} = K_c f^a d^b V \qquad (9\text{–}35)$$

where K_c is, literally, K divided by 33,000 and the velocity exponent c is taken as zero for reasons discussed earlier. Equation (9–35) describes the influence of the parameters on the horsepower demanded at the cutter. An extremely useful quantity is called the *unit power at the cutter* and is defined as the power necessary to remove material at the rate of a cubic inch per minute, or

$$u \text{ hp} = \text{hp}_c/\text{in.}^3 \text{ per minute} = \text{hp}_c/12Vfd \qquad (9\text{–}36)$$

Combining Eqs. (9–35) and (9–36) gives

$$u \text{ hp} = \frac{K_c f^a d^b V}{12Vfd} = C_c f^{a-1} d^{b-1} \qquad (9\text{–}37)$$

As indicated in Eq. (9–28) and Table 9–3, the feed exponent is often about 0.8 while the depth exponent is unity. If, for a given material, *this were the case*, then Eq. (9–37) reduces to

$$u \text{ hp} = C_c f^{-0.2} \qquad (9\text{–}38)$$

so the unit power varies inversely with the feed rate but is *independent* of cutting velocity and depth of cut. The magnitude of C_c is, of course, influenced by *all* of the parameters that affected K in Eq. (9–29). Physically, Eq. (9–38) indicates that it is more efficient from a *power* point of view to cut thick chips, that is, use heavy feed rates, and this is what is done in *roughing* type of cuts, as mentioned at the end of Sec. 9.9. Table 9–4 contains a number of values of unit horsepower at the cutter; in the absence of more complete information, these can be used as good engineering approximations in analyzing power and force predictions.

Three final observations pertain to Eq. (9–38):

1. u hp is independent of velocity if *and only if* the force F_t is independent of velocity [that is, the exponent c is zero in Eq. (9–29)].
2. u hp is independent of the depth of cut if *and only if* the force F_t varies directly with the depth of cut [that is, the exponent b is unity in Eq. (9–29)].
3. The rate of metal removal is reasonably described by $12Vfd$.

Example 9–5

For a certain metal, the magnitude of the tangential (or cutting) force is approximated by

$$F_t = Kf^{0.8}d^{1.0}$$

Suppose that, when $f = 0.008$ ipr and $d = 0.100$ in., F_t equals 400 lbf. How much horsepower would be required at the cutting edge (that is, hp_c) if the cutting conditions were $f = 0.010$, $d = 0.250$, and $V = 100$ fpm?

Solution   Most students first find K via

$$400 = K(0.008)^{0.8}(0.100)^{1.0}$$

TABLE 9–4*   TYPICAL VALUES FOR UNIT HORSEPOWER AT THE CUTTER (u hp) FOR VARIOUS METALS AND MACHINING OPERATIONS

Material	Hardness BHN (R_C or R_b)	Turning	Milling	Drilling
Plain C and low alloy steels	126–162	0.6–0.9	1.2	1.0
	$(30–55)R_c$	1.3–2.0	1.5–2.2	1.0–2.0
Stainless steels	135–275	1.2–1.5	1.1–1.6	1.0–1.4
Free machining steels	118–229	0.36–0.54	——	——
Cast iron-gray and nodular	110–190	0.8–1.0	0.8–1.0	0.8–1.0
	190–320	1.4–1.8	1.6–2.0	1.6–2.0
High-temp alloys (Ni and Co base)	200–360	1.6–2.0	2.0–2.5	1.6–2.0
Aluminum alloys	30–150	0.2–0.3	0.3–0.4	0.2
Magnesium alloys	40–90	0.2	0.2	0.2
Copper and copper alloys	$(20–100\ R_b)$	0.6–1.2	0.6–1.2	0.5–1.1
Leaded brass	30–130	0.2–0.35	——	——

* Original data, revised in the above form, was abstracted from:
1. O. W. Boston, *Manual on Cutting of Metals*, (ASME, 1952), p. 282–295.
2. *Machining Data Handbook,* Metcut Research Associates, Inc., Cincinnati, Ohio, 1966, Section 2, p. 508.
3. ASTME, *Fundamentals of Tool Design* (Englewood Cliffs, N.J.: Prentice-Hall, Inc., 1962), p. 53.

Note that these values will vary with differences in tool signature, sharpness of the cutting edge, and variations in microstructure for a given work material.

that is,

$$K = 400/(0.021)(0.1) = 190,480$$

The new value of F_t is then

$$F_t = 190,480(0.01)^{0.8}(0.250) = 1196 \text{ lbf}$$

Check this, using

$$400/(0.008)^{0.8}(0.1) = K = F_t/(0.01)^{0.8}(0.25)$$

without finding K explicit. Then

$$F_t = 400(0.01/0.008)^{0.8}(0.25/0.10) = 1196 \text{ lbf}$$

Now,

$$hp_c = F_t V/33{,}000 = 1196(100)/33{,}000 = 3.6 \text{ hp}$$

Recalling the earlier comment about the mechanical efficiency of the machine tool being less than 100 percent, consider now how one might determine the size of drive motor needed on a new machine tool. A useful approach has been to introduce a quantity called the tare horsepower, hp_{ta}, which is equal to the total power consumed to run the machine tool while "cutting air," and a measure of mechanical efficiency which describes that part of the total power from the motor that reaches the cutting tool under cutting conditions. In equation form this may be written as

$$hp_m = hp_{ta} + hp_c/\text{eff.} \tag{9–39}$$

where

hp_m = total horsepower delivered by the motor

hp_{ta} = tare horsepower (this may range from 0.5 to 4 hp with larger lathes requiring greater tare hp).

hp_c = horsepower demanded at the cutting tool for the size of cut under consideration [that is, Eq. (9–35)].

eff. = mechanical efficiency (On modern machines with antifriction bearings and efficient drive trains, this is of the order of 95 percent, while with older machines and low efficiency drive trains, it might be a slow as 60 percent).

As a final thought it should be realized that any of the parameters that affect F_t, and are not expressed as direct variables in the form such as Eq. (9–29), will have an effect on hp_c or u hp in a manner in which they influence F_t. Figure 9–41 illustrates such effects.

Example 9–6

For a particular metal, u hp = 0.8 when f = 0.010 ipr, d = 0.100 in., and V = 100 fpm. The cutting force for this metal, $F_t = Kf^{0.85}d^{1.1}V^0$.
If cutting is to be done where f = 0.005 ipr, d = 150 in., and V = 0.150 fpm, determine
1. The value of u hp for these conditions.
2. The size of motor needed if hp_{ta} is 1.4 and an efficiency of 85 percent is reasonable.

Solution

1. u hp $= F_t V/33{,}000(12)Vfd = Kf^{0.85}d^{1.1}V/396{,}000(Vfd)$
so u hp $= Cf^{-0.15}d^{0.1}$, where $C = K/396{,}000$
Now the new u hp can be found by first finding C. From the initial conditions,

$$0.8 = C(0.010)^{-0.15}(0.1)^{0.1}$$

so

$$C = 0.8(0.01)^{+0.15}/(0.1)^{0.1} = 0.505$$

Then, for the new conditions,

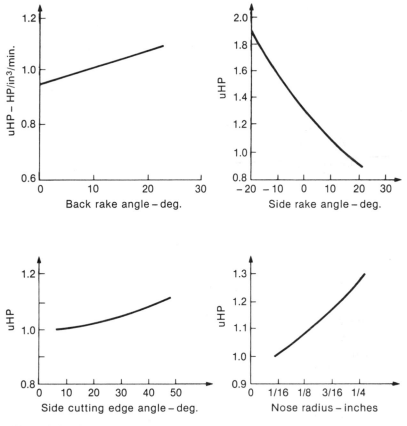

Figure 9–41   Empirical results showing the effect of tool signature on uhp; note that these are relative values.

$$u \text{ hp} = 0.505 \ (0.005)^{-0.15}(0.150)^{0.1}$$

so

$$u \text{ hp} = 0.924$$

Another approach is as follows:

$$u \text{ hp}(f)^{0.15} \ d^{-0.1} = C$$

so

$$0.8(0.01)^{0.15} \ (0.100)^{-0.10} = u \text{ hp}(0.005)^{0.15} \ (0.150)^{-0.1}$$

or the new $u \text{ hp} = 0.8 \ (0.01/0.005)^{0.15}(0.100/0.150)^{-0.1}$
so $u \text{ hp} = 0.924$ *without evaluating C.*
2. With Eq. (9–39)

$$\text{hp}_m = \text{hp}_{ta} + \text{hp}_c/\text{eff.}$$
$$\text{hp}_m = 1.4 + (12Vfd)u\text{hp}/0.85 = 1.4 + 12(150)(0.005)(0.15)(0.924)/0.85$$

so $\text{hp}_m = 1.4 + 1.47 = 2.87$ (that is, a 3-hp motor would just suffice).

Example 9–7

A lathe spindle is driven by a 5-hp motor, the machine efficiency is 80 percent, and the tare hp is 0.5. If, for a metal being turned, the u hp is 1.0, what is the maximum possible rate of metal removal, (RMR)?

Solution   From Eq. (9–39)

$$5 = 0.5 + \text{hp}_c/0.8$$

or

$$\text{hp}_c = 4.5(0.8) = u\ \text{hp}(\text{in.}^3/\text{min})$$

so maximum RMR $= 3.6/1 = 3.6$ in.^{3}/min

9.10.3 Forces and Power in Drilling

It is unnecessary to expend the detail used in the previous sections when analyzing the equivalence in drilling operations since an almost one-to-one approach could be followed. The major components of interest in drilling are the *torque* T_0, which is needed to shear the chip, and the *thrust* T_h, which acts opposite and parallel to the linear drill motion. Note that for this type of operation it is more useful to speak of a torque instead of a cutting force, although they have a corresponding meaning. In fact, we can show this correspondence if we reverse the procedure used earlier by starting with power requirements and note the force breakdown at the cutting edge as shown in Fig. 9–42. There the total force per edge is resolved into three components. The radial component F_r on each edge leads

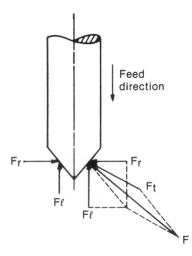

Figure 9–42   Schematic of forces acting on a drill.

to a net balance of zero and may be discarded from further consideration. The *thrust force* T_h is the sum of F_ℓ on each edge, while the force causing the torque acts normal to the edge and is assumed, for simplicity, to be concentrated at the midpoint on each of the two cutting edges (*Note:* this force component is directly analogous to F_t used earlier in single point turning).

Now if attention is concentrated at the cutting edge and if the same definitions apply as in Eq. (9–30),

$$\text{hp}_c = \text{hp}_t + \text{hp}_\ell \tag{9–40}$$

where in drilling, hp_t is demanded by the torque and hp_ℓ by the action of the thrust component. First consider the thrust component, where the symbol T_h will be used as the *total* thrust force opposing motion of the drill into the workpiece (that is, disregard F_ℓ from here on). The velocity in the direction of drill motion is the product of the feed rate in inches per revolution times the rpm; dividing this product *fN* by 12 gives the velocity in feet per minute; thus,

$$\text{hp}_{Th} = T_h fN/12(33,000) \tag{9–41}$$

Now the force producing the torque can be used to give the *torque hp* as

$$\text{hp}_{To} = F_t V/(33,000) \tag{9–42}$$

where the actual torque T_o, in *lb ft*, is

$$T_o = F_t(D)/(4 \times 12) \tag{9–43}$$

and D is the drill diameter in inches. The velocity corresponding to this point at which the force is indicated is

$$V = \pi(DN)/(2)(12) \text{ (in feet/minute)} \tag{9–44}$$

Introducing Eqs. (9–43) and (9–44) into Eq. (9–42) gives

$$\text{hp}_{To} = 48T_o(\pi DN)/(D)(24)(33,000) = (2\pi T_o N)/33,000 \tag{9–45}$$

This is the well-known relation between torque and rpm developed in most sophomore courses in mechanics.

Although we could have introduced Eq. (9–45) directly, it is hoped that the approach used above will point out that the torque effect in drilling is really equivalent to the tangential component in single point turning. Equation (9–40) may be expressed as

$$\text{hp}_c = (2\pi T_o N)/(33,000) + (T_h fN)/(12)(33,000) \tag{9–46}$$

As discussed after Eq. (9–33), although perhaps not as obvious, the portion of the *total horsepower* at *the cutting edge* due to the thrust component (that is, the feeding direction) is minimal as compared with that demanded by the torque. Again, this is because of the velocity or speed ratio and *not* because T_h is negligible. In fact, in every cutting operation, the major power demand comes from the *force component* associated with the *major velocity* involved in the operation and hp_c can, for all practical purposes, be defined in terms of that component only. For drilling, this would reduce Eq. (9–46) to

$$\text{hp}_c = 2\pi T_o N/33{,}000 \tag{9-47}$$

In a manner identical to that which led to Eq. (9–29) for turning, torque could be expressed as a function of machining variables by

$$T_o = C f^m D^n V^p \tag{9-48}$$

Typical values for m, n, and p, for many metals, are 0.8, 1.8, and zero (note that T_o is independent of velocity in general just as F_t was in single point turning). Using these numerical values in Eq. (9–48) and introducing the result into Eq. (9–47) give

$$\text{hp}_c = 2\pi (C f^{0.8} D^{1.8}) N/33{,}000 \tag{9-49}$$

which is analogous to Eq. (9–35). In drilling, the rate of metal removal in cubic inches per minute is the area of the drilled hole times the feed and rpm; that is,

$$\text{RMR} = \pi D^2 f N/4 \tag{9-50}$$

The unit horsepower at the cutter for drilling is obtained by using Eqs. (9–49) and (9–50) to give

$$u \text{ hp} = 2\pi C N f^{0.8} D^{1.8}(4)/33{,}000(\pi)D^2(fN) = C_1 f^{-0.2} D^{-0.2} \tag{9-51}$$

The similarity with Eq. (9–38) is apparent and for those materials having an exponent $n = 2.0$ in Eq. (9–48), the unit horsepower would again be dependent on the feed rate only.

Although the *thrust* horsepower is usually negligible, the thrust *force* itself can be extremely large and must be considered in fixture design and clamping analyses. As a function of the usual parameters, a typical expression for the thrust force would be

$$T_h = C_2 f^{0.8} D^{1.0} \tag{9-52}$$

It cannot be emphasized too strongly that the ratio of chisel edge to drill diameter can have a decided influence on the thrust, and a lesser extent on the torque. For the purposes herein, the typical values shown in Table 9–5 may be used for *first approximations*.

Example 9–8

For a certain drilling setup, the torque (T_0) and thrust (T_h) are expressed by

$$T_o = 600 f^{0.8} D^2 \quad [T_o \text{ in lb ft, } f \text{ in ipr, and } D \text{ (drill diameter) in in.}]$$
$$T_h = 50{,}000 f^{1.0} D^{1.2} \quad (T_h \text{ in lbf})$$

Using a ¾-inch drill, a feed of 0.010 ipr, and an rpm of 600, determine the horsepower demanded by the torque and thrust reactions.

Solution   Torque: Using Eq. (9–45), we have

$$\text{hp}_{T_o} = 2\pi T_o N/33{,}000$$

or

$$\text{hp}_{T_o} = 2\pi (600)(0.010)^{0.8}(0.75)^2(600)/33{,}000 = 0.968 \text{ hp}$$

TABLE 9–5* TYPICAL VALUES FOR THE COEFFICIENTS AND EXPONENTS IN TORQUE AND THRUST RELATIONS IN DRILLING WHERE: TORQUE $(T_0) = C_1 f^m D^n$ and Thrust $(T_h) = C_2 f^x D^y$

Material	T_0 in pound-feet			T_h in pounds force		
	C_1	m	n	C_2	x	y
Aluminum alloys	550	0.83	1.9	48,000	1.1	1.2
Leaded brass	418	0.73	1.9	6,600	0.6	1.0
1020 steel	1590	0.78	1.8	40,000	0.78	1.0
1045 steel	1590	0.78	1.8	42,000	0.78	1.0
4130 steel	1560	0.78	1.8	43,000	0.78	1.0

* Selected values adapted from O. W. Boston *Metal Processing* (New York, John Wiley and Sons, Inc., 1951), pp. 337–338. *Note* that these can vary with the helix angle, the size of the chisel edge, the microstructure of a given work material, and type of cutting fluid. The feed rate f is in ipr and the drill diameter D is in inches.

Thrust: Using Eq. (9–41), we have

$$hp_{Th} = T_h fN/396,000$$

so

$$hp_{Th} = 50,000(0.010)^1(0.75)^{1.2}/396,000 = 0.009 \text{ hp}$$

This illustrates why the hp_c in drilling is, for all practical purposes, due to the *torque* demands as in Eq. (9–47).

Some final observations regarding the thrust force are made with respect to Fig. 9–43. To balance the drill it is essential to grind the point so that the angle and cutting edges are, essentially, evenly divided by the drill axis. This produces equal horizontal components and a hole that is just about equal to the actual drill diameter. The chisel edge is extremely *inefficient* as a cutting edge; rather, the material beneath the chisel edge is forced to flow by plastic deformation. It is for this reason that thinning of the chisel edge can drastically lower the thrust force, or that large holes are often produced *after* an initial small hole is drilled. If the major contact of the chisel edge is avoided, a tremendous

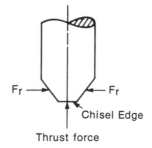

Fr

Fr

Chisel Edge

Thrust force

Figure 9–43 Schematic of forces on the cutting edges and chisel edge of a drill.

decrease in drill thrust follows. If the thrust force is used to hold the work in place on the drill table, *breaking through* of the chisel edge can lead to serious consequences, since this force is released for all practical purposes and the work will spin around with the drill. This is why the holding of a workpiece by hand is to be avoided when one is drilling materials that demand high torque for cutting. So long as the chisel edge exerts its influence, holding the workpiece may appear safe, but once the edge breaks through the bottom, the torque will tend to tear the work away from the operator. However, overthinning of the chisel edge can lead to a point whose structural weakening would necessitate very low feed rates to prevent the point from breaking. In practice, compromises must be made; what is important here is to understand some of the principal reasons why such approaches must be taken.

9.10.4 Closing Comments

If one understands the concepts presented in the previous two sections, then the use of available values of the u hp, can be applied readily to literally any machining operation if one is interested in determining the approximate magnitude of horsepower needed to produce a certain rate of metal removal. In addition, the level of cutting force F_t required can be easily computed once hp_c is found. Unfortunately, the *magnitudes* of the radial and longitudinal forces (which may be called by different names) cannot be deduced from such information. Assuming that such magnitudes are of interest, one must resort to a sensible literature search, hoping that, for the combination of conditions involved, earlier reported findings will be useful.

Example 9–9

A slab of steel, 6 in. wide by 4 in. high, is to be milled with a facing cutter to reduce the 4-in. height to 3.9 in. in a single pass. The 10-in.-diameter cutter contains 16 teeth; an average feed rate of 0.006 inches per tooth is desired, and the cutting velocity is to be about 130 fpm. If the u hp is about 1.2, estimate the magnitude of the cutting or tangential force expected.

Solution   First, find the table feed rate to be set on the machine as found from Eq. (9–1),

$$F = fnN, \text{ where } n = 16 \text{ and } N = 12(130)/\pi 10 = 50 \text{ rpm}$$

then

$$F = (0.006)(16)(50) = 4.8 \text{ in./min}$$
$$\text{RMR(in.}^3/\text{min)} = Fwd, \text{ where } w = 6 \text{ in. and } d = 0.100 \text{ in.}$$

so

$$\text{RMR} = 4.8(6)(0.1) = 2.88 \text{ in.}^3/\text{min}$$

then

$$hp_c = u \text{ hp(RMR)} = 1.2(2.88) = 3.46 = F_tV/33,000$$

so

$$F_t = 33,000 \text{ } hp_c/V = 33,000(3.46)/130 = 877 \text{ lbf}$$

We add a final comment about u hp for grinding operations. Such values are *much* higher than those associated with turning, as shown in Table 9–4. With grinding, values of 5 to 20 are not uncommon and, as related to Eq. (9–38), result because the size of cut (f,d) is exceedingly small. Even though the u hp is quite large, the RMR (in.^{3}/min) is very low, and as a consequence the *cutting* horsepower is also low. Grinding wheels operate at cutting velocities of the order of 5000 ft/min; thus, the torque or force requirements are extremely low. That, as noted earlier, is one of the major reasons why surface finish and dimensional control are superior with grinding as compared with turning, milling, or other *large* chip operations; abrasive operations induce much smaller forces, so deflection of the workpiece and distortion of the work surface are much less severe compared with large chip operations.

9.11 PLASTICS OR POLYMERS

Many of the observations and concepts covered earlier regarding metals find a reasonably direct correspondence with plastics. Rather than repeating the extensive coverage devoted to metals, a concise, qualitative summary, especially of any differences in machining behavior, will be presented. For specific recommendations of such factors as tool signature, size of cut, cutting velocity, and so forth, the reader can consult the text by Kobayashi* as an excellent starting point. In most instances, recommended differences in machining plastics as compared with metals, come about because of the significant differences in many of the properties of these two classes of solids. The following list includes these major differences:

1. Plastics have a much lower elastic modulus than metals. As an example, the modulus of steel is of the order of 100 times that of plastics. Because of this lower rigidity, plastics display greater deflections for certain machining operations; thus *elastic recovery* can cause problems.
2. For the same quantity of heat applied to equal volumes of a plastic and a metal, the temperature rise of the plastic is larger. This may lead to a softening of the plastic and a ''gumminess'' of the workpiece.
3. Properties such as specific heat, thermal conductivity, and coefficient of thermal expansion, are very different for these two classes of solids.
4. Plastics begin to soften at much lower temperatures.
5. Plastics have different toughness-to-strength ratios than metals. This can cause cracking as has occurred when plexiglas is drilled at too high a feed rate.

Because of these differences, cutting conditions that are fully acceptable with metals may have to be altered to provide successful machining of plastics. For example, turning

---

* A. Kobayashi, *Machining of Plastics* (New York: McGraw-Hill Book Co., 1967). That book contains a wealth of information, involving a number of common plastics as work materials.

requires relatively larger rake angles, although an upper or critical value must be maintained; beyond this, a degradation of surface finish occurs. Additionally, large relief angles are needed to minimize rubbing between the tool and finished surface. Such large angles, in both instances, are needed to alleviate the effects which can occur because of the large elastic recovery of the machined surface that results after the tool edge passes by that surface.

Rather serious problems can occur when drilling plastics if improper conditions prevail. As chips are removed along the flutes of the drill, the heat involved in cutting can cause the plastic to soften to the extent that the chips *weld* to the flute surfaces. Not only does this prevent easy removal of subsequent chips, but the necessary torque may reach a level that causes the drill to fail by literally twisting into two pieces. Large helix angles and an effective lubricant reduce the tendency of flute packing and, possibly, drill failure. However, care must be exercised in the use of a fluid. Many plastics are susceptible to structural degradation and, additionally, the phenomenon of *crazing*. A craze is a type of void that is filled with a web-like structure of fibrils. This comment indicates that sound suggestions from those knowledgeable about such chemical reactions must be sought. Due to their low elastic modulus and rather high coefficient of expansion, plastics have, in some cases, ended with drilled holes that are *smaller* than the drill diameter. This may present problems when attempts are made to back out the drill after a relatively deep hole has been machined.

As with metals, both continuous and discontinuous chips can occur; in general, continuous chips provide a better surface finish. However, if excessive heat is induced during cutting, the surface can actually melt; melted material, upon solidification, gives an extremely poor finish.

With regard to the magnitude of the cutting force, trends similar to those found with metals have been observed. Increasing the rake angle tends to reduce this force, whereas a larger size of cut (that is, feed and depth of cut) causes an increase in force.

Finally, with thermoset types of plastics, many of which are composites that contain quite abrasive filler materials, tool wear poses a serious problem. In certain instances, diamond cutting tools must be used for successful performance, although ceramics and carbide tools are often adequate.

The thrust of this section is to indicate that although many of the basic concepts related to the successful machining of metals show a useful correspondence when plastics are used, there can be significant differences. As a final comment, this section does not imply that machining is the operation used to *process* polymers in some initial configuration. Both forming operations (Chapter 8) and casting processes (Chapter 11) are used to produce various *shapes* of polymers. As with metals, machining is often used as a *subsequent* process.

REFERENCES

1. L. E. Doyle, C. A. Keyser, J. L. Leach, G. F. Schrader, and M. B. Singer, *Manufacturing Processes and Materials for Engineers,* 3rd ed. Englewood Cliffs, N.J.: Prentice-Hall, Inc., 1985.
2. E. P. DeGarmo, *Materials and Processes in Manufacturing,* 3rd ed. New York: The Macmillan Co., 1969.
3. H. W. Yankee, *Manufacturing Processes.* Englewood Cliffs, N.J.: Prentice-Hall, Inc., 1979.
4. R. A. Lindberg, *Processes and Materials of Manufacture.* Boston: Allyn and Bacon, Inc., 1964.

PROBLEMS

9–1. A hole of 0.5 in. diameter and one inch depth is to be produced by using a drill of that diameter. If the drill is to operate at 100 rpm and a feed rate of 0.008 in. per revolution, determine
   (a) The machining time (in minutes) after the full diameter is first reached.
   (b) The rate of metal removal when the full diameter is being cut. Answer first in English, then SI, units.

9–2. A hole of 25.4 mm diameter is to be enlarged to 28 mm by boring. The single point boring tool is to operate at a feed rate of 0.076 mm/rev and the piece rotates at 250 rpm. If the length of the hole is 48 mm, find
   (a) The machining time.
   (b) The rate of metal removal

9–3. A plate having dimensions of 18 in. length, 6 in. width and 2 in. thickness is to be reduced to 1.75 in. thickness by using a shaper. The total length of stroke is set for 20 in., allowing a 1-in. overtravel at each end of the 18-in. dimension. If the *average* cutting velocity is to be 60 ft per min and the feed rate is set for 0.015 in. per stroke, find
   (a) The machining time to shape the full surface.
   (b) The rate of metal removal in both English and SI units.

9–4. For a particular work material, tool material, tool signature and cutting fluid combination, experiments indicate that for turning,

$$VT^{0.1}f^{0.7}d^{0.4} = 2.0$$

Assume that the desired tool life T is 30 minutes and the depth of cut d is 0.250 in. All parameters carry English units.
   (a) If a feed rate is specified as 0.008 ipr, what cutting velocity V should be specified?
   (b) What rate of metal removal results in a?
   (c) If a feed rate of 0.005 ipr is used instead, what value of V should be used?
   (d) What is the RMR for these new conditions?

9–5. Tool life information for a certain set of cutting conditions indicate
   (a) $VT^{0.25} = 250$ when $f = 0.010$ ipr and $d = 0.200$ in.
   (b) $VT^{0.25}f^{0.6}d^{0.35} = K$

A 5-in. diameter bar of steel is to be reduced to 4.4 in. in a single turning pass, with a feed rate of 0.010 ipr. If a 40-min tool life is desired, what cutting velocity should be used?

9–6. Machining studies for a certain work-tool material combination show that

$V_{60} = 1.6f^{-0.65}d^{-0.25}$ (V_{60} = velocity for a 60-min tool life)
$VT^{0.1} = 90$ when $f = 0.008$ ipr and $d = 0.150$ in.

If an operation will involve a feed of 0.012 ipr and depth of cut of 0.250, what cutting velocity should be used if the tool life is to be 25 minutes?

9–7. In turning, tool life can be described in terms of cutting velocity via $VT^n = C$, whereas the metal removal rate, RMR, comes from $12Vfd$. Develop an expression to express tool life in terms of cubic inches removed per tool failure. This should be expressed explicitly as a function of the parameters V, f, d, and C.

9–8. High-speed steel tools are used to turn steel bars from an outer diameter of two inches to 1.75 in. using a feed rate of 0.020 ipr. After turning two full bars at 300 rpm, a tool fails, whereas at 235 rpm, nine bars are machined before tool failure. The length of each bar is 10 in.
Sintered carbide tools are to replace the HSS tools for this operation, and the velocity for a 60-min. tool life (that is, V_{60}) for carbides is 2.5 times V_{60} for HSS. An appropriate tool life equation for carbides is $VT^{0.23} = C$.
If carbides are to turn 50 full bars before failing, what rpm should be used if the same feed rate and depth of cut are maintained?

9–9. Large propeller shafts are to be turned on a lathe having 20 hp (maximum). The tare hp is 1.5 and the efficiency is 85 percent. Estimates of the *maximum* level of cutting conditions in any one combination are

$$f = 0.024 \text{ ipr}, \qquad d = 0.250 \text{ in.}, \qquad V = 150 \text{ fpm}$$

For the tool-work combination, it is known that,
(a) $F_c = K_1 f^{0.7} d^{0.85}$
(b) $u \text{ hp} = 1.0$ when $f = 0.008$ ipr and $d = 0.050$ in.

Does the lathe possess adequate horsepower? Show supporting calculations.

9–10. Pertinent drilling information indicates

$$T_o = 4180 f^{0.7} d^{1.9}$$

where T_o is in lb-ft, f in ipr, and d, the drill diameter, in inches.

$$T_h = 6940 f^{0.6} d^{1.0}$$

where T_h is in lbf and f and d as above.

$$VT^{0.15} f^{0.65} d^{0.95} = 4$$

where V is in ft/min, T is in min, and f and d are as above.

A drill press to be used possesses a drive motor of 7.5 hp, an efficiency of 90 percent, and negligible tare hp. The operation will employ a ½-in.-diameter drill, and the desired tool life is 45 min. Two feeds, 0.010 and 0.020 ipr, are to be checked in the following analyses.
(a) Including *all* contributions to power demands, is adequate horsepower available for both feed rates?

(b) Regardless of your answer in part a, which combination of f, d, and V gives the highest RMR *if* adequate power were available?

9–11. Milling is to be used to reduce a plate from 305 mm length by 152 mm width by 75 mm height to 305 mm by 152 mm by 70 mm in a single pass. The 250-mm-diameter cutter has 20 teeth, and the *average* feed rate is to be 0.200 mm/tooth. A reasonable cutting velocity is about 0.5 m/s. If the unit horsepower for this operation is 0.9 hp/in.^{3}/min, determine the magnitude of the cutting force.

9–12. A company must purchase a new lathe for turning large drive shafts. Both roughing and finishing cuts will be performed using the following conditions:

Finish turning: $f = 0.010$ ipr, $d = 0.100$ in., $V = 300$ fpm
Rough turning: $f = 0.030$ ipr, $d = 0.350$ in., $V = 100$ fpm

When the same material was machined on an existing lathe, the motor indicated the following horsepower demands:
(a) $hp_m = 1.5$ when $f = 0.010$ ipr, $d = 0.050$ in., $V = 150$ fpm
(b) $hp_m = 2.5$ when $f = 0.010$ ipr, $d = 0.100$ in., $V = 150$ fpm
(c) $hp_m = 4.0$ when $f = 0.020$ ipr, $d = 0.100$ in., $V = 150$ fpm

During these three tests a tare horsepower of 0.5 and machine efficiency of 90 percent were noted. It is also known that

$$VT^{0.2} = 400 \quad \text{and} \quad VT^{0.2}f^{0.55}d^{0.2} = K$$

pertain for this situation.

The manufacturer of the lathe to be purchased estimates that the tare hp will be about 5 while the efficiency should be around 95 percent. What size or capacity motor (that is, horsepower) should be specified?

_10_
joining
processes _____

10.1 INTRODUCTION

The topics discussed in this chapter include the chemistry and mechanics of the following processes:

1. Welding and other processes that join parts by localized melting of the parent and filler metals.
2. Brazing and soldering, processes which bond solids by filling the joint with molten filler material, without melting the parent metal.
3. Cold bonding.
4. Diffusion bonding, sintering, and coalescence of ceramic, metal, and plastic powder particles.
5. Gluing and adhesive bonding.

Joining and bonding technologies are too numerous, too complex, and too varied to adequately describe in a general textbook. Furthermore, new developments are appearing at a great rate, so any short description of the subject would soon be out of date. Much information of the type needed by manufacturing engineers, namely, the cost effectiveness of these processes, the skills required to operate the machinery, and the production rate of the machinery, can be found in publications of the American Society for Metals and the Society of Manufacturing Engineering, as well as in the sales brochures and catalogs of equipment suppliers. The necessary arts and skills for training purposes are described in the well-illustrated instruction manuals and textbooks of industrial arts schools.

But reliable information on the *chemistry* and *mechanics* of many of the processes and on the mechanical strength of the product is generally not available. Most new developments of old processes and most new processes contain some proprietary or *secret* aspects, and many products are advertised with exaggerated claims of superiority over other processes, either in terms of low cost, ease of processing, or high product quality. A common practice in industry, in order to learn the real capability of processes, is to send parts to several purveyors of *new* processes for evaluation purposes. Should one or two of the processes be found useful, the next step is to determine whether the technology of the candidate processes can be successfully applied where needed. This depends on the expertise and resources of the vendors, which can best be estimated by asking questions on the fundamentals of joining and coating. Thus the emphasis in this chapter will be on the atomic aspects of bonding or joining of separate bodies, followed by a discussion of the problems and properties of the bonded system. With this approach all processes, old and new, may be better understood.

10.2 JOINING PROCESSES:
AN OVERVIEW

Few products are made of a single piece; rather, most products are made of several pieces, which are then joined together in some way. One class of joining, which is not discussed in this book, is by the use of *mechanical* devices such as bolts or screw threads, rivets, stakes, bands, clips, cast-in inserts, press-fits, and so forth. The most important disadvantage of mechanical joints is their high cost, since the surfaces to be joined must usually be made fairly precisely, holes must often be provided, and several operations are required to effect the joining operation. The major advantage of mechanical bonding is that the strength of the joint can be *predicted* quite reliably through the use of design handbooks or handbooks of mechanical fasteners. Fewer data are available for the joining processes discussed below, mostly because the strength of the joint is strongly dependent on how the process was done. Thus, where reliability of a small production run of products is more important than economy, mechanical joining will often be specified.

The economic advantage of *adhesive* and *cohesive* bonded joints derive from the fact that imprecise surfaces can be used, and the processes are fast. (*Adhesion* is defined as bonding between dissimilar materials, and *cohesion* is defined as bonding between pieces of identical materials.) Their major disadvantage is that the joined parts are not readily separable for maintenance or other use. Think how awkward photography would be if the back of a camera had to be unsoldered and resoldered whenever a film cartridge is to be changed. However, for many other products, permanent joints are acceptable or even preferred. Examples are floor covering, pages in books, auto body parts, and electrical connectors.

Bonding processes themselves can introduce problems which require special care to minimize. For example, polymeric adhesive compounds produce stresses when they contract at a different rate and amount than does the substrate. This will produce distortion of the finished part, particularly if the cross section is thin. Welding processes also produce distortion and leave residual stresses, because they progressively melt and solidify small amounts of the part interfaces as the heat source moves along. All heat sources for the welding processes cause these problems; these include the electron and ion beams, light beams including lasers, electrical resistance, electric arc, plasmas, and hot gases. A second problem is entrained gases. For example, if an adhesive supply contains excessive amounts of gases and solvents from the mixing and formulation processes, these may produce bubbles and other voids in the solidified joint. Fusion welding processes that operate in air or under water also entrain oxygen and water vapor into the molten metal, thereby forming oxides and other defects.

A third problem with bonding is seen in welding, where there is often the problem of undesirable metallurgical phase change in the parent metal. There is always a region extending down into the substrate beneath the weld, known as the heat affected zone (HAZ). The temperatures in the HAZ will over age and soften a hardened aluminum, and the fast cooling that follows heating may be fast enough to produce brittle martensite in alloyed steel. Many welded joints will fail in the HAZ rather than *at the original interface*.

There are, therefore, two aspects of the integrity of a bonded joint, namely, the nature of the bond interface or thin region between the two separate parts, and the properties of the near surface region within the bodies near the joint. Both will now be discussed.

10.3 THE NATURE OF INTERFACE BONDS

One fact of nature is that if two atoms are brought close enough to each other, they will bond together. The spacing between atoms in the solid state and the strength of the bond depends on the electron structure of the atoms. For the primary solid bonding systems, namely, the ionic, the covalent, and the metallic systems, the operative distance is about 0.3 nm and the strength is very high, in the range of 1 to 3 GPa. For the vanderWaals and dispersion forces the distance is somewhat greater and the bond is weaker, of the order of 0.1 to 0.2 GPa.

If any two atoms will readily bond together, then two clean and flat solids, each being composed of atoms, will also bond together by mere *contact,* that is, without heat of pressure. This does not accord with common experience, however. Two *clean* coins, for example, are readily separated even though they are vigorously rubbed together. But the same two coins, when thoroughly cleaned in the vacuum of space, will bond or weld together by mere contact. It should be apparent that there is a difference between cleanness as commonly defined and the cleanness required to effect strong bonding. A perspective on cleanness will be given after the further consideration of bonding energy.

It is useful to view the bonding between atoms in terms of energy. For example, in order to form a crack in a solid, that is, in order to separate atomic bonds, external energy must be applied. This amount of energy (ignoring plastic flow at the crack tip) is the order of one MJ per square meter of new surface as measured by the methods of fracture toughness. Healing of such a crack involves the liberation of the same amount of energy, either as mechanical energy or as thermal energy.

Another method of accounting for energy of bonding is to measure the energy of adsorption of gases on solids. This begins by grinding a brittle material into a fine powder in a vacuum; the fine powder has a large ratio of surface area to volume. When a gas is admitted into the enclosure, the surfaces become covered with one or several layers of gas, and the temperature of the powder increases. The source of the heating is the high energy state of the free surface in the vacuum. When the gas adsorbs on the surface, the energy state of the gas and solid surface combination has a lower potential for bonding to yet other gases than did the solid surface alone. The energy difference is *given off* as heat, and this heat of adsorption is unique to each combination of gas and solid surface. For some combinations the gas can be driven off the solid by heating in a vacuum (note that most ''vacuum'' is only a reduced gas pressure, not the complete absence of gas). This is called *physical adsorption,* for which the heat of reaction is in the range of 0.05 to 0.5 eV (or from 4 to 40 kJ) per mol of the adsorbed gas (see Ref. [1]).

Very few surfaces remain truly clean, even if they can ever be made clean in some way. If a square meter of surface begins in the perfectly clean state in air at atmospheric pressure, it will be bombarded by gas molecules at a rate $\emptyset_0$ that can be estimated from the equation (Pirani):

$$\emptyset_0 = \frac{4.7 \times 10^{28}P}{\sqrt{(MT)}} \text{ molecules per second} \qquad (10–1)$$

where P = pressure in Pa (1 atm. $\approx 10^5$Pa)
$\quad\ M$ = molecular weight (big molecules move slowly)
$\quad\ T$ = temperature in degrees Kelvin

This is a very fast rate as compared with the fact that a square meter of surface can hold only about 6.2×10^{18} molecules of nitrogen. Actually the entire surface does not become covered at once. If a molecule strikes a surface but only remains attached at a site not previously occupied, then the rate of surface coverage at any instant is

$$\emptyset = \emptyset_0(1 - \emptyset) \qquad (10–2)$$

where $\emptyset$ is the fraction of uncovered surface. $\emptyset$ also equals N/N_0, where N is the number of molecules that had previously settled on the surface and N_0 is the maximum number that can occupy a unit of area. Then $\emptyset = (dN/dt) = N_0(d\emptyset/dt)$. Substitution yields $\emptyset_0(1 - \emptyset) = N_0(d\emptyset/dt)$, for which the solution is

$$\text{In } (1 - \emptyset) = -(\emptyset_0/N_0)t \qquad (10–3)$$

Calculations for nitrogen at 250°F (120°C) in a vacuum of 10^{-6} Torr. (1.33×10^{-4} Pa) show that the surface is covered progressively over several seconds of time, as shown in Table 10–1. These data have been verified in adhesion tests in vacuum. Atmospheric pressure is about 7.5×10^8 as dense as the pressure used in the calculation, and thus the rate of molecular bombardment of gases on a surface would be 7.5×10^8 as great were it not for fact that a molecule cannot travel a distance equal to its mean free path without colliding with others in a dense atmosphere. We can only estimate, therefore, that the rate

TABLE 10–1 RATE OF COVERAGE OF A SOLID SURFACE
BY ADSORBING GAS

% of Surface Covered	Time, Seconds	
	in vacuum of 1.33×10^{-4}Pa	at atmospheric pressure
25	.8	multiply
50	1.7	the
75	3.5	previous
90	6.0	column
95	7.5	by
99	12.0	10^{-8}

of coverage of a surface in atmosphere is about 10^8 that shown in the center column of Table 10–1. These are very short times, hardly perceived in ordinary experience.

The above calculation applies to the first adsorbed layer (excluding oxide layers for purposes of discussion) of gases, including water vapor. Subsequent layers condense more slowly, depending on the temperature and humidity. Once covered, the surfaces are not in a sufficiently high state of energy to bond firmly to other solid atoms. This layer of gas, which is about 80 percent as dense as liquid, is then the first fundamental problem to overcome in the joining of solid surfaces together.

Many adsorbed gases react chemically with the substrate to form a new compound, the most common of which is a metal oxide. These chemical reactions, which occur after physical adsorption, usually liberate between 10 and 100 Kcal/mol, and more. The total reaction is referred to as *chemical adsorption,* or *chemisorption,* and it is not reversible. Because the energy of chemisorption is very much higher than for physical adsorption, an oxide usually forms quickly in air. But the rate of oxidation is limited by the diffusion of ions and vacancies through the oxide film. The adsorption of additional gases upon the oxide film takes place at a higher rate than that gas can be diffused into the oxide. Thus the oxides are also covered with physically adsorbed layers of gas. Both layers are between 2 to 10 nm thick on technologically clean surfaces (see Ref. [2]).

Although the adsorbed films form quickly, there is a limit to their mechanical durability. Rubbing two solids together displaces small regions of the films and allows some microscopic solid bonding. In fact, the measurable resistance to sliding, called *friction,* involves some solid bonding. However, though locally strong, there are very few bond sites, and the total joint strength is hardly measurable in directions normal to surface. The adsorbed gases are one reason, but another is the nonconformity of two touching surfaces. This uneven topography usually consists of several dimensional scales, ranging from the microscopic to the visible or macroscopic level. The smaller scale protruberances or *asperities* produce *roughness,* whereas the larger-scale features are called *waviness.*

The consequence of roughness and/or waviness is that contact between surfaces is made on only a few widely separated *points*, which in reality are very small areas. As combinations of normal and shear stresses are applied to these small areas by external loading, without sliding, the stress state rises, plastic flow begins, and the sizes of the small contact areas increase. When the external loading is decreased, the stored elastic energy beneath the contact areas produces an elastic recovery to a less deformed state, resulting in a decrease in the size of the small areas of contact. All of this usually occurs on the scale of the asperities.

During sliding some of the adsorbed film will be displaced from the contact regions into the *valleys* between. They remain in the system, limiting the amount of surface that can be *cleaned* and bonded by rubbing. But even in those regions that were bonded, much of the bonding would be fractured by elastic recovery when the stresses on the contacting region are reduced, leaving a scarcely measurable adhesion.

We can estimate the amount of technological surface in actual contact if we assume that the contact pressure in isolated asperity contact areas is about three times the yield strength of the materials in the asperities. If one cube of lime-soda glass (density of 2.2 gm/cc, YS of 150 MPa) were placed upon another, each of dimension 100 mm, the apparent area of contact would be 10^4 mm^2, but the accumulated minute contact area would be only 2.2 mm^2, a factor of 4500 less. Sliding of one glass surface over another will produce fracturing, but with metals the asperities will plastically deform into conformity with each other to increase the bond area.

In vacuum, practical weld strengths can be achieved by sliding, and such a process is known as *cold welding*. In air some truly cold (ambient temperature) welding can be achieved by squeezing two parts together so that the surfaces expand to expose areas upon which there is little adsorbed gas. A more practical process is to rub two surfaces together repeatedly or continuously until the majority of the surfaces have been plastically deformed. This reduces the effects of local elastic recovery. In addition, if the rubbing heats the surfaces, some physically adsorbed gases are removed and the yield strength of the materials is reduced, thereby enhancing the possibility of plastic flow. This process is often called *friction welding*.

So far the discussion has focused on the bonding of solids. Liquids also bond or attach to solids. This phenomenon is called *wetting*. As in the case of solid bonding, there are degrees or strengths of wetting, but these quantities are not expressed in units of stress. One way to express the tendency for a drop of liquid to spread over, or wet, a flat horizontal solid surface is in terms of a contact angle (after T. Young). For each combination of liquid and solid there is a specific contact angle β. Figure 10–1 shows how the contact angle is measured. Table 10–2 gives the values of β for four fluids on glass. Water with detergent is seen to wet glass better than does tap water, but the alcohol wets completely. In the same way a glue or adhesive compound may wet surfaces to different degrees, with consequences on the quality of the bonded joint; molten aluminum may be deposited on a solid surface of aluminum, but it may not attach well. In the latter case the

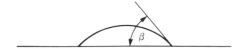

Figure 10–1   The measurement of contact angle.

TABLE 10–2 THE CONTACT ANGLE FOR FOUR FLUIDS

Liquid	β
tap water	110°
water with detergent	80°
furfural	30°
isopropyl alcohol	<1°

oxide on the solid aluminum impedes wetting of liquid metal to the solid with the effect of preventing good welding. With time, and in inert atmosphere, the liquid-covered oxide dissolves away, or with sufficient heating the oxide will melt and float away so that good wetting may occur. Thus wetting is a dynamic event.

In summary, the state of matter on earth is such that separate solid bodies do not readily bond together. This is fortunate because if human bodies were to stick to whatever they touch, all matter would soon become one coherent (or incoherent) mass. Joining and coating therefore require an intervention in the natural order, by introducing some special effort. These special efforts in the form of manufacturing processes will now be discussed.

10.4 WELDING

Fusion welding begins by placing two solid parts near to each other. Recall that two surfaces are never able to contact each other at every point though clamped tightly together. But in practice large gaps (of the order of millimeters) often exist between surfaces. For fusion welding, the acceptable range of gap size is about one-fourth to one-half of the diameter of the *welding rod* used, which in turn is usually no larger than the thickness of the thinnest member being welded. These ranges are wide and depend upon the circumstance and the experience of the people doing the work.

Welding proceeds by melting a small volume of metal at the surface of each part. If these two pools of molten metal can be contained, that is, prevented from falling away, they may mix together. If they then solidify, a solid bridge between the two parts has been effected, and the parts have been joined or welded together. Very localized fusing of two parts together is called *tack welding* and is sometimes done in preparation for a final weld. The final weld will usually, but not always, require the addition of *filler material* because of the gap that usually exists between surfaces to be joined. Filler materials can be of the same composition as the metal parts, or perhaps a material that will add some desirable property to the joint.

Welding methods are often classified by the heat source used to melt metal; there are five major heat sources. One is a flame, which is achieved by the oxidation of the carbon and hydrogen in fuel gases. Carbon releases 60,900 J/gm of heat, and hydrogen releases 254,600 J/gm, but these are awkward fuels to use by themselves. Combinations or compounds of carbon and hydrogen, such as butane (C_4H_{10}), propane (C_3H_8), and acetylene (C_2H_2) are easier to handle and will produce hot enough flames for melting most materials. The hottest flame is achieved by using oxygen with acetylene, which produces 27,720 J/gm. This flame, if adjusted properly, will just melt a small piece of tungsten (MP = 3410 ± 20°C). If acetylene or any other gas is burned with air, the flame temperatures are a few hundred degrees cooler than when burning oxygen, because of the need to heat up the large fraction of useless nitrogen in air. It should be noted that the combustion of fuel gases produces CO_2 and H_2O as well as complexes of nitrogen or other gases in the vicinity of the flame, some of which become entrained into the molten weld metal.

   Another common heat source is the *electric arc*. An arc is a high-density flow of electrons, across a voltage, from a cathode to the anode, with a flow of ions in the opposite direction. Welding is done at voltages between 30 and 75 volts with arc lengths of about 2 to 10 mm. The technology of arc welding, incidentally, also includes getting the arc started. An arc would not spontaneously begin by applying 75 volts over a gap of 2 mm. An arc is therefore "struck" by momentarily touching the cathode to the anode, which induces a flow of electrons. After electron flow begins a useful arc can be maintained, and the arc may *carry* several hundred amperes.

   The actual temperature of the arc is a fictitious property, but its effect may be estimated in terms of a temperature. An arc in air has a *surface temperature* of the order of 10,000°K, while one drawn between iron electrodes, as in the majority of welding, has a *surface temperature* of about 6,000°K, because the iron vapor has higher electrical conductivity than does air. (The arc loses about 20 percent of its heat by radiation when welding is done with a short arc and a current of 100 A.) A metal arc is therefore sufficient to melt all materials by radiation and convection alone, but electron bombardment of the anode also heats the surface. The cathode is correspondingly cooled by evaporation of electrons. In arc welding the workpiece is usually the anode, upon which the impinging power density is about 10^7 to 10^8 W/m^2, and the filler material becomes the cathode. Industrially, this arrangement is called *straight polarity*. In some cases, such as when thin sheet is welded, *reverse polarity* is used to prevent melting the sheet before the filler material is melted.

   In and around the arc there is a large magnetic field which causes a flow of gases and molten metal to attain velocities of the order of hundreds of meters a second, generally from the cathode to the anode. Some of the metal will be heated to the point of evaporation, which, because of expansion, produces a net flow of material out of the arc. Up to 10 percent of the metal may be lost in this way, and it condenses as dust on everything in the vicinity of the welding operation. In addition, there is a *spatter* of liquid metal out of the arc. Between 10 and 20 percent of the filler metal is lost from the weld by spatter, and this metal is distributed around the weld as moderately firmly attached spheres.

   The evaporation of metals proceed somewhat selectively according to the boiling point of the metal. A table of melting points and boiling points of various metals is given in Table 10–3. The end result of selective boiling of metal is to change the composition of the weld.

   In practice, arc welding involves a number of choices in order to achieve the desired joint properties most reliably. These include:

TABLE 10–3 THE MELTING POINTS AND BOILING POINTS OF SOME ELEMENTS AND OXIDES AT ATMOSPHERIC PRESSURE

Element	Density g/cc	MP°C	BP°C	Th.cond.	Oxide ΔH, cal/m	MP°C	BP°C
Ag (silver)	10.50	961.9	2212	4.29(W/cmK)	$Ag_2°$ $-7,740$	230	
Al (aluminum)	2.70	660.4	2467	2.36	Al_2O_3 $-404,080$	2072	2980
Au (gold)	19.32	1064	2807	3.19			
Be (beryllium)	1.85	1278	2970	2.18			
B (boron)	2.34	2079	2550(s*)	0.32			
Cd (cadmium)	8.65	321	765	≈ 0.9			
C (carbon)	1.8–2.3	3550	3367(s*)	0.01 – 26			
(diamond)	3.15–3.53						
Cr (chromium)	7.19	1857	2672	0.97	Cr_2O_3 $-274,670$	2266	4000
Co (cobalt)	8.9	1495	2870	1.05			
Cu (copper)	8.96	1083	2567	4.03	CuO $-37,740$	1326	
					Cu_2O $-40,550$	1235	1800
Fe (iron)	7.87	1535	2750	0.87	FeO $-65,320$	1369	
					Fe_3O_4 $-268,310$	1594	
					Fe_2O_3 $-200,000$	1565	
Mg (magnesium)	1.74	648.4	1090	1.57	MgO $-144,090$	2852	3600
Mn (manganese)	7.3	1244	1962	0.08			
Mo (molybdenum)	10.22	2617	4612	1.39			
Ni (nickel)	8.9	1453	2732	0.94	NiO $-57,640$	1984	
Pb (lead)	11.35	327.5	1740	0.36	PbO $-52,800$	886	
Si (silicon)	2.33	1410	2355	1.68	SiO_2 $-210,070$	1723	2230
Sn (tin)	5.75	231.97	2270	0.5–0.7	SnO_2 $-68,600$	1630	1800(s*)
Ta (tantalum)	16.65	2996	5425	0.57	Ta_2O_5 $-492,790$	1872	
Ti (titanium)	4.54	1660	3278	0.23	TiO_2 $-125,010$	1825	
V (vanadium)	6.11	1890	3380	0.31			
W (tungsten)	19.3	3410	5660	1.77	WO_3 $-201,180$	1473	
Zn (zinc)	7.13	419.6	907	1.17	ZnO $-84,670$	1975	
Zr (zirconium)	6.5	1852	4377	0.23	ZrO_2 $-262,980$	2715	

*sublimes

1. Selecting the method by which the weld joint will be protected from the effects of entrained gas bubbles and inclusions, and the system for preventing gases and inclusions in the weld.
2. Selecting a type of filler material.
3. Setting the velocity of welding.
4. Setting the voltage and the limiting current provided by the power supply for the operation.

Protection against gas entrainment may be done either by keeping harmful gases out of the weld region, or by removing whatever gases and products of reaction enter the weld metal. The first may be done with the use of an inert gas such as argon. (Nitrogen is not sufficiently inert in the welding environment, and the other noble gases are very expensive.) The flow rate of inert gas must be controlled to keep harmful gases away without carrying excessive amounts of heat away. In some cases a small amount of readily ionizable gas must accompany the inert gas to control ionization of gas in the plasma or arc. Oxygen is one such gas, and it *stabilizes* the arc. A stable arc is one in which the region on the cathode, from which electrons depart, and the region on the anode, at which they bombard, do not dance about wildly, making welding difficult or impossible to do. But primarily, the inert gas minimizes the amount of oxides and nitrides that can form in the weld metal.

The second method of control is to feed a solid substance, containing several complex compounds, into the arc area. These compounds are called *fluxes,* and they provide some combination of inert gases when heated, some atomic elements to lower the electrical resistance of the gases in the arc, and some atomic elements to add desirable properties to the weld joint. Such compounds are often contained in a coating on the *electrode* or filler material for hand-welding operations. In the older automatic filler-wire feed machinery, the fluxes are in a granular form that drop around and cover the moving arc. The arc is submerged under the flux, and the method is appropriately referred to as *submerged arc welding.* Granules of flux may be applied overhead or to vertical surfaces by blowing them into the arc, but it is expensive. Newer automatic machinery uses hollow filler-wires or tubes, filled with flux, which are more useful for all positions of welding.

Filler materials may be steel, brass, aluminum, or other materials. It is not necessary to use a filler material of the same composition as the pieces to be joined. Joint composition may be adjusted by adding alloys to the fluxes as well as to the filler metal. Thus for welding steels and irons, a low-cost filler material may be used. Chromium may be added to the flux to impart some corrosion resistance to the weld; nickel may be added to retain ductility of the weld joint at low temperature, and particles of tungsten carbide (WC) may be added to impart wear resistance to the weld.

Engineers in manufacturing are usually concerned with both the cost to weld and the final properties of the weld. For both purposes the arc voltage, arc current, and velocity of arc movement must be selected. High arc power (voltage × current) heats and melts large quantities of metal in a short time, and high velocity distributes the power over a large area. There are practical limits imposed on velocity by the ability of people or machinery to properly direct the path of the weld operation, but there are no real limits on the arc power.

Joint strength is achieved by control of *weld penetration* and to a lesser extent by control of *bead height.* Both of these terms can be defined with the aid of Fig. 10–2. The penetration is the depth to which the plates shown in Fig. 10–2 are fused together. To achieve this penetration a section of each plate is melted, and the size of the molten pool is determined by the rate of heat input versus the rate of heat conduction into the plates. The heat enters at the surface, by radiation and convection, and also by electron bombardment if the plates are the anode or positive pole.

Theoretical equations relating penetration to the arc power, arc length, velocity of welding, and polarity of the arc are very cumbersome. It is thus more common to develop empirical equations by measuring the penetration that results from exposing a surface to a moving arc of varying powers. The results will vary over a wide range, depending on the polarity of the arc if it has DC applied, and whether the arc is a metal arc or one drawn

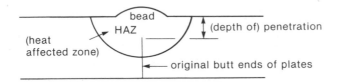

Figure 10–2   Sketch of the cross section of a weld, joining two thick plates.

by a tungsten or carbon electrode. The common form of such equations is

$$p = k\frac{E^a I^b}{v^c} \tag{10–4}$$

where E is the voltage across the arc during welding (usually in the range of 25 to 75 volts, *not* the open circuit voltage), I is the current in amperes through the arc (usually in the range of 75 to 1000 amperes, not the limiting current set on the machine), and v is the welding speed or velocity (usually in the range of ⅓ to 5 meters per minute). The exponents and the constant k are characteristics of the metal as well as the process, and the functions of the thickness of the plates being welded, the diameter of the filler rod, the atomic elements passing through the arc, the arc length, and other parameters. Typically for welding steel at moderately high speeds, where p is to be given in mm, $k \approx 10^{-4}$, $a \approx 0.5$, $b \approx 1.5$, and $c \approx 0.33$.

Precise values for careful prediction of welding costs should be obtained from handbooks on welding or should be obtained by tests using the actual production equipment. The performance of various combinations of equipment and welding materials vary as much as 5 to 1. Modern production equipment for arc welding no longer depends on the characteristics of generators and transformers. Rather, the equipment can produce virtually any desired electrical waveform, such as pulses of various voltages, polarities, and durations, depending on the need. Each waveform will have different equations for penetration and bead buildup. The manufacturer of the equipment and vendors of welding supplies have some data of this type available.

The bead height is as difficult to predict as the penetration. Again, these should be empirically determined if this dimension is important. Generally, the bead height is about one-fourth of the penetration, for metal arc welding with straight polarity. For reverse polarity the bead height approaches one-half of the penetration, and for welding with an alternating voltage the bead height is halfway between one-fourth and one-half.

Three other heat sources are available beside the electric arc and gas flame: these are the *plasma-arc*, the *laser*, and the *electron beam*. The plasma-arc system passes gas mixtures (for example, 75% A_2 + 25% H_2 or other combinations with helium or nitrogen) through or near an arc, where the gas is ionized. The gas absorbs most of the energy in the arc. This stream of ionized gas can be directed toward any cool surface, including a nonconducting material, where the ionized gas evolves heat and reverts to molecular gas.

Lasers produce electromagnetic radiation which is coherent and therefore can be propagated as nearly parallel beams. Radiation that impinges on a solid surface is reflected, absorbed, and transmitted in various proportions according to the *optical* properties of the target substance and the wavelength of the radiation. A common type of laser is the helium-neon laser, which produces red light of 632.8 nm wavelength, and another is the argon laser, which produces blue light of 488 nm and green light of 525 nm. These do not heat surfaces very efficiently. Infrared (invisible) radiation is better in that it is mostly absorbed by metals, which causes heating. The CO_2 laser, which produces infrared radiation in the range of 1000 nm or 1 μm, is the best commercial type for heating most solid surfaces.

The electron *beam* is noncoherent radiation of wavelengths in the range of 0.01 to 10 nm. Because of its short wavelength, over 98 percent of the radiation is absorbed when the electrons *bombard* an anode. Electrons are emitted from a large surface of heated thorium-tungsten or other source. They are directed toward another surface, and given energy, by applying several thousand volts between the source, which is the cathode, and the target, which is the anode. (Higher voltages produce X-rays.) The beams can be focused by electric or magnetic fields upon a small region on the work surface.

It is difficult to rate these sources relative to the others in terms of their temperatures, but figures are available on the maximum practical power density available on the surfaces upon which these heat sources impinge. For the plasma it is between 10^9 and 10^{10} W/m^2; for the CO_2 laser it is between 10^{10} and 10^{11} W/m^2, and for the electron beam it is between 10^{11} and 10^{12} W/m^2. The rate of heating for each of these sources can be increased at will, but a limit is reached when the molten pool is yet very shallow when the surface of the metal evaporates.

The plasma is a stream of ions, which revert to molecular gases when cooled by the target surface. The gases can be inert, which expel oxidizing gases, but unfortunately they can also become entrained into a weld and produce bubbles. Both the electron beam and the laser beam are "clean," and both beams can be shaped and focused precisely and quickly, but both are expensive. The laser is very *inefficient* in the use of energy, using only about six percent of the power taken from the electrical source. The electron beam is about 75 percent efficient, but it must operate in a vacuum of about 10^{-3} to 10^{-4} Torr. (0.1 to 0.01 Pa). The ordinary arc welder is also about 75 percent efficient in terms of the use of power to effect deposition of the weld metal. Pulsed arc welders are somewhat more efficient, whereas the plasma arc system is somewhat less efficient.

In principle, if the edges of two parts are melted progressively and allowed to fuse together, eventually the two parts should be joined. However, a major problem arises because of the progressive nature of solidification. This may be seen by holding a bar by the ends in a very stiff vise, as shown in Fig. 10–3. If the rod is heated, with no heat

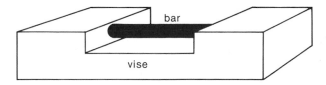

Figure 10–3 Sketch of a bar held in a vise.

transfer to the vise, the rod will expand by an amount equal to $e = \propto \Delta T$, where $\propto$ is the thermal coefficient of expansion and ΔT is the temperature rise. If the vise resists this expansion completely, as if it were rigid, the rod will not change in length. Rather, its expansion would be exactly countered by a compressive strain, and the stress in the rod would rise to

$$s = eE = E \propto \Delta T \tag{10–5}$$

The linear increase in stress as a function of temperature rise is shown graphically in Figure 10–4 as line segment a. If the bar is cooled from any value of ΔT along line a, the stress in the bar will return to zero. But if heating continues to a temperature at which the yield strength of the material in the rod is reducing, the change in stress then follows a different course, as shown by line segment b. In this region the bar is still expanding, but it is also being *plastically* compressed. If now the temperature decreases after plastic flow occurs, the stresses will again decrease linearly with temperature, but will follow curve c in Fig. 10–4. If the rod were originally loosely inserted into the vise, it will separate from the vise at the temperature at which the stress becomes zero, and perhaps fall out. If the rod had been bonded to the surface of the vise, further cooling would develop a tensile stress at the interface between the rod and the vise as shown by line d in Fig. 10–4.

The rod in the vise is a useful analogy for a nugget of molten metal in a weld area if the rod is heated to the melting point. The stresses in the rod are seen to pass through a maximum and decrease to zero at MP. Solidification and cooling now involve contraction only and the (attached) rod is left in a state of tensile stress, as shown by line e. These stresses cannot increase linearly because temperature affects the flow stress of the material. Thus line e is curved as shown. The stress state now remains or *resides* in the rod, and it is referred to as a *residual stress*. The magnitude of the residual stress, in the case of a rigid vise, is seen to be dependent on the amount of differential heating and on the properties of the material of which the rod is made. In the more practical elastic case, the vise would have strained (open) so as to reduce the stresses in the rod upon heating. In a weld there is no sharp dividing line between the heated and upheated regions. Thus it is best to simply recognize that the residual stresses arise from differential strains that arise from local heating, as well as from localized phase changes and from local plastic flow due to mechanical applied stresses.

There are two practical considerations that arise from the inevitability of localized

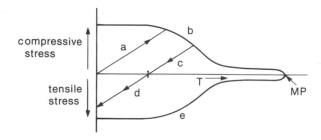

Figure 10–4   Stresses in a bar that is held in a vise and heated.

and differential strains due to conventional welding. The first is that these strains can exceed the fracture strain of such brittle materials as cast iron, primary martensite, and ceramic materials. These materials are then likely to crack during cooling of the weld, wherever they are located, in the HAZ, outside of the HAZ, or in the bead. There are several ways to prevent such cracking. These include:

1. Weld with a filler material that has a yield strength significantly lower than the fracture strength of the joined materials.
2. Heat a large area around the joint before welding so that the stress gradients from the weld region outward are small, that is, heat the vise in the above example.
3. In the case of steel, heat the area around the weld above the M_s (martensite start) temperature to prevent the formation of the brittle martensite when the weld cools (see Chapter 7).

The second problem is that welded parts and structures inevitably become distorted during welding, and distort again when machined or heated after welding. Figure 10–5(a) shows an example of two plates, perpendicular to each other, and welded in the corner. A molten bead will have been placed in the corner and then cooled. During cooling it contracts. If the plates are held rigidly, the finished and cooled weld will be in a state of tensile residual stresses. If the plates are not held in place, the solidifying bead will alter the angle between the plates by as much as 2 to 4 deg. In fact, if the rigidly held plates were released after welding the angle would again change, but by a smaller amount. One solution to the distortion problem is to plastically bend the joint to an angle of 92 to 94 deg and release it. Elastic recovery may return the angle to 90 deg. A better solution is to set the unwelded parts to an angle of 92 to 94 deg, and then weld without restraint. Contraction of the bead will pull the parts to an angle near 90 deg. The proper angles for setting the parts can be calculated if enough of the relevant details of the welding process are known, but it is more quickly and accurately done experimentally.

The above example shows distortion across the *width* of the weld. There is also distortion due to tensile residual stresses along the length of the weld. This can be seen by passing an arc along the edge of a steel bar. The result will be a curved bar, as shown in Fig. 10–6. The radius of curvature can be estimated by assuming that the arc has melted the top layer of the bar of height t, to a depth *of penetration p,* and that the molten layer has contracted when it solidified and cooled. If the bar were held straight during this contraction, then the top layer would have a tensile stress in it, and the lower part of the bar would have a compressive stress so that the forces in the entire bar would be equal. Calculation of thermal contraction and the observation that welding with brittle filler materials causes cracking in the weld beads suggests that the strain in the top layer is in

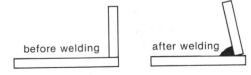

before welding   after welding   **Figure 10–5**   Two plates welded at an angle.

Figure 10–6 A bar, bowed by welding all along the top edge.

the range of 0.001 to 0.003. Thus for convenience, let us assume that the tensile stress in the top layer is equal to the yield strength S_y of the metal, acting at a distance $p/2$ from the top face. The force exerted by the top layer, per unit width of the bar is $F_t = S_y p$. A moment M of magnitude $S_y p$ times the moment arm, $t/2 - p/2$, is required to hold the bar straight. Thus, $M = S_y p(t - p)/2$. If the external moment is removed, the bar will bend to a radius R, which may be calculated from $R = EI/M$. I, per unit width of the bar, is $t^3/12$ so that

$$R = \frac{Et^3}{6S_y p(t - p)} \tag{10–6}$$

It may be seen that where p is zero, and where $p = t$, R is infinite; that is, the bar is straight. R is minimum when $t = p/2$, or when the depth of penetration is half the height of the bar. For this condition, and for a material where $E/S_y \approx 500$, $R \approx 333t$. Thus for a bar of $t = 1$ cm, $R \approx 3.33$ m. By this method one could make a hoop of steel 6.66 m in diameter. The same can be done by localized heating above the recrystallization temperature without melting. By this method large beams may be made curved for bridge decks, and by this method a moderately bent beam may be straightened.

In the design of welds there should be a strategy for accommodating the distortions that occur. As mentioned above, distortion can be minimized by beginning with parts that are oriented with a negative distortion. Welding will simply straighten the parts, but leave residual stresses in them. If now material is removed from one face of the welded part, for example, by grinding, the residual stress distribution will be altered and the part will distort again. The distortion of welded structures such as ships can be minimized by assuring that the welding along the left side of the ship keeps pace with the welding along the right side. A bent part can be straightened by plastic bending in the opposite direction, or it may be straightened by *peening* on the surface containing the tensile stress that is the cause of the undesirable bend.

10.5 "CUTTING" OR PARTING BY FOCUSED HEAT SOURCES

All of the heat sources that will melt metal quickly can be used for "cutting" as well as for joining. The simplest approach is simply to melt a slot from a piece of metal to make two parts.

The cutting of steel may be done with a special oxyacetylene torch by heating an edge of the steel to about 850°C and then directing a stream of oxygen toward the heated region. As the iron oxidizes, enough heat is liberated to heat up the layer beneath.

The oxide and some metal melt and are blown away by the oxygen stream, and the process is self-sustaining. In fact, the original flame can be shut off, leaving only a stream of oxygen to accomplish the cutting. This process depends primarily on two factors; the heat of oxidation must be high, and the thermal conductivity of the metal must be low. Few metals fulfill these conditions as well as does iron, as may be seen from Table 10–3. For some that do not, iron powder can be injected into the gas or plasma stream to aid the process. It should, in principle, be possible to cut cast iron in this manner as well, but the flakes of graphite (Chapter 11) deflect the oxidizing stream away from the metal.

10.6 BRAZING AND SOLDERING

In many cases, metals and ceramic materials can be joined without melting the parent material. The casting of a soft metal into the gap between two parts can produce a joint provided the soft metal fills the gap, wets the parent surfaces, and finally bonds to the parent materials. Gap filling is best achieved by heating the filler metal to a temperature where it has low viscosity.

Wetting is achieved by cleaning the parent surfaces, usually with fluxes. A flux typically contains some elements or compounds that will, with the oxide of the parent material, form a new compound that has a lower melting temperature than that of the filler metal. This new compound will usually have a lower density than that of the joint metal, so it will float on the molten metal, which now readily wets the solids.

The success of cleaning can readily be checked by observing the contact angle between a drop of molten filler metal and each of the surfaces to be joined, as shown in Fig. 10–1. Or, if the parts have been joined but later they become easily separated, one can observe the extent to which the filler metal was drawn into the gap. The extent to which the filler metal *should have been drawn* into the gap can be estimated by using the equations for calculating the height of rise of a liquid in a vertical capillary with one end set in the liquid. Examples of good and poor wetting are shown in Fig. 10–7. For the

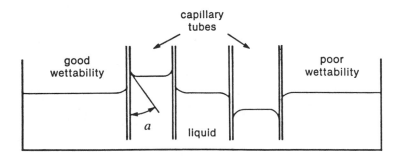

Figure 10–7   Sketch of the effect of wetting forces on liquid height in a small tube.

case of liquid rising in the tube, the lifting force F of the surface film is the vertical component of the *surface-tension* force T acting around the inner surface of the tube, or $F = \pi \, dT \cos a$. (Here a is proportional to the contact angle of a drop on a flat surface.) This is countered by the weight of the fluid, $\pi \, d^2 h \rho g / 2$, where ρ is the density of the fluid and g is a gravitational constant. Thus

$$h = \frac{4T \cos a}{d \rho g} \tag{10–8}$$

It may be seen that for poor wettability, that is, where $a > 90$ deg, the fluid level in the capillary will be lower than that of the outside level, and in joining, the filler metal will resist entering the gap.

Good wetting and a strong bond are effected where some constituent in the filler metal is soluble in some constituent in the parent surface. For example, only very small amounts of lead will dissolve in iron, and thus lead does not wet an iron surface very well. But lead is a very useful and low-cost constituent in solder. To make a versatile solder, therefore, tin, antimony, silver, copper, and indium are usually included. None of these metals have oxides that interfere with the joining process. Aluminum would seem useful because of its low cost and low melting temperature, but its oxide forms very quickly and it is hard to remove.

Brazing and soldering are different from each other in that brazing uses a predominantly copper filler alloy, whereas soldering uses filler alloys made chiefly of lead and tin. Usually the parts to be joined are materials that are much stronger than the filler metal. Though the filler metals are soft and weak, a joint may be made very strong by effecting a proper geometry. The three geometric variables that control strength are joint contact area, thickness of filler, and whether the stress applied to the joint is a shear stress or normal stress. Figure 10–8 shows two ways to join parts, by a butt joint or by a lap joint. It can be seen that the load-carrying cross section of the filler material is restricted in the butt joint, but potentially unlimited in the lap joint. The lap joint could therefore be made stronger than the butt joint, but it has two disadvantages. The large step is rather conspicuous, and a large tensile force will cause bending, because the forces in each part are acting in different planes. Both of these problems can be overcome by a double lap joint as shown in Fig. 10–8. Another difference between the two is that the butt joint applies a tensile stress upon the filler metal, whereas the lap joint applies a shear stress. Shear strength is usually about half the tensile strength of metals, but in addition, a thin joint in tension can benefit from *triaxiality*.

Triaxiality can be shown in the sketch of Fig. 10–9, where a soft solid joins two much stronger solids. The parts are loaded in tension, and the stress in the joint is largely normal (that is, little imposed shear). As the tensile load on the parts increases, the stress state in the joint approaches the yield strength of the filler material. The filler material will

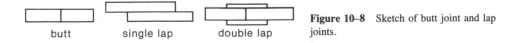

butt   single lap   double lap

Figure 10–8   Sketch of butt joint and lap joints.

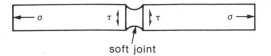

Figure 10-9   Triaxiality effects in a thin joint.

soft joint

deform plastically with the effect of diminishing its horizontal dimension. However, the parent material will constrain the filler material from changing dimension by exerting an outwardly directed shear on the surfaces of the filler material as shown. This has the effect of imposing a horizontal tension on the filler material. By the principles of plasticity, additional vertical strain in the filler material requires a stress equal to the plastic flow stress *plus* a stress approximately equal to the horizontally imposed stress. The thinner the joint, the more complete the constraint, and thus thin joints appear to have high strength. Generally, a joint of approximately 10^{-4} mm shows the highest strength. Perhaps thinner films would be even stronger, but at this dimension it is difficult to determine what the film thickness is. There is little commercial interest in very thin films because of the high precision required in making the adjoining parts.

Soldering in particular, but brazing also, may be done by mass production methods because of the use of low-melting-point fillers. One can assemble complex structures, wires, brackets, and the like with flux and filler material properly placed. The entire assembly can then be heated in a furnace, or in a large flame shield, or by induction, or by infrared radiation. Alternatively, one can heat each joint in the assembly in turn by a fast-moving laser beam or an electron beam, programmed to focus briefly on each joint. These technologies are readily automated, but inspection is not as readily automated.

10.7 SINTERING, DIFFUSION BONDING, HOT-ISOSTATIC COALESCENCE

A growing technique for making solid parts is to press and coalesce powders into structural shapes. Parts ranging from styrofoam cups to roller bearing races are made in this way. Powders range in size from about 1 μm upward, and virtually every material is available in this form. Whereas in former days one chose materials from among those microstructures available on phase diagrams, it is now possible to mix a variety of powders together to achieve a wide range of properties. This is sometimes called making parts by *nonstochiometric chemistry*. The variations are infinite in number, far more than with the choices from phase diagrams.

The success of powder technology lies in compacting of the powder to near full density, and diffusion of the particles together. In the compacting operation the major problem is to avoid density gradients. Because of friction between the dies and the powder it is difficult to achieve uniform compaction. For example, if a long tube is filled with powder, and then if a rod were inserted into each end of the tube to compact the powder, the powder would be more dense at the ends and least dense in the middle. More uniform compacting, with lower applied force, can be done if some oil is mixed with the powder. Good compaction can also be aided by die design. Generally, the best part shape is such that, for example, if a die plunger moves 1 cm, it should compact the powder everywhere by the same volume fraction. But if a powder were compressed from all sides rather than from one side, the powder would be still more dense, and there would be fewer planes of weakness that arise from shearing the powder during uniaxial pressing.

Pressing a material from all sides is very difficult to do. One could press a small amount uniaxially, on three orthogonal surfaces in turn, but at some risk of fracture. In practice a powder is first compacted a small amount, usually uniaxially, to develop some *strength*. Then it is encased in a fluid-tight container, inserted into a chamber, and pressurized hydrostatically.

Compressed powder has little physical strength. The strength of this powder is known as *green strength*. A useful increase in powder strength can be achieved by heating without hydrostatic pressure or with very little pressure. This process joint is called *sintering*. The powder particles are bonded together over a fraction of their surfaces, as shown in Fig. 10–10(a). The maximum strength of a powder is developed by heating powder particles while under pressure, so that they conform to each other and bond together. This is known as *hot isostatic pressing*, or HIP. When well done the particles are completely bonded, leaving porosity of less than one percent, as shown in Fig. 10–10(b). The pressures for this process range from 100 MPa to 300 MPa, and temperatures as high as 2000°C can be reached.

An important attribute of the finished powder product is the remaining porosity, or its converse, density. Almost any density may be achieved, up to about 99.999 percent of the theoretical density. This is done by heating to about 50°C above the solidus temperature of some major constituent in the powder and maintaining a hydrostatic pressure of the order of 250 MPa while hot. An interesting mechanism of bonding is seen in the processing of Si_3N_4 powder. If a small amount of oxygen is admitted into the enclosure, an oxide forms in conjunction with the Si_3N_4 to form a low-melting-point compound, which holds the powder together quite well.

10.8 GLUING AND ADHESIVE BONDING

Adhesive bonding is done by filling the gap between two parts with a liquid which will solidify by one method or other. There are two important aspects to the strength of adhesively bonded joints, and these are the strength of the material in the joint and the strength of the interface.

a. High porosity

b. Low porosity

Figure 10–10   Sintered and HIP'd particles.

The strength of the material in the joint depends on the material, the geometry of the joint, and the methods used to apply the adhesive material, as discussed previously in soldering. There are many types of adhesives, and most of them are too difficult to describe by molecule structure and reaction mechanisms. The manufacturers of adhesives have been vigorous in marketing their products, so they supply detailed information on the properties, curing methods, and precautions to observe when using adhesives. The major generic types of adhesives are listed below.

Major types of adhesives and some of their properties

1. *Epoxies* are liquids that polymerize, or cure into a *thermosetting* solid, to hardness up to 90 ShoreD. They are available with a wide range of cure rates. Some are available in two "parts" which must be thoroughly mixed, and as soon as they are mixed, curing proceeds, with some heat evolution. Some are available in one part, that is, where the two parts are already together but formulated to cure very slowly at 20°C. These must be heated in order to cure in a few minutes. Epoxies are the strongest and most reliable of the adhesives and are available in forms that maintain strength up to 250°C. They are not recommended for joining plastics, because they attack the polymers chemically.

2. *Acrylic adhesives* are available in two major forms:
   a. The *anaerobic* form in the liquid state is prevented from curing when oxygen "terminates" or prevents free radical polymerization or hardening. Curing occurs when the normal exchange of oxygen becomes one way, namely evolution only. This occurs by containing some of the liquid between impermeable solids and exposing very little of the surface of the liquid to oxygen. This process is hastened by heating the liquid. Curing also requires the presence of metal ions, which are not available in most plastics. Where metal ions are not available, such as on surfaces, a "primer" can be applied to the surface (of the parts to be bonded) to supply the needed ions.
   b. The *cyanoacrylates* are thermoplastic; that is, they soften at elevated temperatures and become brittle at low temperature. The ethyl form is made to bond plastics and rubber, and the methyl form is made for bonding metals. These soften in the range of 50°C to 100°C. There is also an allyl monomer which retains strength up to about 220°C. All of these adhesives polymerize by an ionic mechanism initiated by weak basic ions, water being the most common. Adsorbed water on all surfaces is one source of initiator.

3. The *urethanes* are as strong as epoxies, and have greater toughness than many adhesives. However, they are formed by the reaction of a hydroxyl group and an isocyanate group, which produces toxic fumes.

4. The *silicones* are weak but resilient. They consist of Si and O in the chain. Some cure by liberating acetic acid and should therefore not be used on zinc, copper, or concrete. Others liberate the less reactive alcohols or amides, these are more expensive but are more widely useful.

5. The *methacrylate* adhesives are as good as the epoxies and cyanoacrylates, and they have the advantage of bonding plastics together.

An important part of the technology of adhesives is the design of the joint. Generally, polymeric adhesives should not be used to fill large gaps, one reason being that low-viscosity liquids will not stay in large gaps. Another reason, at least for the acrylic adhesives, is that curing is initiated at the interface of the liquid adhesive and the parts being bonded. A thick joint requires much time to cure, and a thick joint is exposed to oxygen over more of its surface than is a thin joint; this inhibits some curing. For the adhesives that evolve volatile chemicals during curing, the diffusion distance to the outer surface limits the rate of full cure. For the adhesives that cure by combination of their constituents, these considerations are not important.

Most of the adhesives provide joint strength of the order of 10 to 20 MPa, which usually refers to the shear strength. The joint strength in simple tension may be about half the shear strength, and the peel and/or cleavage strength is considerably less. Joints are usually more durable where one of the bodies is made of a material of low elastic modulus. Adhesively bonded joints will usually fail in the adhesive if it is inadequately cured; otherwise the interface usually fails. It is useful to visualize the nature of the bond sequence through the interface.

Virtually all metal surfaces are covered with an oxide. Metals such as aluminum, titanium, and chromium containing steels have relatively thin (3–5 nm) and adherent oxides. Iron, copper, silver, and many others have thicker (5–10 nm) and weaker oxides. Some oxides crumble and become detached when they are very thick. These should be removed before an adhesive is applied. However, an oxide layer will inevitably form, and most such layers are invisible. The instructions to remove oxides must refer to the loosely attached oxide. Next upon the oxide in a clean atmosphere is a layer of adsorbed water, as mentioned above, about 10 nm thick. In the presence of humans and machinery, however, most surfaces will also accumulate oils and greases. These may form under or over the water film. The major difficulty with these contaminants is that they constitute a thick fluid layer, which diminishes joint strength. The oily films are usually removed by a basic chemical solution, usually sodium compounds. These include carbonates, bicarbonates, sesquicarbonates, hydroxides, tripolyphosphates, tetrapyrophosphates, metasilicates, and orthosilicates.

The purpose in cleaning is to reduce the thickness of the fluid layer. If a liquid wets two parallel flat plates, the force required to pull the plates apart normally is inversely related to the thickness of the fluid film. At the edge of the plate there will be a meniscus of liquid, curved inward if the liquid wets the plate material. Surface tension in the fluid surface will tend to produce a flat meniscus, and thus a curved meniscus implies that there is a higher pressure outside of the mensicus than within the liquid. We may estimate this pressure difference from the simpler case of the pressure difference between the inside and outside of a liquid drop or a bubble of gas in a liquid. The equation for this pressure differential is $p = 2\gamma/R$. Here γ is the surface energy and R is the radius of the drop or bubble. In the case of liquid between two flat plates the separation of the plates h is about $2R$. The differential pressure from outside to inside the liquid, acting over the plates, will produce a force F needed to separate the plates and the magnitude, $F = 2A\gamma/h$. If $\gamma = 30$ dynes/cm, $A = 1$ cm^2, and if there is a thick contaminant film about $h = 10^{-5}$ cm thick, then $F \approx 6 \times 10^4$ dynes, or about 60 newtons of force. But if the surface is "clean," that is, if the adsorbed gas film is $h = 10$ nm (10^{-6} cm) and the gap between two such adsorbed films on two surfaces is filled with a hardened polymer, the force required to separate the surfaces normally—that is, to draw the adsorbed water inward—will be 600 N/cm^2 or about 600 MPa. This is the approximate strength of many adhesive joints. It can readily be seen that an additional contaminant film would lower the joint strength.

Admittedly, most adhesively bonded joints do not fail in tension, but rather in shear. In this case the fluid films might flow and readily fail. However, in a thin joint the asperities of the surfaces, that is, the surface roughness, will be much higher than the thickness of adsorbed water, thus preventing easy shear.

REFERENCES

1. A. H. Cottrell, *The Mechanical Properties of Matter*. New York: John Wiley and Sons, 1964.
2. L. Holland, *The Properties of Glass Surfaces*. Chapman and Hall, 1964.
3. R. Pirani and R. Yarwood, *Principles of Vacuum Engineering*. Reinhold, 1961.

PROBLEMS

10–1. Assume a cube of steel of 100-mm dimension, set upon a steel table. Both have a YS of 80 MPa. Calculate the real area of contact between the two due to the weight of the cube alone and due to the addition of your own weight.

10–2. If one second of time is required to melt a steel surface with an electric spark, and 10 seconds are required when an oxyacetylene torch is used, what is the approximate energy density on the surface due to the latter?

10–3. Calculate the penetration due to arc welding steel at a velocity of 3 mm/s.

10–4. Sketch the effect of an elastic vise instead of a rigid vise on a figure of the type of Fig. 10–6.

10–5. With reference to the data in Table 10–2, how far would each of these liquids rise in a capillary tube of 0.5 mm diameter?

11
casting _____

In the most general sense, casting includes all processes where a material in liquid form is introduced into a cavity (or mold) so that upon complete solidification, the surfaces of the new solid product have the opposite shape of the mold. Consider the pouring of water into an ice-cube tray; the cavity is basically a cubical hole and upon solidification, a solid cube of ice has formed. This is the simplest example of casting that involves all of the major fundamentals involved, and shows that the process entails a liquid-to-solid transformation to produce the desired shape. Thus, unlike the processes called forming and machining (cutting) discussed in earlier chapters, where end products are produced from one solid state to another, casting usually involves more complex possibilities with regard to the microstructures that may result upon solidification.

Although certain processes which use the word *molding* fit the definition of casting, others do not, and this could lead to confusion. For our purposes, any molding process that involves conversion from the liquid to solid will fall under the heading of casting. Those that are processed from one solid to another are classified differently. In this way, the fundamental nature of the process takes precedence over the name given to the process itself.

Most of the *analyses* developed for those processes that involve the altering of a starting solid into some other solid configuration were based upon the subject of solid mechanics, and applied forces or induced stresses were of primary concern. In contrast, the major disciplines involved in casting include fluid mechanics, heat transfer, concepts of solidification, and thermal contraction. Consider an example of involving *most* of these topics. A hot fluid is being pumped through a pipe made of stainless steel, and the temperature of the fluid at various locations along the pipe is to be predicted. If such factors as the thermal conductivity of the steel, the rate of flow, and the likely effects of both conduction and convection are known, a reasonably straightforward analysis can be made by using an appropriate background in fluid mechanics and heat transfer. Of course, the geometry involved is relatively simple, and solidification is not considered. The great majority of cast products are far more complicated in shape, and even those who possess advanced technical backgrounds in the topics of fluid flow and heat transfer would have a difficult time to provide a similar, straightforward analysis concerning temperature gradients. Because of this, we intend to cover such situations from a physical rather than a mathematical point of view. After all, manufacturing engineers are not expected, for example, to be involved with details of mold design; rather, they are often confronted with the processing of cast products that have been produced by others.

Here we distinguish between two broad categories of cast products. The first includes those that are produced to a particular shape that is as *close* to the desired product configuration as possible. As an example, the crankshafts shown in Fig. 11–1 would require only a limited number of subsequent grinding operations to become a finished product. Similarly, if one considers a carburetor body, which has an extremely complicated shape, only some final machining operations are needed to produce the finished part. For products like these, it is *always* desirable to first cast the part with as little *excess* material as possible, that is, to leave only such excess as is needed to be certain that final dimensions can be achieved by subsequent finishing operations. The second category

Figure 11–1   Photograph of several cast crankshafts.

includes those materials that are cast to an initially simple shape (often called an *ingot*) which is then subsequently processed by other methods to produce various forms that are then used in other processing modes. For example, a cubical billet of steel may be cast via a series of operations, as shown in Fig. 11–2. It is then subjected to one of a number of possible forming operations that lead to a wide variety of shapes such as flat sheets, round bars, wire, and the like; note the great variety of possibilities shown on Fig. 11–2. These operations produce a *wrought* structure which possesses a greater uniformity of finer grains, far fewer internal voids, and greater ductility than did the initial cast structure. Thus, ductile steel sheets used in subsequent forming operations were *initially* cast to some very different shape; this is true for the many shapes used in a variety of other processing operations. In this context, it could be said that casting is the truly primary process, since forming and cutting operations are performed on parts that were initially cast to shape. Continuous casting, a more recent process, is discussed in Sec. 11.8.

11.2 ASPECTS OF THE BASIC PROCESS

Consider Fig. 11–3, where a simple cube is to be produced; for now let us pay no attention to the material which forms the mold. Since the solidified product must be eventually removed from the mold, the latter is constructed of two separate parts (*cope* and *drag*) which meet along the *parting line* when assembled. The molten material is fed into the cavity via the *pouring basin*, from which it travels through a channel (called a *runner*), fills the *mold cavity* itself, and then proceeds to fill a *riser*. (This simply provides a reservoir of molten material, whose purpose is discussed shortly.) Using this simple illustration, we now address the variety of occurrences that take place prior to, during, and after final solidification results.

First, the molten material must possess adequate *fluidity** so as to completely fill all cavity of the system before solidification begins anywhere. The term fluidity involves the combined effects of the fluid flow *and* heat transfer aspects involved. If, for example, the molten metal had the consistency of molasses as compared with water, the flow rate into the mold would be extremely slow. As it contacts the walls of the runner and early sections of the mold cavity, heat transfer into the various contact surfaces begins, and the temperature of the leading zone of molten material decreases. Once this reaches the freezing temperature, solidification begins; if the flow rate is low enough (because of inadequate fluidity), the leading zone solidifies across an entire section, thereby preventing further flow. As a consequence, the entire cavity, from pouring basin to riser, would never be fully filled and an incomplete casting would result. Techniques involving initially higher fluid temperature, mold materials of lower thermal conductivity, better mold design, and, in some cases, additions of particular elements into the initial fluid may be used to alleviate this problem. Some of these, however, can introduce other problems if not handled properly. For example, increasing the fluid temperature will increase the

---

* Particular test methods have been developed to measure the relative fluidity of molten materials. See R. A. Flinn, *Fundamentals of Metal Casting* (Reading, Mass.: Addison-Wesley, 1963), pp. 87–95.

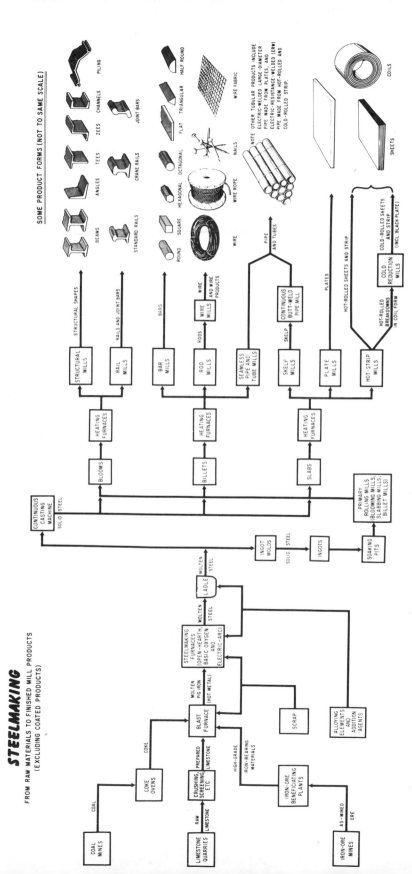

Figure 11-2 Flow diagram showing the principal process steps involved in converting raw materials into the major product forms, excluding coated products.

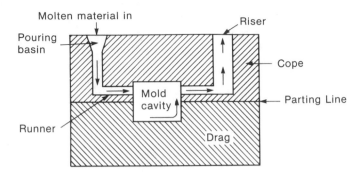

Figure 11–3   Sketch of the major components related to a simple casting.

solubility of the liquid for gases, such as hydrogen and oxygen, and will produce a greater amount of *porosity* (discussed later) in the solid product. With certain metals, the addition of silicon leads to improved fluidity *but* can cause subsequent problems in machining operations, since the presence of silicon oxide, which is quite abrasive, leads to higher wear rates of cutting tools. These examples indicate that the manner by which fluidity is controlled cannot be wholly arbitrary.

Now assume that the entire mold has been completely filled and the cubical cavity (still molten) begins to cool down. Some thermal contraction begins, and the volume of the molten cube decreases. To compensate for this, the riser, if properly designed and located, then *feeds* molten metal to the cubical cavity to compensate for such shrinkage.* As solidification of the cube continues, the change from a liquid to solid causes additional contraction due to the latent heat of fusion. (In the liquid form we are at a higher energy level as compared with the solid form, so as solidification occurs, and is accompanied by a volume decrease, heat is given off.) Finally, when the entire cube has solidified and cools to room temperature, further solid-state contraction occurs. Near the end of full solidification, the last regions of liquid that solidify must undergo a volume decrease. As a consequence, small voids or holes result. They may form within the part or on the surface and can be reduced to acceptable minimum sizes through proper casting design or with the use of *chills;* both are discussed later.

11.3 TYPES OF MOLDS AND MOLD MATERIALS

11.3.1 Expendable Molds

As the name implies, molds of this category are broken apart to remove the cast part; thus, they are expendable. The most widely used material for such molds is *sand,* which is

---

* The size of the riser is crucial here, since if a section *across* the riser freezes too soon, it can no longer supply molten metal to the cavity. A good rule to observe is never to try to feed thick sections through thin sections.

usually mixed with clay, filler material such as cereals, and water. The combined effects of clay and water provide the binding and necessary strength (to maintain a desired shape of cavity) which would be absent if only dry sand were used. Cereals, as they burn off, provide one method to allow gases, air, or even steam to permeate through the mold and away from the casting as it solidifies. Both the size and shape of the grains of sand influence the surface finish of the end product. In Fig. 11–4, a schematic of the interface between a number of grains and the molten material is shown. As heat is transferred from the molten material, the grains tend to expand; if they were packed together as indicated, they would tend to heave (much like a concrete highway without adequate expansion joints). The surface finish that results is directly influenced by such action. Using grains of finer size and adequate filler material, which after burning off leaves spaces for freer expansion between adjacent grains, can alleviate this problem and improve surface finish. In a comparative sense, however, typical sand castings produce the poorest surface finish of all casting processes. If this poses a problem, such surfaces can always be machined to produce smoother surfaces; of course, this introduces added cost. There are, however, several advantages in using sand. It is inexpensive and plentiful, and, because it can be exposed to high temperatures without gross deterioration, practically any material can be sand cast. In addition, there is really no restriction on the size of the part that can be made by this method.*

Shell molding utilizes fine-grained sand that is mixed with and bound by a thermosetting resin. In essence, the pattern (usually metal) of the part to be cast is coated with the sand-resin mixture and cured at an elevated temperature to produce a thin shell which is then simply lifted away from the pattern. Making two such shells (each effectively duplicates half of the full pattern) and then clamping them together produce a cavity that duplicates the outer shape of the initial pattern. As compared with sand castings, parts made by shell molding produce much better surface finish and less dimensional variation (that is, closer tolerances). These characteristics result because the finish of the shell cavity itself is much smoother than cavities produced in sand castings.

Plaster molding involves the use of plaster of Paris and additives such as asbestos and silica, which improve the strength of the base material. These ingredients are mixed with water to form an almost pastelike composition, which is then poured over the pattern of the part to be cast. Once the mold material sets to a reasonably hardened consistency,

---

* Sand casting is often subdivided into three categories, namely green-sand mold, skin-dried mold, and dry-sand mold.

Grains at face
of cavity prior
to pouring.

Grains expand when heated
by molten material but due
to resistance from adjacent
expanding grains, some heave
so interface contour is not
as smooth and surface finish
reflects this contour.

Figure 11–4 Schematic illustrating how expansion of sand grains causes heaving, which results in surface variations of the cast part.

the two halves (similar to those discussed under shell molding) are stripped from the pattern and dried at an elevated temperature to provide adequate strength. The halves are assembled together, and the complete internal cavity produced is then ready for accepting the molten material. Although such molds produce castings of good surface finish, close tolerance variation, and excellent detail in terms of surface reproduction, they cannot be used with molten materials having high melting temperatures, since the plaster mold will rapidly degrade at temperatures in excess of about 1200°C. This problem can be avoided if the mold material contains certain ceramics instead of plaster of Paris. Procedures involved to produce two halves of the mold are quite similar to that discussed under plaster molding. Because the major ingredients differ, however, the latter process is called *ceramic-mold casting*.

With *investment casting,* also called the *lost-wax process,* the pattern is made of wax, which is dipped into a mixture of refractory material (such as fine silica) and liquid. The wax is coated by this *slurry* (that is, silica plus liquid) and when the casting has dried, redipping follows. This is continued until the full coating possesses adequate thickness to provide the necessary strength for handling. After heating the entire unit at a few hundred °F in an inverted position, the wax melts and runs out of the mold (thus the lost-wax concept), which is then heated to between one and two thousand °F. This removes any fluid from the mold material, which in turn, hardens and sets. In general, excellent reproduction of surface details, good surface finish, and close dimensional control result with parts cast by this method. Unlike plaster molds, the materials used to form the mold by investment casting can handle materials of higher melting temperatures.

To summarize the major concepts in this section, a satisfactory mold material must be capable of receiving the molten material and permit it to solidify without the mold's degrading. Thus the melting temperature of the mold must be greater than that of the material to be cast. In addition, the smoother the internal surfaces of the mold cavity, the better will be the surface finish and dimensional control of the cast product.

11.3.2 Permanent or Reusable Molds

When cast parts are to be produced in large volumes, it is beneficial, where possible, to use molds that need not be broken up to remove the casting. Instead, such molds can be used over and over; thus the word *permanent* is applied. As compared with those discussed in Sec. 11.3.1, permanent molds provide much greater production rates. Again, the mold material must possess the ability to avoid degradation from contact with the molten material to be cast, and thus such molds are made from various metals, refractory materials, or graphite; the choice to a large extent depends upon the temperature of the molten material as it is introduced into the mold cavity.

Permanent-mold casting is one category of such processes. In essence, the mold halves are clamped together and the molten material is introduced into the cavity; flow occurs by the effect of gravity. After solidification, the halves are separated and the casting is removed, often with the aid of ejector pins. When metal molds are used, the cooling rates involved are much faster then those associated with the methods discussed in Sec. 11.3.1, since the thermal conductivity of metals is much higher than sand, plaster of Paris, and the like, so the resulting microstructure of the cast product can differ accordingly. Besides high production rates, this process produces a product having a good surface finish and close dimensional control. Of course, the initial cost of equipment is higher than expendable molds, but if the quantity of production is great enough, the initial investment can be justified economically. It is also noted that permanent molds do not permit the permeation of gases through the mold walls as do sand molds, so special venting techniques must be used when such molds are designed.

The second major category is called *die-casting,* and the principal difference between this process and permanent-mold casting is that the molten material is fed into the mold cavity under a pressure (rather than gravity). Depending upon the size of the cavity and the type of molten material, pressures up to thousands of pounds per square inch may be required, and the resulting forces acting on the cavity walls tend to separate the die halves. Consequently, high clamping forces must be applied to prevent such separation; this is not as serious a problem with permanent-mold castings. Die casting of relatively low-melting-point materials is usually referred to as *hot-chamber* die-casting; as illustrated in Fig. 11–5, pressures of a few thousand psi are typical. With higher-melting-point materials, the process is called *cold-chamber* die-casting; this is shown in Fig. 11–6, and pressure requirements are perhaps ten times higher than those used with the hot-chamber process. As with permanent-mold casting, high production rates, good surface finish, and close dimensional control result with die-casting. Because cavities are filled more quickly, due to the pressure, die casting is somewhat faster.

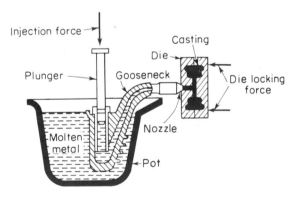

Figure 11–5   Schematic of hot-chamber die casting.

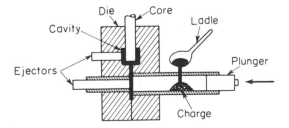

Figure 11–6   Schematic of cold-chamber die casting.

11.4 CASTING OF METALS

As *pure* metals are poured into a mold cavity, nucleation of solid grains begins along the interface of the liquid and mold walls. Conversion of material from liquid to solid occurs at a constant temperature, since the composition of any remaining liquid at any instant remains essentially constant. Thus, from the start to the end of full solidification, liquid and solid phases of the same composition coexist throughout. In most cases, a layer of fine, equiaxed grains first forms in the region of the interface, where the rate of heat transfer is highest and there is little resistance to uniform grain growth from the inner liquid. Further solidification generally produces large *columnar dendrites* which, as they grow away from the equiaxed region, find no resistance to growth towards the center. Because adjacent dendrites tend to interfere with each other's lateral growth, a type of *preferred* orientation results in the formation of long, columnar grains. Figure 11–7 illustrates a typical cast structure of a pure metal.

With *alloys,* solidification from beginning to end does not occur at a constant temperature. Consider two elements called *A* and *B,* where the melting temperature of *A* is substantially higher than *B*. Further assume that these elements are fully soluble in each other in both the liquid and subsequent solid states and that we begin with 50 kg of each element as solid bars. As the melting temperature of *B* is reached, that mass will liquify first; eventually at a higher temperature, *A* becomes liquid and the two elements form a homogeneous solution of liquid. As the temperature is lowered, solidification begins, and the first solid to form will be richer in *A* (the higher-melting-temperature element). In a sense, this is the exact opposite of what happened when the two solid elements were heated, since *B* melted first. As discussed in Chapter 7, if the cooling rate is slow enough (that is, nearly equilibrium cooling) as more solid solution forms, the effect of diffusion will produce a homogeneous structure whose *composition* at any time will show a lowering of element *A* by percent. Thus when the entire mass has solidified, its composition will be 50 percent *A* and 50 percent *B*. But to arrive at this final structure demands a cooling rate low enough to allow diffusion to be effective. Many times when alloys are cast, the initial solid forms at a rate of cooling that is much faster than that called equilibrium; in addition, the cooling rate varies widely from surface to center. As a consequence, the composition of grains in regions adjacent to the surface can be quite different from those nearer the center; in addition, the composition of grains themselves vary from surface to center. This is called *coring* or a cored grain structure. Figure 11–8

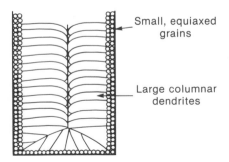

Small, equiaxed grains

Large columnar dendrites

Figure 11–7 Schematic of the cast structure of a pure metal showing equiaxed grains at the surface and columnar dendrites in the interior.

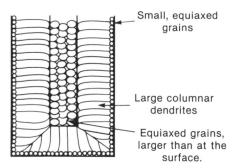

Small, equiaxed grains

Large columnar dendrites

Equiaxed grains, larger than at the surface.

Figure 11–8   Same as Fig. 11–7 except for an alloy.

is a schematic of an alloy casting; it is noted that the central region contains some equiaxed grains and lesser-sized dendrites on the whole as compared with pure metals. These differences arise because (unlike the freezing of pure metals, where only the *masses* of solid and liquid change during solidification) as alloys solidify, not only do the masses of solid and liquid change, but the *compositions* of the two phases change as well. Several methods can be used to decrease the nonhomogeneity of cast metal alloys. One is to use an *additive* (such as sodium in aluminum alloys or ferrosilicon in cast irons) which induces nucleation in regions across the section rather than preferentially at the mold walls. The avoidance of columnar dendrites and a more uniform grain size result. To produce a more uniform structure in terms of *composition,* reheating of the casting to elevated temperatures for relatively lengthy time (often called *soaking*) accelerates diffusion effects in the solid state. A more homogeneous structure results and is somewhat similar to that expected *if* equilibrium cooling rates had prevailed during initial solidification.

11.5 CASTING OF POLYMERS

Because of the names given to many processes that produce polymers in the form of rods, sheets, and so forth, strict categorization poses some problems. For example, *extrusion* sounds like a forming process, and to a major extent it is; many thermoplastics are processed this way. Solid pellets of the material are fed from a hopper into a chamber containing a heating zone which softens the pellets. Often a screw-type conveyer then applies pressure to the softened pellets and forces them through a die whose particular configuration governs the shape of the outlet product. Since the pellets, although heated, are not poured into a mold cavity to solidify, this can't be called a casting process as we have defined it. Instead it best falls under the category of a forming process. Figure 11–9 shows the essentials of this process.

Another widely used process, using either thermosets or thermoplastics, is called *injection molding;* this has all of the characteristics of a casting process and is so defined here. Pellets (or granules) are fed from a hopper to a heating chamber, where they become molten. Under pressure, either from a plunger or a screw drive, the melt is forced to flow into a cavity (usually a split mold). Upon solidification, the mold is opened and the part

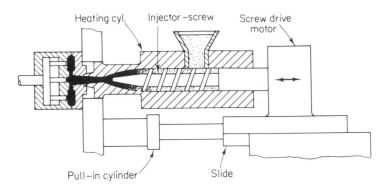

Heating cyl.   Injector-screw   Screw drive motor

Pull-in cylinder   Slide

Figure 11-9 Essentials of screw extrusion.

is removed; often ejector pins assist in separating the part from the split mold. This process is quite similar to die casting, especially the hot-chamber type; Figure 11-10 is a schematic showing various details of this process.

Rotational molding, used to produce relatively large, hollow parts such as drums or tanks, is another process that poses difficulties in terms of categorization. A two-piece mold, usually metal, is designed in such a way that after the mold is clamped to form a unit, it can be rotated about two axes perpendicular to each other. The polymeric material, either in the form of a fine powder or liquid, is introduced into the mold cavity, which itself is heated. As rotation commences, particles are thrown against the mold walls by centrifugal force, whereupon they fully melt and coat the cavity wall. After cooling to form a solid, the part is then removed. Since this process essentially involves the cooling of a molten material to produce a shape governed by a cavity, we classify it as a casting process.

Both thermoplastic and thermoset materials are produced to shapes such as rods or tubes, with techniques similar to those discussed in Sec. 11.4. With thermoplastics, the monomer is mixed with ingredients such as additives, heated to a molten condition, and poured into a mold to solidify. Items such as wheels and gears are processed in this way. Procedures used with thermosets are similar.

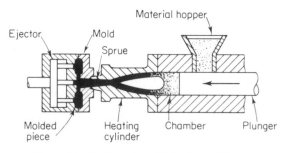

Material hopper

Ejector   Mold   Sprue

Molded piece   Heating cylinder   Chamber   Plunger

Figure 11-10 Essentials of injection molding.

11.6 POTENTIAL DEFECTS AND POSSIBLE CURES

Undesirable occurrences such as holes and cracks can result in castings for several reasons. Since the solubility of *gases* is much greater in a liquid than solid material, upon solidification from the melt, such gases can end up forming small holes. Hydrogen, nitrogen, and oxygen typify gases that can cause problems. Because the solid structure contains these porelike holes, the term *porosity* is often used to describe this action; in effect, the entire cross section is not sound. This type of porosity can be alleviated by providing a means for these gas *bubbles* to vent to and from the surface of the melt. Another technique is to *purge* the melt. This involves flushing the melt with an inert gas such as argon or helium. There, for example, hydrogen in atomic form diffuses into the bubbles of inert gases, which are then carried to the surface and out of the molten material. Pouring in a vacuum is done on occasion; this reduces the amount of dissolved gases but requires special equipment.

Another, often severe effect of porosity occurs in the internal region of the molten material where the last bit of metal solidifies; this was mentioned in Sec. 11.2. Since the last liquid to solidify must undergo a volume contraction, a void, hole, or porous region results. Figure 11–11 is an illustration; this is called *shrinkage*. This type of porosity can be reduced or avoided by improved casting design or the use of *chills*. In either case, the fundamental problem to be solved involves a higher rate of heat transfer from those sections which would solidify last (that is, reduce temperature gradients). As shown in Fig. 11–12, a metal chill, which could be a small piece of steel, is located in a sand mold *before* pouring proceeds. When the cavity is filled, the largest section now cools at a faster rate than it would if the chill were not present; the rate of heat transfer through the steel is much greater than through the sand mold. This tendency to reduce the volume of *hot spots* leads to more uniform cooling across the section, thereby lessening the size of the porous region that would normally occur. Using chills against external surfaces is also helpful. A simple illustration of improved casting design is given in Fig. 11–13; the radial shape leads to a more uniform cross section, thereby reducing temperature gradients

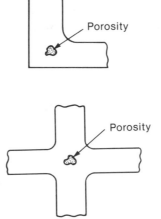

Figure 11–11   Illustration of shrinkage cavities.

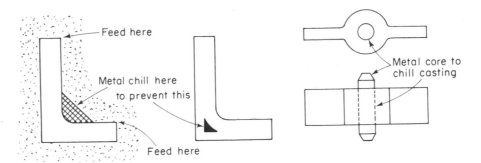

Figure 11–12 Examples of the use of chills.

across the corner. In general, except in large ingots, porosity due to gases leads to small, spherical holes, whereas shrinkage causes larger voids with rougher surfaces.

Another possible defect of this type is called *interdendritic* shrinkage. In regions where dendrites have formed as three-dimensional solids containing branchlike growth, adjacent dendrites eventually interlock and trap still molten material within the interlocked branches. As this material solidifies and shrinks, small voids or holes result, since additional liquid cannot be supplied to compensate for such shrinkage. This type of porosity, in terms of pore size, is generally quite small and in many cases is not as serious as the voids left by the type of shrinkage cavities discussed above. Of course, any type of porosity is undesirable from a strength viewpoint, since the effective area supporting loads or stresses is smaller than that based upon external dimensions. Stress concentration effects may also be more severe. Also, if the cast product is to contain a gas or fluid under pressure, there is a greater problem with porous structures, since the pressurized gas will have a greater tendency to permeate through the porous structure compared with one that contains few if any pores.

Hot tearing is another serious defect that often occurs; in essence, an actual crack (or hot tear) results in the cast part. Figure 11–14 indicates how this comes about. There the cavity, shaped like an I-beam, has ends of somewhat larger thickness than the center portion. As solidification occurs, the center section will completely solidify before the ends do. With further cooling, the center begins to undergo thermal contraction, but

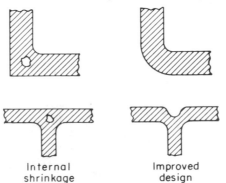

Internal
shrinkage

Improved
design

Figure 11–13 Design alterations to reduce or eliminate shrinkage.

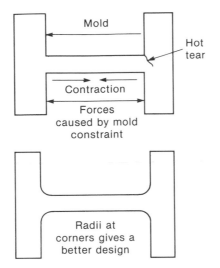

Figure 11–14   Schematics to illustrate the possible cause of hot tearing and an improved design to reduce such a possibility.

constraint from the mold walls prevents free contraction and induces forces to act against the ends as shown. If these forces reach a high enough level, the induced stresses cause a hot tear or crack to occur as indicated. Such a result must obviously be avoided. Better casting design, as shown in Fig. 11–14, can alleviate this problem. Another solution is to literally break up the mold once the entire *outer surface* has solidified. Thus, removing the ability of the mold to cause excessive constraint negates the possibilty of hot tearing. Finally, a change of material is another possibility; using a material that can tolerate greater strain to fracture could solve this problem.

   Although numerous other kinds of defects (mainly at the surface) can be categorized, those we have discussed are often the most serious and, for purposes of illustration, provide an adequate introduction to this topic.

11.7 RESIDUAL STRESSES

Whether this topic should fall under Sec. 11.6 or be treated separately is a matter of preference. Since cast products will *nearly always* contain residual stresses and in many instances perform acceptably, we treat this separately from the topic of defects. The primary cause of residual stresses is the nonuniform thermal gradients that typify the process of solidification of a molten material. Consider Fig. 11–15, which illustrates a component consisting of two thick round selections (top and bottom) connected by a thin-walled annular section. Because of the different thicknesses, the annular section will solidify first, and then attempt to undergo solid-state thermal contraction as the temperature decreases. Uninhibited contraction is restrained by the thicker sections. Now the top section will fully solidify next and as it begins to shrink in the radial direction (due to contraction in the solid state), it applies radial forces to the upper portion of the annular section. A similar sequence of events follows as the bottom (thickest) section

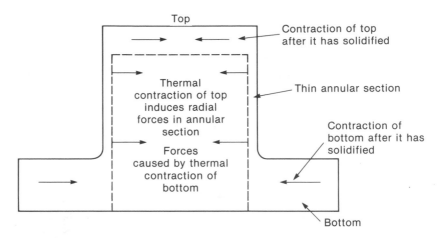

Figure 11–15 Illustration of a cast component which may contain residual stresses after solidification.

solidifies and contracts, except that the lower section of the annular section experiences forces induced by the contraction of the bottom section. This complicated series of events can produce residual stresses of varying magnitudes in different regions of the final casting.* There are several potentially serious consequences that can result. First, when this component is placed in service, the applied loads, forces, torques, pressures, and the like induce stresses that are superimposed upon the residual ones, so the onset of yielding, necking, or fracture could occur at lower levels of *applied* stresses than would happen if the residual stresses were not present. Another detrimental effect can result if such a casting were subjected to subsequent machining operations, since they induce forces (thus, stresses) as the cutting action takes place. For example, suppose the full height of the annular specimen is to be turned † to produce a particular dimension and improved surface finish. When the stresses caused by the turning operation are superimposed upon the residual stresses, fracture in the form of a crack could occur around the periphery, where the annular section meets the thickest section at the bottom. Several solutions could alleviate these potential problems. First, the casting could be subjected to a stress relief by proper heat treatment; this entails an added cost. Next, the thickness of the annular section could be increased; this would lessen the degree of thermal differences as the various sections solidify and undergo thermal contraction. As a consequence, the magnitudes of residual stresses would be lowered. Of course, this design change increases the weight of the casting; this can be reduced by machining off some of the material considered as excess. However, this increases material costs and adds a cost for machining. Finally, a substitute material possessing greater inherent strength and resistance to crack formation and propagation could be considered. Among

---

* If they lead to hot tears as discussed in Sec. 11.6, the stresses are effectively relieved *but* a cracked part results.

† Discussed in Chapter 9.

other problems that may result because of residual stresses are distortion that accompanies subsequent machining, difficulty in holding desired tolerances, and long-term dimensional instability.

11.8 BATCH VERSUS CONTINUOUS CASTING

Although a variety of metallic alloys are cast to nearly final shape, we include this section to illustrate two broadly different techniques that are used to produce cast structures that are further processed to a variety of shapes such as flat sheets, round bars, and the like. In addition, the discussion is restricted to low-carbon steels; although this is arbitrary, it is noted that the tonnage of such alloys used for industrial purposes makes them one of the most widely used materials.

Batch casting involves the production of an individual ingot cast in a large mold. Once solidified, the ingot can then be subjected to a variety of operations, discussed earlier in connection with Fig. 11–2. The types of steels of concern here are broadly classified as *rimmed* or *killed*.* Rimmed steels are not deoxidized before solidification occurs, and as the ingot freezes, dissolved carbon and oxygen react to cause a violent evolution of CO bubbles; this is the *rimming* effect, and it leads to a stirring of the molten metal. One of the consequences is the segregation of carbon towards the center of the ingot. The net result is that the surface regions of the ingot contain a carbon content lower than the overall average composition. If processed into sheets, for example, the rolled sheet will likewise have surfaces of low carbon that tend to be free of carbide particles; in essence, those regions are nearly pure iron.

With *killed* steels, elements such as aluminum and silicon are added to the melt. They combine more readily with oxygen than does carbon, thereby forming oxides (that is, the steel has been *deoxidized*) and the rimming action is prevented. Solidification occurs without violent bubbling; hence the term "killed." Not only is segregation of carbon to the center avoided, but the oxides often float to the top of the melt. As a result, the composition across the ingot is quite uniform and largely free of porosity. One drawback is the extremely large shrinkage cavity (called a *pipe*) that forms at the top of the ingot; this must be cut off or *cropped* before further processing, which leads to a larger material loss than that connected with rimmed steels; Fig. 11–16 illustrates these differences.

A more recent method for producing various shapes involves *continuous casting*. Figure 11–17 illustrates different methods that are used. Essentially, the molten metal flows through cooling chambers that act as molds. With proper cooling, a solid skin forms and acts to support the partially cast structure which, with further cooling, completely solidifies. As it exits in a solid form, it can be cut to desired lengths or subjected directly to forming operations that produce shapes of desired contours. Compared with ingot

---

* *Semi-killed* is a further subdivision that falls between these two. These are partially deoxidized and have less porosity than rimmed steel and little or no pipe as with killed steels. Rimmed steels are often "capped" to control the rimming action and reduce the degree of porosity that would otherwise result.

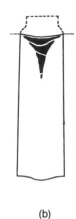

Figure 11–16   Sketch of the cross sections of ingots made from (a) rimmed and (b) killed steels.

(a)   (b)

casting and subsequent processing, this technique provides far greater production rates. Of course, the specialized equipment is expensive.

11.9 CAST IRONS

In our experience, many students who have completed an introductory course in materials science have heard of *cast irons* but have decided misconceptions about these metals. First, they seem to think that these are the only metal alloys that are cast to some shape and, in addition, that these alloys are inherently brittle. From what has already been discussed in this chapter, the first idea is clearly wrong, but the second needs some attention. Cast irons are basically alloys of iron and carbon; however, they differ from steels in two major aspects regarding *composition*. While steels seldom contain carbon in excess of one percent, cast irons, as a class, contain carbon from about two up to four percent. In addition, and of extreme importance, while silicon is usually controlled in steels at low levels (~0.2 percent), it is deliberately added in cast irons in amounts up to 3.0 percent.

There are three *primary* types of cast iron; they are white, gray, and nodular. The term "primary" as used here indicates that these irons can be produced directly from the melt. In Sec. 7.4.1, various steel microconstituents like ferrite and pearlite were discussed. These same micros occur in cast irons, and their relative amounts depend upon both the initial solidification rates and any subsequent solid-state heat treatment. In cast irons there is the additional possibility of cementite (Fe_3C) *dissociating* into iron and free carbon (*graphite*). A combination of a *slow* cooling rate and a *high* percentage of silicon provides a greater chance for this dissociation to occur.

White cast iron (white iron) contains relatively lower carbon and silicon than do the others, and to produce this material requires relatively high cooling rates. As a consequence, no free graphite results, and all of the excess carbon (other than the small amount in any ferrite) exists in the *combined* form of cementite (Θ) In general, the structure of most white irons usually consists of pearlite and rather large masses of free cementite. Because the carbon content is higher than with typical steels, the *amount* of

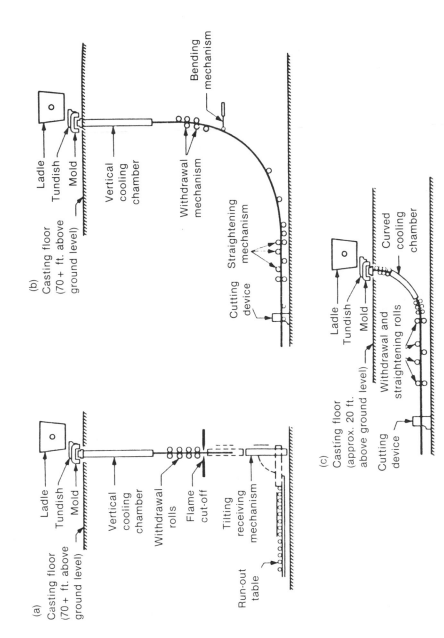

Figure 11–17 Three methods used in continuous casting of steel.

(a)
Casting floor
(70 + ft. above
ground level)

Ladle
Tundish
Mold

Vertical
cooling
chamber

Withdrawal
rolls

Flame
cut-off

Tilting
receiving
mechanism

Run-out
table

(b)
Casting floor
(70 + ft. above
ground level)

Ladle
Tundish
Mold

Vertical
cooling
chamber

Withdrawal
mechanism

Bending
mechanism

Straightening
mechanism

Cutting
device

(c)
Casting floor
(approx. 20 ft.
above ground level)

Ladle
Tundish
Mold

Curved
cooling
chamber

Withdrawal and
straightening rolls

Cutting
device

free cementite is much greater. Since Θ is extremely hard but quite brittle, the ductility of white cast iron, as measured in a tensile test, is negligible; it does, however, provide excellent wear resistance.

Gray cast iron (gray iron) contains larger percentages of carbon and silicon than does white. As these metals solidify and cool to room temperature, a large amount of dissociation occurs and free graphite, in the form of sharp-ended *flakes,* forms part of the final structure. These flakes act as stress risers such that the stresses at the flake tips are much higher than the nominal stress (that is, load divided by area). The net result is that gray irons display little tensile ductility. If these metals are given subsequent heat treatments, it is possible to reduce the *size* of the flakes, which increases the tensile strength but, since the *shape* of the free graphite remains as sharp-ended flakes, ductility cannot be improved to any meaningful degree. In any event these cast irons are substantially stronger in compression than in tension, since compressive loading tends to ''close'' the flakes and the stress-raising effect is avoided.

Nodular cast iron (also called *ductile* or *spheroidal*) has compositions similar to gray iron but also contains the elements cerium and/or magnesium. These elements influence the formation of free graphite to produce spheres or nodules, rather than flakes. Since the stress concentration effects are greatly reduced, nodular irons display much greater tensile ductility than does gray iron; reduction of area up to 25 percent is possible.

Figure 11–18 shows typical photomicrographs of these three cast irons; note the obvious difference in the shape of the graphite in the gray and nodular irons. We add here that, in a comparative sense, the gray and nodular irons involve much larger percentages of silicon and slower cooling rates as compared with white iron. In fact, if one wished to produce a structure of white iron, the cooling rate *must* be fast enough to prevent dissociation of iron carbide. Consequently, white iron cannot be fully produced throughout thick sections.

For completeness, *malleable cast iron* is included and is referred to as a *secondary* type. It is classified this way because, unlike the others, it *cannot* be produced directly from the melt. One must form white iron (note that this therefore limits the section sizes of malleable iron). The white iron is then heated to about 1750°F for up to two days and then slowly cooled to room temperature. At the elevated temperature the massive regions of cementite dissociate to form free graphite that has a shape which is unlike that in gray or nodular. Further dissociation can occur when the austenite attempts to transform to pearlite at the lower critical temperature. With slow enough cooling rates, the cementite (which must exist if pearlite is to result) may break down into ferrite and additional free graphite. Such a structure would produce a *ferritic malleable iron,* as shown in Fig. 11–19. Although greatest ductility can be produced with a ferritic nodular iron, malleable irons possess greater ductility than either the inherently brittle gray or white irons. It is noted, however, that the use of malleable irons is on the decrease, while nodular irons have replaced them in many applications. This is probably due to two factors. First, much thicker section sizes can be made directly from the melt with nodular irons; in contrast, since white irons are restricted to thinner sizes, due to the need of higher cooling rates, malleable irons are so restricted. Next, the need to heat treat white irons at elevated temperatures for rather long time periods

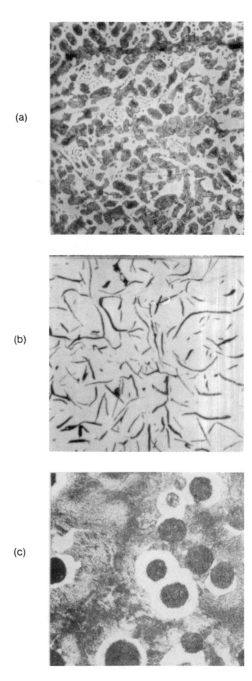

Figure 11–18 Photomicrographs of (a) white, (b) gray, and (c) nodular cast iron.

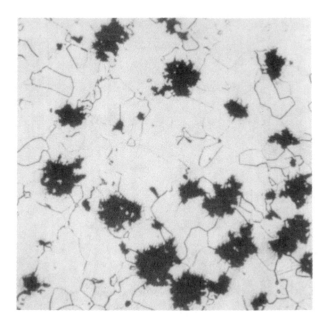

Figure 11–19   Photomicrograph of ferritic malleable cast iron.

adds a cost in producing malleable irons that is usually higher than similar costs associated with the nodular type.

   Compacted graphite iron has a composition similar to nodular iron but lower amounts of magnesium. Probably because of lower magnesium, the shape of the graphite in these irons is not as spheroidal as in nodular irons, nor are the flakes that form as sharp-tipped as in gray irons. Instead the graphite tends to form as a type of flake with somewhat rounded tips. As a result, the mechanical properties of compacted graphite irons fall between those displayed by gray and nodular irons.

11.10 DIRECTIONAL SOLIDIFICATION

There are several important industrial applications that require the preferential solidification in a single direction; in essence, single crystals are produced so that the external surface is the entire grain boundary. Semiconductor materials, such as germanium, are made by the *crystal-pulling* method illustrated in Fig. 11–20. There, the molten metal is made to contact a small seed in the form of a single crystal. As solidification occurs at the interface, the seed is rotated and slowly pulled away from the melt, and the new solid duplicates the structure of the original seed. Alloying elements, called *dopants,* are sometimes included in the liquid metal to produce special properties.

   A second method, called the *floating-zone* technique, is used with higher-melting-temperature materials such as silicon. There, a piece of polycrystalline material contacts

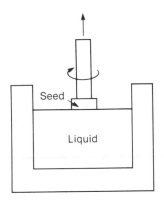

Figure 11–20   Schematic of the crystal-pulling method to produce single crystals.

a single crystal of the material, and then a radio-frequency (RF) heat source is applied while moving upward at a slow rate. In effect, the polycrystalline rod is converted to an ever-increasing length of a single crystal. The net result is a single crystal having a single orientation; see Fig. 11–21. The large single crystal is then sliced to provide thin wafers that are the *chips* used in many electronic devices.

Another important application of directional solidification is in the manufacture of turbine blades used in high-temperature devices such as jet engines. There, by carefully controlling the rate of heat transfer from the molten metal, solidification occurs in a unidirectional manner. As shown in Fig. 11–22, long, parallel grain boundaries depicting columnar grains can result. It is also possible to produce the entire blade as a single crystal whose only grain boundary is the external surface. Such blades provide superior high-temperature properties (such as resistance to creep and thermal shock) compared with conventionally cast blades shown in the same figure.

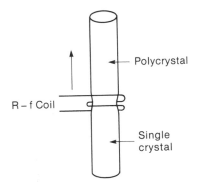

Figure 11–21   Schematic of the floating-zone method to produce a single crystal.

ADVANCES IN TURBINE AIRFOIL MATERIALS

Conventional casting   **Columnar grain**   **Single crystal**

(a)   (b)   (c)

Figure 11–22   Turbine blades produced by three methods: (a) conventionally cast, (b) directionally solidified to produce columnar grains, and (c) a single crystal.

11.11 *SOLIDIFICATION TIME*

From a heat transfer analysis,* the time for solidification of a molten metal may be given as

$$t = B(V/A)^2 \qquad\qquad (11\text{–}1)$$

where t is the solidification time, V is the volume of the shape of the part, A is the total *surface* area of the part, and B is called the *mold constant*. As such, B is a function of such parameters as density and specific heat of the metal as well as the thermal expansion and

* See, e.g., R. A. Flinn, *Fundamentals of Metal Casting,* pp. 32–38.

thermal conductivity of the mold material. Work by Chvorinov* showed a good agreement between Eq. (11–1) and experimental results for shapes of relatively simple geometries. It is best to consider this equation as a reasonable guide that shows a general tendency, since the freezing times of more complex shapes and certain metals cannot be analyzed adequately with such a simple equation.

PROBLEMS

11–1. Figure P11–1 shows a cavity in the form of a tapered plate where the direction of the flow of the molten material is indicated. After solidification is complete, voids along the centerline result, as shown in that figure. Is this defect due to a lack of adequate fluidity, poor mold design, or a combination of both. What do you recommend should be done to avoid this *centerline shrinkage*?

11–2. A tee section, shown in Fig. P11–2, is to be made by sand casting a metal. The only concern here has to do with the most sensible location of the riser, noting the different dimensions of the cavity. Where would you suggest the riser be located? Explain why.

11–3. Coarse-sized grains of sand are used in a particular sand mold. Suppose that at one location of the mold surface, two adjacent grains are in contact and that the constraint from regions adjacent to these grains is rather large. The cavity is then filled with a molten metal. Discuss what happens as the grains become heated and what result is likely to occur on the surface in that region when solidification takes place. How might this result differ if very fine grains of sand had been used?

11–4. In terms of fundamental considerations, why might shell molding be used instead of plaster molding?

11–5. A particular metal component may be produced by die casting or sand casting. Regarding the grain size at or near the surface of the part after solidification, what differences are apt to be found with parts cast by these two methods? Why does this occur?

11–6. Dendrites are found in many castings. Why does such grain formation result and why is this often disadvantageous?

11–7. Alloys of copper and nickel are completely soluble in the liquid or solid state for any composition; consider an alloy of half of each element by mass as shown in Fig. P11–7. Under *equilibrium* cooling, an α solid solution of 50 Cu − 50 Ni results in all individual

---

* N. Chvorinov, *Proceedings of the Institution of British Foundry,* **32** (1938–39), p. 229.

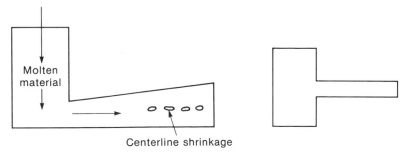

Figure P11–1   **Figure P11–2**

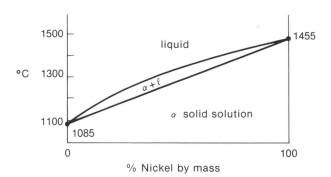

Figure P11–7

grains. If this molten mass were poured into a mold cavity, the composition of solid grains is found to vary from one location to another. Why does this occur?

11–8. If you section a cast structure, you are apt to see a number of voids in the interior. How would you distinguish between those due to gas porosity as compared with shrinkage cavities?

11–9. Figure P11–9 shows a section of an angle plate that is to be cast.
   (a) At which of the lettered locations would a *hot spot* be most likely to occur. That is the region to solidify last.
   (b) Assuming that an unsatisfactory casting results, specify any changes you would recommend to alleviate this problem.

11–10. The boxlike shape shown in Fig. P11–10 is to be produced with a sand mold.
   (a) At which location is a hot tear most likely to occur?
   (b) Explain what might be done to avoid that possibility.

11–11. Concerning residual stresses in castings,
   (a) Explain why they are usually disadvantageous.
   (b) Explain how they might be handled (that is, avoided or removed) after solidification.

11–12. Both pearlitic white and pearlitic gray cast iron are essentially brittle.
   (a) With regard to the microstructure of each, why does this brittleness occur?
   (b) Using only heat treatments (that is, you *can't* change the compositions), specify how you could bring about a *decided* increase in tensile ductility of these two structures.

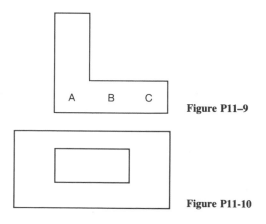

Figure P11–9

Figure P11-10

11–13. Why might ferritic nodular cast iron be chosen over ferritic malleable iron for a certain part? Consider all possible reasons.

11–14. You are to compare the solidification *times* for a *square* cross section whose height is three times its lateral dimension with a *cylindrical* section of the same height whose diameter equals the side of the square section. Repeat the above if the *volumes* and heights of these two shapes are equal. Equation (11–1) should be used.

12
surface
processes _____

12.1 INTRODUCTION

There are relatively few products on the market that are single components and made of homogeneous materials. Examples include nails, cups made of foamed styrene, concrete blocks, steel beams, and rope. It is instructive to visit a shopping center to see how few such products there are.

The great majority of products are assemblies of two or more obvious and separable components, each selected to fulfill some of the desired attributes of the assembly. For example, a durable shoe is, in essence, a composite structure consisting of a wear-resisting sole attached to a flexible upper segment. The versatility of such products is limited only by the imagination of the designer's knowledge of materials and knowledge of ways to attach the separable parts together. The availability of such products is limited by economics, however, mostly by the high cost of joining materials together. Thus there have always been efforts to achieve desirable properties in single components by making the "surface" different from the substrate. The substrate is usually expected to provide mechanical strength, ductility, conductivity, and several other functions. The surface is expected to perform very different functions, namely, to resist wear and corrosion, and to have an acceptable appearance, among other things. This chapter discusses surface processing, where the intent is to achieve properties different from that provided by the substrate. This chapter does not include methods of surface finishing for achieving texture or topography, but it does include such surface finishing processes as painting.

Surface processes can be classified in terms of *surface treatment, surface modification,* and *surface coating.* Short examples in each of these groups are listed below, with longer discussions following:

332

1. *Surface treatments* are the processes by which surface properties are changed separately from that of the substrate. Perhaps the most common example is found in steel. A piece of 10100 steel can be annealed throughout to achieve a hardness of 250 VPN (Vickers Pyramid Number). The surface can then be heated to 730°C by a flame or a laser to some shallow depth and cooled quickly to produce martensite of 800VPN hardness. There is no change in chemistry, only a difference in hardness due to heat treatment.

2. *Surface modification* processes are those that change the chemistry of the surface to some shallow depth, ranging from one μm to about three mm. One old method adds carbon to the austenitic form of low carbon steel, by diffusion. When the entire part is then cooled quickly (quenched in water), the substrate remains tough and the surface becomes hard because of the difference in carbon content. A newer method implants nitrogen and other ions into metals with the effect of distorting the lattice structure near the surface, thereby hardening it.

3. *Surface coating* processes *build up* the dimension of some region of a surface. All types of metals, polymers, and ceramics are used as coatings, and they are applied to all types of substrates.

Surface processes are many and varied, and are applicable to virtually all materials. Data on prices and properties for purposes of evaluating these processes cannot be put into a convenient table; available information for specific production problems should be obtained from vendors of the machinery and supplies available for such processes. Unfortunately, surface processes are often advertised in the same manner as one hears for laundry soap, including testimonials from shop foremen and sundry purchasing agents. An interested process engineer should assess processes by testing them on actual production materials. Before such tests, however, it is well to become aware of the fundamental events that take place in each process. These are described in the next sections.

12.2 SURFACE TREATMENTS

Virtually all processes that change bulk properties will also change only the surface properties, if properly applied. The properties of some materials are changed by heat treatment; the properties of others may best be changed by plastic flow. A partial list of the surface treatments is given in two groups, namely, those that use heat and those that plastically deform.

1. Heat treatment is effected by heating at any convenient rate, but by cooling at controlled rates. The major heat sources are listed below in order of potential increasing surface heating *rate*. The higher the rate of heating, the thinner will be the heated layer, where the goal is to reach some specific surface temperature. A thick layer will resist wear and indentation longer than will a thin layer, but a thin layer will produce less part distortion than does a thick layer. Note that processes are often given names that only partially describe what takes place. For example,

laser hardening of steel implies that a laser hardens steel. In fact, the laser only heats the steel, after which fast cooling (usually in water) effects the hardening.

a. *Flame hardening* uses a gas-fired flame, usually oxygen-acetylene, propane, or other high-temperature fuel. This process can be quickly installed, but it is not as readily automated as some others, and it can not be focused upon very small regions on a surface.

b. *Induction hardening* is done by placing a metal into a loosely fitting coil, which is cooled by water and in which an alternating high current (60 Hz up to radio frequency) flows. The current in the coil induces a magnetic field in the metal, which because of magnetic reluctance causes heating in the metal, mostly in the surface at the higher frequencies. The coil current is shut off and cooling water is applied to the part at the appropriate time. This process is clean and readily automated, but it is restricted in its ability to heat specific regions on a surface.

c. *Laser hardening* uses a laser for heating a surface. The usual wavelength is in the infrared, in the range longer than 1000 nm or 1 μm. The CO_2 laser is commonly used. A laser system is expensive to install, but the beam is easily steered or directed along any path on a surface by automatic control of mirrors, even into regions that are out of sight.

d. *Electron beam hardening* uses a stream of electrons to heat a surface. The electron accelerating voltage is usually held below 25 kV when X rays are to be avoided. The beam can be steered by magnetic lens but only in line of sight. Conventional electron beam systems require that the part being processed should be placed into and removed from a vacuum chamber. This usually requires some time and skill to operate and obviates the use of fluids to cool a heated part. At higher cost, one can purchase an electron beam system which directs a beam from the vacuum enclosure through an orifice into the atmosphere where part handling and cooling can be done conveniently. This beam cannot be steered, and thus the part must be moved about under the beam.

Where cooling of a surface is required, after heating, in order to cause a phase change, it may be necessary to do so by quenching in liquid or by spraying liquid on the hot surface. However, a very thin *layer* of heated material will also cool quickly by conduction to the substrate, if the temperature gradient and the thermal conductivity are high enough. For example, the conduction cooling that follows heating by a laser or by the electron beam can be sufficient to produce martensite in 1040 steel, but this will not occur when the surface is heated by a flame.

2. Some plastic flow processes include the following:

a. *Burnishing* involves pressing and sliding a hardened sphere or (usually) roller against the surface to be hardened. It is a rather crude process which can leave a severely damaged surface. Lubrication reduces the damage.

b. *Peening* is done either with a heavy tool that strikes and plastically indents a surface, usually repeatedly, or by small particles that are flung against a surface with sufficient momentum to plastically dent the surface. The latter is called *shot peening* if the particles are metal of the size of ballistic shot. The velocity of shot or other particles may be as high as 35 m/s: it is, therefore, a very noisy and dangerous process.

c. Skin pass rolling is done with spheres or (usually) rollers, of a diameter and loading such that the surface to be hardened is plastically indented to a small depth. Large rolls will plastically deform thin plate or sheet throughout the thickness, but skin pass rolling can be controlled to plastically deform to shallow depths.

The local plastic flow that occurs in these processes expands an element of material laterally and "thins it," with the effect of developing a compressive residual stress in the surface. A bar that has been shot peened, for example, will bow so that the peened surface will be on the *outer* radius.

The hardness of a surface that has been severely plastically deformed depends on the original ductility of material. Generally, the hardness of an annealed material can be increased by local indentation by a percentage that is twice that of the percentage reduction of area of an annealed specimen in a tensile test.

12.3 SURFACE MODIFICATION PROCESSES

Surface modification processes are those that change the chemistry of existing materials in the surface of the original material. These include the following:

1. *Carburizing* is done to increase the carbon content of steel. The maximum hardness of a piece of steel is related to the carbon content. For structural purposes a steel of less than 0.4 percent carbon is desired for toughness, but for wear resistance and indentation resistance a carbon content of about 1 percent is desired. The carbon content of steel can be increased only when the steel is in the austenitic or face-centered cubic state where the maximum solubility of carbon is about 2 percent (at 1130°C: see Chapter 7). Thus when steel is heated in an atmosphere rich in carbon, some of the carbon will diffuse into the steel. A carbonaceous atmosphere is achieved by using CO, by burning fuel gas with inadequate O_2, or by heating chips of gray cast iron (which usually contain over 2.5 percent carbon). A very rich carbonaceous atmosphere will usually produce a steep gradient of carbon content in the part, which results in large stress gradients and possible cracking during heat treatments. A lean atmosphere adds carbon slowly. The proper depth and thickness of carburized layer is controlled by temperature and atmosphere. However, precautions must always be taken to prevent oxidation, hydrogen diffusion, grain growth of the steel, and undesirable migration of alloying elements in the steel. Carburizing layers of any thickness can be obtained, but the usual thickness is in the range of 1 to 3 mm.

2. *Carbonitriding* may be done either in a gas atmosphere of ammonia diluted with other gas, or it may be done by inserting a piece of steel into a *salt bath,* which is a molten cyanide salt or compound. The cyanide supplies both carbon and nitrogen for diffusion into iron, which itself must be in the austenitic state. The role of the

carbon is described above. The nitrogen that diffuses into the steel forms nitrides with such alloys as aluminum, chromium, molybdenum, vanadium, and nickel, producing a hardness between 900 and 1000 VPN.

3. *Ion implantation* is done in a vacuum of the order of 10 mPa. Many types of ions may be inserted into a wide range of surface materials in this process, but the easiest to describe is nitrogen in iron. Nitrogen gas is ionized in an electric field gradient of 10^5 volts/mm. The ions are then propelled to a high velocity in a field of the order of 100 KeV toward an iron surface held electrically negative. The usual area rate of impingement of ions is of the order of 10^{15}/mm^2. As ions enter the iron surface, several iron atoms are evaporated from the surface, and a channel of atoms is displaced to receive the nitrogen. The nitrogen concentration builds up to about 15 to 20 atomic percent with a peak concentration at a depth of about 0.7 mm.

   An implanted surface is in a compressive state of stress, which will usually increase the fatigue life of the surface. The surface is also harder, but very thin. Implantation affects the corrosion properties of metals and increases wear resistance for some forms of mild wear.

12.4 COATING PROCESSES

A very significant industry has developed which offers as many as 60 coating processes. Most of the processes can be broadly classified as given below. No attempt is made to name the processes, because in most cases the process is named after the machine that applies the coating, or is given the name of the inventor. In the following paragraphs several processes will be described in terms that will lead to an understanding of the vital information an engineer needs concerning a process, namely, the quality of the product. Information on cost must be obtained from the suppliers of coating service. There are very many suppliers, ranging from the substantial industries to the part-time home-based operation. The broad categories of processes include the following major processes:

1. Weld *overlaying* is done with all of the heat sources mentioned above, but most often by arc and by gas flame. Welding produces very strongly adhering layers, which may be built up to any desired thickness. For corrosion resistance the *filler* or coating material may be a stainless steel, and for wear resistance the *filler* may incorporate nitrides and carbides. Soil-engaging plow points and mining equipment are often coated with steel filler materials containing particles of two forms of tungsten carbide, WC and W$_2$C, which have a hardness of the order of 1800 VPN.

2. Spraying of molten and semi-molten metals and ceramics is done in air or in *vacuum*. The durability of the product depends strongly on the strength of the bond between the coating and the substrate, which in turn depends on how much of the adsorbed gases, oxides, and contaminants found on all commercial surfaces are removed or displaced so that the sprayed material can bond to the substrate of the target material. Several processes are described:

a. Molten metal, usually aluminum, is sprayed to coat steel pipe and tanks exposed to weather and to coat engine exhaust systems. The metal doubtless begins to travel from the "gun" to the target in the molten state, but some of the droplets cool to the 2-phase region of the equilibrium diagram before they reach the target. This transaction is not instantaneous, because a phase change entails the evolution of some heat. In any case the spray travels at various speeds, usually less than 30 m/s. If the spray is solid, the particles would bounce off the target. Liquid would wet a solid surface and solidify, but 2-phase droplets partially flatten against the target surface and remain attached partly by wetting forces due to the liquid phase of the spray. A "wet" snow ball hurled against a wall behaves the same way. Upon solidification some other bonding mechanisms must be involved, however. Recall that all solid surfaces are covered with adsorbed gases. The hot sprayed metal, upon striking the target surface, will cause desorption of some of the water. A bond is therefore effected between the sprayed metal and the oxide on the metal substrate. Later the sprayed metal contracts and doubtless produces high residual stresses at the bond interfaces, which will limit the adhesive strength of the film to the substrate. But practically, sprayed coatings are fairly durable against very mild abrasion. Their effectiveness against corrosion depends on their continuity. Here again, one can pile drop upon drop from the spray, but the drops must fit tightly together to prevent the incursion of acids and other corrosive substances. Each drop will bond to another through an oxide film, and there will be high residual stresses because of differential contraction from one drop to another.

b. The coating of surfaces for wear resistance is a fast growing industry. One process uses a spray which is produced by feeding a powder into the flame of a gas fired or plasma torch. The powder can be a mixture of dozens of available metals, ceramics, and intermetallic compounds, selected both for cost and effectiveness for resisting wear. The spray velocity is in the range of 150 to 500 m/s, and the sprayed material reaches the target surface again in the semi-molten state. The firmness of attachment, or stress to separate the coating is of the order of 70 MPa, which is adequate for many tasks, but not for severe abrasion. One process achieves a velocity as high as 1300 m/s of particle impingement, by detonation of a fuel gas in a tube containing a powder of the coating material. The high-velocity particles from such a device apparently remove a large amount of adsorbed water and other contaminants. Perhaps there is also an effective packing of particles in the layers of coating. This type of coating appears to have a strength of attachment in excess of 140 MPa, which makes it much more suitable than other processes for abrasion and erosion resistance.

c. Paints and polymers are in a class of coatings usually used for appearance and also for mild corrosion protection, but not for wear resistance. These materials are applied to a surface by spraying, wiping, or rolling of liquid. For effective bonding the surface to be coated must be clean and the liquid coating must wet the solid surface. The coating is then expected to solidify, either by the evaporation of a solvent or thinner from the coating, or by other mechanisms of polymerization of the molecules.

d. Surfaces can be coated by *electroplating,* usually in the range from 0.5 μm to about 0.25 mm thick. The common coatings are chromium, nickel, copper, zinc, cadmium, tin, and molybdenum. Some coatings are hard and provide wear resistance. Some are soft and provide protection against scuffing, while others are well suited to protection against corrosion. The process is done in an acid (electrolyte) bath containing a salt of the metal to be plated (for example, a nitrate, a sulfate, or others). A few volts are applied with the part to be plated as the cathode (−). The plating ion concentration, the bath temperature, and the applied voltage must be carefully controlled to avoid poor adhesion of plating to the substrate, spongy plating, or large crystals in the plating. Overvoltage must be avoided because it produces hydrogen, which usually embrittles some metal. In addition, since the plating thickness is proportional to the current density, some care must be taken in part design, anode geometry, and shielding to make the plating of the proper thickness in all areas.

e. *Electroless plating* is a process that is named such because it was developed to overcome some of the difficulties of electroplating. Coatings of nickel-phosphorous or nickel-boron alloys may be applied to a wide range of metals and alloys. Plating occurs by hydrogenation of a solution of nickel hypophosphite, usually available commercially with proprietary buffers and reducing agents. Coatings of any thickness can be applied. The applied coating has a hardness of ≈500 VPN, and the hardness increases to ≈900 VPN when heated to 400°C for one hour.

f. *Impregnated* coatings are not strictly coatings but are usually classified as such. They are formed by direct contact of the surface to be coated with a solid, liquid, or gas of the desired element. An alloy forms in the surface of the part to be coated, which has different properties than that of the substrate. The catalog of such processes is large, including *calorizing* (Al), *carburizing* (C), *chromizing* (Cr), *siliconizing* (Si), *stannizing* (Sn), and *sherardizing* (Zn).

g. Another process that is not strictly a coating involves the melting of a thin layer of a metal part, and then sprinkling TiC or other hard compounds into the molten layer. Upon solidification the TiC becomes firmly bonded and serves to increase wear resistance.

h. *Physical vapor deposition* (PVD) is a process that is done in a vacuum of about 10 mPa. The coating material is heated and evaporated (boiled). This vapor fills the enclosure and condenses on cooled surfaces, including the part to be coated. Coatings of any thickness up to about 100 μm may be applied. The adhesion to the surface (often called the substrate) depends on the cleanliness of the surface, but PVD coatings are readily rubbed off unless the coated part has been heated for some time, allowing diffusion of some of the coating into the oxide on the part surface.

i. *Chemical vapor deposition* (CVD) takes place in a "vacuum" of about 10 to 100 mPa. The enclosure also contains a gas, which includes ions of the type to be deposited on the part surface. There usually is sufficient chemical reaction of the coating with the part to effect a bond. Chemical reaction occurs at the surface of the base metal M′, with deposition of the coating metal M. There are three types of reactions:

(1) When the coating medium or vapor is a chloride (for example),

$$MCl_2 + M' >< M + M'Cl_2$$

(2) By catalytic reduction of the chloride at the base metal surface when the treating atmosphere contains hydrogen

$$MCl_2 + H_2 <> M + 2HCl$$

(3) By thermal decomposition of the chloride vapor at the base metal

$$MCl_2 <> M + Cl_2$$

The last reaction appears the simplest, but thermodynamically it is often not possible or very economic. Specialists in these processes should be consulted on such detail.

PROBLEMS

12–1. List 20 items each of the three classifications, surface treated, surface modified, and surface coated among consumer products.

12–2. List 10 items in hardware stores that are single-component items and made of one homogeneous material.

12–3. If austenite of 0.4 percent carbon will transform to martensite at a cooling rate of 600°C/s, what should the minimum temperature gradient be in a bar of steel so that conduction heat transfer will accomplish the formation of martensite?

FOREWORD TO PART C

Here we attempt to tie in costing with particular processes. In Chapter 13, examples are given to illustrate how one can proceed to determine the cost of specific processing operations. This involves the individual costs of labor, materials, overhead, and the like. Attention is also paid to the manner by which such costs might be reduced. In Chapter 14, the idea of integrating design and the method of processing to make the designed part is put forth. A natural consequence of this integration leads to the need to consider alternative methods of processing. In essence, different methods are often available, and a comparison of the costs of each will often dictate which should be used. Two major points are stressed here. First, the most economical method may depend on the use of equipment that is not available, so a decision involving the expenditure of money as an investment in capital equipment enters the picture. Such decisions go well beyond the jurisdiction of the manufacturing engineer; all he or she can do is to present the most accurate cost factors possible, and then it is up to upper management to decide if new investments are to be made. The second major observation leads to the general conclusion that there is no absolutely "best" method that provides greatest economy. The quantity or number of parts to be produced almost always plays a key role in deciding which of several methods is most economical.

_13_
costing
of
manufactured
parts

13.1 INTRODUCTION

This chapter concentrates on procedures for estimating the costs to manufacture parts, restricting our attention to material, labor, tooling, and what is loosely known as overhead. Four methods are discussed, with some industrial examples given.

1. Simple calculations are made of the costs in _cost centers,_ where a prescribed manufacturing technique is employed to produce or process a component. The effects of varying the production parameters (speeds of processing, for example) in such cost centers are then investigated, leading to considerations of optimizing the processing conditions through that cost center.
2. The cost of cutting materials with a tool is calculated, and found to be dependent on a balance between the high cost in labor when cutting slowly, and the high cost in tooling when cutting fast.
3. The economics of alternative methods of manufacturing the same shape in the same materials are studied, particularly comparing the modern metal forming process which produces _near net shape,_ with the traditional metal removal methods.
4. Costs of competitive routes to manufacture either the _same_ overall design of article, or articles of _different_ design that will all perform the same function, are illustrated by examples.

Some of the examples below use data obtained several years ago, without converting old data to present-day values. It is a useful exercise for the student to insert only one or two modern cost figures into the given examples in order to gain experience in predicting the effect of possible future variations in prices. For products made in other countries one must consider different divisions of costs of material, labor, and overhead in different countries; also, there are different tax laws, direct and indirect government subsidies, and so on across the world.

13.2 PROCESS PLANS: COST CENTERS

The cost of producing a component is usually estimated from process plans. The complete plan is a detailed list of *what* equipment and material is to be used, and *how much* equipment time, material, and man-hours are required to produce the designed product (or perhaps to perform a service for another cost center). The standards for many products are usually set by industrial engineers who have data both from handbooks and from previous experience of the company. They decide what production rate is reasonable to achieve with the equipment available, with the level of workforce skills, and with the quality of material currently available.

Note that estimates are usually developed for the *existing* available equipment and the way it has been used in the past, and not necessarily for the optimum equipment and tooling. There is wisdom in this practice in some instances, but sometimes it is vital for economic survival to apply specialized knowledge of materials and processes to develop new ways to make old products.

As an example of a process plan for a simple cost center, consider the production of steel plate in a hot rolling mill. The function of the cost center is typically to convert slabs 100 to 150 mm thick into plate 12 mm thick and then to shear the plate into 15-m lengths. After some years of experience, manufacturers develop *cost standards* for the amount of material required to produce a particular amount of prime or salable product, allowing for scrap and scale (mostly iron oxide), and for the equipment time or man-hours required to produce that amount of salable product. Slightly different standards will apply for different width of plates; the wastage in scrap and scale is a smaller proportion, the wider and thicker the plate. There is about 18 percent wastage for 750 mm wide × 12 mm thick hot rolled plates, but only some 6 percent wastage for a 2 m wide × 12 mm thick plates. For the narrow plate 1.18 tons of ingoing slab yields 1 ton of salable plate, whereas 1.06 tons are needed to yield 1 ton of wide plate. Once standards are established, the quantities actually used are compared with the standard quantities. In this way unified comparisons may be made day by day, or between several mills. Note that although the material standards are expressed here in tons, there is often some benefit in using a dollar-equivalent value, particularly if two mills in different parts of the country use different sources for raw materials and the prices vary somewhat independently.

No cost center is in production 100 percent of the time. In the case of a rolling mill, for example, *downtime* or *outage* occurs when rolls must be changed, or when there are mechanical and electrical breakdowns. Operators are allowed time for personal needs and rest time, and there will be time involved for setting the job up, in starting up, and in finishing the process. These factors may have a negligible effect in some cases, but a significant influence on the overall costing in others. Setup time in complicated machining operations or in welding is usually relatively greater than in other operations such as plate rolling.

Production standards for a rolling mill may either be expressed as allowed hours per ton of salable product (no credit is given for the production of scrap) or, alternatively, its reciprocal, which is production rates (tons per hour). Then the *production performance* of the cost center can be assessed in terms of the ratio of the *earned standard hours* (that is, the standard allowed hours/ton multiplied by the quantity produced) divided by the actual number of hours taken to do the job. Thus, for some given plate thickness and width, if the standard operating practice is 0.007 hour/ton, it should take $(0.007 \times 500) = 3.5$ hours to roll 500 tons of salable plate. If it actually takes only 3.0 hours to do the job, the "production performance" is given by the quotient of $3.5/3.0 = 117$ percent.

To determine the standard product cost, the value of the starting material is added to the cost of the labor required to perform the operation. Taking the 1.18 tons of slab at \$105/ton to yield 1 ton of salable rolled product the gross material cost is $105 \times 1.18 = \$123.90$. But the scrap and (oxide) scale from the 1.18 tons can be recovered and sold for \$12.00, so the net material cost is \$111.90 to produce 1 ton of salable plate. For labor, if the standard operating time is 0.007 hour/ton, and the cost of several employees operating the mill is \$245 per hour, then that cost is $\$245 \times 0.007 = \$1.72$ per ton. The product cost is therefore $\$111.90 + \$1.72 = \$113.62$ per ton of salable product. Note that labor is only about 1.5 percent of the direct cost of the product. A rolling operation is therefore not *labor intensive* but is *capital intensive*.

To this, one can add the overhead charges (both direct and indirect) associated with the product. For convenience, these are often quoted as a percentage of the labor costs of a batch (assuming some standard production plan), but note that things like factory rent, interest on money borrowed to buy machines, and the like remain fixed when a plant is not operating at full capacity; these moneys have to be found even if there is zero production.

13.3 MACHINING ECONOMICS

In previous chapters, the words *desired tool life* were used, and it is now appropriate to discuss why one particular tool life might be better than another. Practical interest is centered around one of two viewpoints, namely,

1. The combination of parameters (cutting conditions) that leads to a tool life associated with the minimum *cost per piece*.
2. That tool life which leads to the maximum production rate *or* minimum *time per piece*.

At first glance, the above distinction may seem contradictory, but as we will now show, the conditions leading to greatest economy are *never* equivalent to those providing maximum production.

First, consider the viewpoint of cost per piece. The total cost in *dollars per piece* may be designated as C_p and is defined as follows*:

$$C_p = C_i + C_m + C_c + C_g + C_s + C_r \qquad (13\text{--}1)$$

where

C_p = total cost in dollars per piece.
C_i = idle cost. This includes loading and unloading of the workpiece, plus any other machine handling time.
C_m = cutting or machining cost. This involves *only* the time (and, therefore, the cost) when material is being machined or cut.
C_c = tool changing cost. In essence this involves the time when a tool must be replaced with a freshly ground tool after it has been used to machine a number of pieces. As such, it must be prorated over the number of pieces produced prior to the change.
C_g = tool cost per grind. This includes the cost to regrind a worn tool (the original grind is included) and involves the depreciation of the initial cost of the cutting tool.
C_s = setup cost. This involves the cost to get the machine tool ready for operation. If handled in this manner, this cost is prorated over the total number of pieces produced on the existing setup.
C_r = cost of raw material.

Note that most authors ignore C_r and C_s, while others combine C_c and C_g. Our preference is to break the cost C_p down into as many specific items as possible.

Now consider each of the components individually. On any operation, noncutting time costs money, since operators are being paid hourly and since the department in which the operation is carried out must have some type of *burden*† or *overhead* rate that must be accounted for. Let R_m be the sum of the operator's rate *and* burden rate of the department; its units will be dollars per *minute*. Suppose that the total idle time t_i is the number of minutes required to handle the loading and unloading of the workpiece and, perhaps, alter the feed rate or spindle speed setting of the machine. Then

$$C_i = t_i R_m \qquad (13\text{--}2)$$

The machining cost is simply the product of R_m and the time it takes to complete the cutting operation; this is t_m and its units are minutes. Then

$$C_m = t_m R_m \qquad (13\text{--}3)$$

If it is assumed that the same operator changes tools when a worn tool must be replaced, the total cost of changing the tool is simply the product of the tool changing time t_c in minutes and R_m. This total cost must be distributed over the total number of pieces that were produced during the life of the tool. Now the number of pieces produced per tool failure (that is, up to the time the tool is changed) is simply the tool life in minutes divided

---

* *Each component* has the units of dollars per piece.
† Burden includes maintenance, depreciation, and indirect labor. Data on these items are usually readily available in a given company. Details may be handled differently, *but all industrial* organizations utilize some factor of this type.

by the time in minutes required to machine one piece. In essence, then, we are assuming that a fixed *fraction* of the tool is *used up* to machine each piece. Thus if the tool life T were 60 minutes and the machining time per piece t_m were 2 minutes, then 30 pieces would be machined before tool changing would occur and $\frac{1}{30}$ of the total tool changing cost would be attributed to each piece. In symbolic form:

$$C_c = t_c R_m (t_m/T) \tag{13-4}$$

With regard to the tool grinding cost, it is best to think of this in terms of a cutting tool whether it has one edge or many. One aspect of this cost is that associated with the actual grinding of the tool; this requires so many minutes t_g. Since hourly operator and overhead rates may be different for grinding than for the machine operator and his department, the symbol R_g will be used, although its components have the same meaning as used with R_m.

In addition to the actual cost to grind a tool, the original cost of the cutting tool should be prorated over the number of cuts the tool makes, rather than to regard it as part of overhead costs. Each time the tool is reground, some of it is used up; if the initial tool cost is divided by the number of permissible grinds, then this cost D_g added to the actual grinding cost gives the total cost involved to replace a worn tool. Again, as with Eq. (13-4), this cost must be sensibly distributed over the number of pieces produced per tool failure, so

$$C_g = [t_g R_g + D_g](t_m/T) \tag{13-5}$$

Note that the terms inside the bracket can be considered as the cost in dollars associated with a ground tool (that is, \$/tool) where the ratio t_m/T is the fraction of a tool used per piece; thus, the final units are still in dollars per piece.

With regard to the setup cost, consider that it might cost one hundred dollars to assemble and set up all components required to complete a certain operation. Say that after ten thousand parts are produced, the existing setup is broken down and the machine tool prepared for a new operation. The setup time t_s multiplied by the labor and overhead rate, say R_s, would lead to the one-hundred-dollar cost. If that is divided by the *total* pieces machined on that setup N_s, the setup cost per piece becomes

$$C_s = \frac{t_s R_s}{N_s} \tag{13-6}$$

The raw material cost per piece C_r is self-evident. Inserting Eqs. (13-2) through (13-6) into Eq. (13-1), we obtain

$$C_p = t_i R_m + t_m R_m + t_c R_m (t_m/T) + [t_g R_g + D_g](t_m/T) + \frac{t_s R_s}{N_s} + C_r \tag{13-7}$$

Consider now a turning operation where

$$t_m = L/fN \tag{13-8}$$

With $N = 12V/\pi D$ and $VT^n = C$ or $T = (C/V)^{1/n}$ substituted into Eq. (13-7), we have

$$C_p = t_i R_m + R_m(\pi DLV^{-1})/12f + t_c R_m(\pi DLV^{1/n-1})/12fC^{1/n}$$
$$+ [t_g R_g + D_g](\pi DLV^{1/n-1})/12fC^{1/n} + t_s R_s/N_s + C_r \qquad (13\text{–}9)$$

To find the cutting velocity that will yield a *minimum cost per piece*, we perform the operation $\partial C_p/\partial V = 0$. This leads to the following form for this velocity V_{cm}

$$V_{cm} = (CR_m{}^n)/[(1/n - 1)(t_c R_m + t_g R_g + D_g)]^n \qquad (13\text{–}10)$$

Since $VT^n = C$, Eq. (13–10) may be easily revised to give the *tool life* for minimum cost per piece. This is T_{cm} or

$$T_{cm} = (1/n - 1)(t_c R_m + t_g R_g + D_g)/R_m \qquad (13\text{–}11)$$

If we are interested in maximum production rate (*minimum time per piece*), the expression equivalent to Eq. (13–1) is

$$t_p = t_i + t_m + t_c(t_m/T) + t_s \qquad (13\text{–}12)$$

Note that the terms associated with grinding and raw material in Eq. (13–1) have no counterparts in Eq. (13–12). Taking $\partial t_p/\partial V = 0$, one finds

$$V_{tm} = C/[(1/n - 1)t_c]^n \qquad (13\text{–}13)$$

and

$$T_{tm} = (1/n - 1)t_c \qquad (13\text{–}14)$$

Note that in Eq. (13–10) if the cost connected with grinding approaches zero, V_{cm} approaches V_{tm}. A similar comment applies regarding T_{cm} and T_{tm}. Of course, tools and grinding are never free, and that is the key reason why the most economical tool life (or velocity) will not also provide the maximum production rate. Figure (13–1) shows a qualitative plot of the components for minimum cost per piece. Since Eq. (13–14) will *always* predict a shorter tool life than will Eq. (13–11). The optimum velocity and cost for maximum production will be higher than that for the minimum cost condition.

Now a practical word about the most sensible use of the ideas and results just presented. Whether we use Eq. (13–11) or Eq. (13–14), it is usually a relatively simple

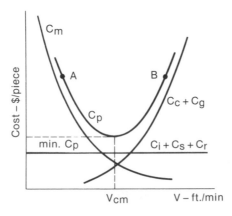

Figure 13–1 Components in the analysis of minimum cost in cutting.

task to obtain the necessary *standards* related to t_c, R_m, t_g, R_g, and D_g for a given operation, and a reasonable estimate can be made for the exponent n since much data are available. However, the machine operator does not set the tool life; rather he sets the rpm which produces a cutting velocity. Thus it is Eqs. (13–10) and (13–13) that would be most useful *if* the value for the constant C were available.* Often, this is not the case and instead of making wild guesses, it is best to use the actual production machine as an experimental tool while it is physically producing parts. For example, suppose that the value of T from Eq. (13–11) is found to be 75 minutes. By starting with some reasonable velocity, parts can be produced and the tool life of the operation can be found. Suppose the tool lasts only 30 minutes. A modest lowering of spindle speed should be made (remember how sensitive tool life is to small changes in velocity) and the new tool should be timed to failure. In a few such tests one can find the most appropriate spindle speed available that will produce a tool life close to the desired 75 minutes. Because many machine tools *do not* possess variable speed spindle drives, but rather a number of discrete values of the spindle rpm, it may be impossible to obtain the exact velocity to produce the desired tool life. Instead, one wants to get as *close* to the optimum speed as possible and not be operating at speeds that are much too high or low, as indicated by points A and B on Fig. (13–1). After all, the exactness of the value used for n is not without flaws, and the reality of this fact should be remembered at all times.

Note from Fig. (13–1) those components that are unaffected by changes in velocity but which must be considered in Eqs. (13–7) or (13–9) when the *actual cost* per piece is desired. Finally, if more than one tool cuts at the same time, the *tool life for minimum cost* increases in comparison to that which prevails for one tool cutting. In effect, the total tool cost increases with multitool operations and, as we see in Eq. (13–11), this necessitates longer values of T_{cm}. Although such operations are difficult to analyze directly, the above principle will hold and should be kept in mind.

Example 13–1
————————

1. The cutting velocities for minimum *cost* per piece and minimum *time* per piece are to be determined for a turning operation where a feed rate of 0.008 ipr and depth of cut of 0.075 in. are involved and high-speed steel tools are to be used. For this operation, the following are known:

> Machine operator rate = $13.00 per hour
> Machine department overhead = $15.00 per hour
> Tool grinding rate = $10.00 per hour
> Grinding department overhead = $12.00 per hour
> Cost of a 0.5 in. by 0.5 in. by 3 in. long HSS tool = $ 3.75
> Number of regrinds per tool = 15
> Tool regrinding time = 3 minutes
> Tool changing time = 2 minutes
> Tool life exponent = 0.10

---

* This would come from an equation such as Eq. (9–8) for the particular feed and depth *being used*. Alternatively, Eq. (9–17) may be used.

$VT^{0.1} = 175$

Idle time $t_i = 1.5$ minutes

Solution   For minimum cost per piece, Eq. (13–10) is used. There,

$R_m = (13.0 + 15.00)/60 = 0.467$ \$/min
$t_c = 2$ min, $t_g = 3$ min
$R_g = 3(10.00 + 12.00)/60 = \$1.10$/edge
$D_g = 3.75/15 = \$0.25$/edge,

so

$$V_{cm} = \frac{175\,(0.467)^{0.1}}{[\{(1/0.1) - 1\}(2 \times 0.467 + 3(1.10) + 0.25)]^{0.1}} = 112 \text{ fpm}$$

For minimum time per piece, Eq. (13–13) is used, so

$$V_{tm} = 175/[((1/0.1) - 1)2]^{0.1} = 131 \text{ fpm}$$

2. Now suppose a *throwaway* sintered carbide tool is used. One type is a thin blank of carbide that is ½ in. by ½ in. square and ⅛ in. thick.* At each corner, a nose radius is usually ground, so with four corners and two faces (top and bottom), eight cutting edges are provided by this tool as purchased. The blank is clamped in a special tool holder such that the form of the holder produces the tool signature. When one cutting edge fails, the blank is unclamped, rotated 90 degrees to provide a fresh cutting edge, and then reclamped. After all eight edges are used up, the blank is *literally* thrown away—thus the name. Note there is *no* regrinding done on such a tool. Here, the machine operator and burden rates are the same as in part 1 and

Cost of throwaway = \$2.50
Number of edges = 8
Tool changing time = 0.75 min

$$VT^{0.2} = 400$$
$$R_m = 0.467 \text{ \$/min as before}$$
$$t_c = 0.75 \text{ min}$$
$$t_g = 0, R_g = 0$$
$$D_g = \$2.50/8 = 0.3125 \text{ \$/edge}$$

Then

$$V_{cm} = \frac{400\,(0.467)^{0.2}}{[\{(1/0.2) - 1\}(0.75 \times 0.467 + 0.3125)]^{0.2}} = 283 \text{ fpm}$$

and

$$V_{tm} = \frac{400}{[\{(1/0.2) - 1\}(0.75)]^{0.2}} = 321 \text{ fpm}$$

---

* Such throwaway tools are also made in triangular and circular cross sections.

The fact that the velocities using carbide tools are much higher than their HSS counterparts indicates why production rates (or RMR) are much greater when carbides are used. Note that the cost of the special toolholder is included in the overhead rate in this analysis. Since such a toolholder can be used for many years, it is simplest to handle it in this way. The same comments apply to the toolholder used with the high-speed tools.

Example 13-2

Using the necessary information from Example 13–1, determine the minimum cost in dollars/piece for both situations if

C_r (material cost/piece) = $0.85
C_s (setup cost/piece) = $0.05
D (work diameter) = 2.5 in.
L (length of cut) = 10 in.

Solution 1. For HSS tools:

$$N = 12V_{cm}/\pi D = 12(112)/2.5\pi = 171 \text{ rpm}$$
$$(T_{cm})^{0.1} = 175/V_{cm} = 175/112 = 86.74 \text{ min}$$

With Eqs. (13–1) or (13–7),

$C_i = t_i R_m = (1.5)(0.467) = \$0.7005$
$C_m = t_m R_m = 10(0.467)/(0.008)(171) = LR_m/fN = \$3.4137$
$C_c = t_c R_m (t_m/T) = 2(0.467)(10)/(0.008)(171)(86.74) = \$0.0787$
$C_g = [t_g R_g + D_g](t_m/T) = [3(1.1) + 0.25](10/(0.008)(171)(86.74) = \$0.2992$
$C_s = \$0.05$
$C_r = \$0.85$
$C_p = \$0.7005 + \$3.4137 + \$0.0787 + \$0.2992 + \$0.05 + \$0.85$
$C_p = \$5.39 \text{ per piece}$

2. For the throwaway carbide:

Only C_m, C_c, and C_g differ from part 1, since they are the only terms influenced by the different velocities.

$$N = 12(283)/2.5(\pi) = 432 \text{ rpm}$$
$$T^{0.2} = 400/283 \text{ or } T = 5.64 \text{ min}$$
$C_p = 10(0.467)/(0.008)(432) = \$1.3513$
$C_c = 0.75(0.467)(10)/(0.008)(432)(5.64) = \$0.1797$
$C_g = (2.5/8)(10)/(0.008)(432)(5.64) = \$0.1603$
So $C_p = 0.7005 + 1.3513 + 0.1797 + 0.1603 + 0.05 + 0.85$

or

$$C_p = \$3.29 \text{ per piece}$$

Note that with HSS tools there is a higher C_m (machining) and higher C_p (grinding), but a lower C_c (tool changing).

13.4 COSTS OF JOINING METHODS

13.4.1 Arc Welding Costs

Arc welding processes are fairly consistent, in that the time and material required to produce a given length of weld, by a given welding method, is reasonably constant between different jobs. Thus it is easy to establish standard costs. The cost of producing a weld is essentially that of materials, labor, and overhead. Material costs in welding include the cost of electrodes,* fluxes, shielding gases, and other consumables. The cost of electricity for arc welding is either included with the "consumables" or it may be included in the overhead charges, depending on the accounting procedures of different companies. (About 1.75 kWhr of electricity is required to deposit one pound of weld, that is, 3.85 kWhr/kg).

The cost of welding is conveniently expressed in terms of dollars per unit run of weld (formerly per foot, nowadays per meter). Thus we need to know

1. The time to produce a given length of weld (which, with the labor rate, gives the labor cost per length).
2. The amount of electrode used together with the electrode cost per unit weight (formerly per pound, nowadays per kilogram) to give the electrode cost.
3. The quantity of flux and/or shielding gas required, together with relevant unit costs (if the flux is separate from the electrode).
4. The overhead rate, expressed usually as a percentage of the labor cost. As with other metal working processes, it is difficult to make generalizations on overhead as it depends on local accounting methods, but figures ranging over 50 percent to 400 percent of the labor cost per job are found between different organizations.

In these types of calculations it is necessary to bring in the efficiency in the use of materials and labor:

1. Not all the electrode is actually deposited as weld metal; some electrode weight is lost in spatter and vaporization. Some portion of the weight of electrode for manual welding is the flux, and the unused or "stub ends" of these electrodes are discarded. The "deposition efficiency" is the percentage of the weight of original electrode actually deposited as weld metal. Typical values of the factor for various welding processes are

Stick-electrode welding   65 percent
Self-shielded flux-cored welding   82 percent

---

* Most arc welding processes supply filler metal in wire form, where the arc is formed between the filler wire and the surface to be welded. In manual arc welding the wire is available in short rods or sticks, frequently coated with *solid flux, and usually referred to as electrodes.*

Arc welding with shielding gas   92 percent
Submerged arc welding   100 percent

2. An operator does not, in practice, weld all the time that he is paid. This leads to the so-called "operating factor" defined simply as the ratio of the arcing time to the total time for which the welder is being paid. It is expressed as a percentage usually, but also as a decimal (<1.0) in the costing calculations which follow. The operating factor depends on the particular welding process being employed, and also on the management and organization of the fabrication shop. The usual range is between 20 percent and 60 percent, with even lower percentages being found for constructional welding in the field and some higher values for automated welding in the shop. Reasons for the different values are as follows:

a. *Manual welding.* The operating factors range from 10 percent to about 45 percent; 65 percent represents the extremes of human achievement. In addition to the usual personal breaks, there are additional breaks caused by setting up equipment, preparing the job for welding, shifting of the working positions of the operator, changing electrodes, and chipping off slag (flux) to check the weld or to prepare for a second welding pass. Since deslagging can take much time, there are considerable advantages in using electrodes of easy slag detachability.

b. *Semi-automatic welding,* for which the typical range of operating factors would be 25 percent to 60 percent. Generally, the biggest advantage of the semi-automatic process is the absence of electrode changing and reduced need to shift operator position. These factors will increase the operating factor by between 5 percent and 15 percent. For many applications, a higher welding current and increased travel speed can be utilized; with bare wire inert gas welding, there is no deslagging to be done, which can significantly improve the operating factor.

c. *Automatic welding.* With fully mechanized automatic welding, it is possible to work at an operating factor of over 90 percent, provided the setup time and deslagging times are very small.

In all of these processes, poor quality welding reduces the operating factor considerably. Poor welds are repaired by grinding away sections of defective weld and rewelding, which requires much time and is chargeable to the cost center.

For common designs of weld in most metals, handbooks and sales and product brochures list the time required to run the weld, the cross-sectional area of the weld, and hence the weight of electrode required to make the run, and the quantity of solid flux and cover gas that will be required for the job. Where data are not available, the deposition rate may have to be taken on test welds.

Calculation of cost may be done with the following equations and the symbols listed in Table 13–1.

cost of electrode/ft of weld $= (WE)(CE)$
cost of flux/ft of weld $= (WF)(CF)$
cost of gas/ft of weld $= (VG)(CG)$

TABLE 13–1   Symbols used in calculations of the cost to weld.*

$$
\begin{aligned}
A &= \text{Cross-sectional area of the weld, (in.}^2) \\
CE &= \text{Cost of electrode (or filler wire), (\$/lb)} \\
CF &= \text{Cost of flux, (\$/lb)} \\
CG &= \text{Cost of shielding gas, (\$/cu ft)} \\
CL &= \text{Cost of labor and overhead, (\$/ft or weld)} \\
CM &= \text{Cost of consumable welding materials, (\$/ft of weld)} \\
CR &= \text{Cost of labor and overhead, (\$/hr)} \\
D &= \text{Deposition rate, (lb/hr)} \\
DE &= \text{Deposition efficiency, weight ratio of weld metal deposited to electrode consumed} \\
OF &= \text{Operating factor, ratio of productive time to total time required by the welder} \\
OH &= \text{Overhead} \\
M &= \text{Melt-off rate of the electrode, (lb/hr)} \\
S &= \text{Speed of electrode travel, (ft/hr)} \\
T &= \text{Time required to weld, (hr/ft)} \\
VG &= \text{Volume of shielding gas required per linear foot of weld, (ft}^3\text{/ft), or (ft}^3\text{/hr)} \\
WE &= \text{Weight of electrode (or filler wire), (lb/ft)} \\
WF &= \text{Weight of flux required, (lb/ft)} \\
WW &= \text{Weight of weld metal deposited, (lb/ft)}
\end{aligned}
$$

* Taken from the *Lincoln Electric Welding Handbook*.

so that the materials cost per foot run of weld is*

$$CM = (WE)(CE) + (WF)(CF) + (VG)(CG) \qquad (13\text{–}15)$$

It is convenient to combine the labor rate and overhead into one parameter CR (when overhead is based on labor charges), in which case

$$CL = (T)(CR)/(OF) \qquad (13\text{–}16)$$

where T is the time required to run a foot of weld and OF is the operating factor. We note that $T = 1/S$, where S is the welding speed. Thus as an alternative expression,

$$CL = (CR)/S(OF) \qquad (13\text{–}17)$$

It is important to be aware that in the case of multiple-pass welds, the total time required to complete the joint may be obtained from the "average speed," which is the *harmonic mean* of successive passes, that is,

$$S = \frac{n}{1/S_1 + 1/S_2 + \ldots} \qquad (13\text{–}18)$$

where $S_1, S_2, \ldots$ are the speeds of the first, second, $\ldots$ welding passes, after n total passes over the same length of run. Hence

$$T = \left(\frac{1}{S_1} + \frac{1}{S_2} + \ldots \frac{1}{S_n}\right) \qquad (13\text{–}19)$$

*Note that the term *(VG)(CG)* would be zero in submerged-arc or self-shielded electrode welding, since no shielding gas is used. Similarly, the term *(WF)(CF)* would be zero with all welding processes other than submerged-arc. Note also that, as an alternative, *(VG)* may be given as cubic feet per hour rather than per foot of weld.

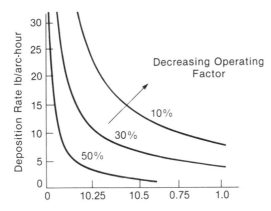

Figure 13-2 The relationship between metal deposition rate and labor cost.

Owing to the inverse relation between welding speed and time to effect a weld, the relation between metal deposition rate and labor costs takes the form shown in Fig. 13–2; the deposition rate (which depends directly on welding current and electrode size) is given by $S(WW)$. At high deposition rates, the labor costs for each pound of weld metal deposited are relatively small; at low deposition rates the labor costs are relatively higher, and the effect of a low operating factor becomes important.

Although the amounts of consumables required to produce a given weld will vary with different processes, it turns out that the cost of filler metal does not vary widely between different welding processes when the deposition efficiency is taken into account. Table 13–2 shows the cost of consumables per foot of weld making a ¼-in. fillet by various methods, and it is seen that all are about 5 cents/foot, in 1973 prices. Consequently, estimations of *comparative* welding costs may be quickly made by using Eq. (13–17) relating to labor, overheads, welding speed, and operating factor.

Equations are also available to calculate costs. We may take as an example a ¼-in. horizontal fillet weld, using a ⁷/₃₂ in., E7024 electrode. For a speed of 85 ft per hour and

TABLE 13–2 ESTIMATED COST* OF CONSUMABLES FOR FOUR WELDING PROCESSES

Process	Type of Electrode	Electrode Cost $/lb	Deposition Efficiency	Cost of Deposited Metal	
			%	$/lb	$/ft
Shielded metal arc	E7024	0.191	65	0.29	0.045
Self-shielded flux-cored	E70T-G	0.315	82	0.38	0.059
Submerged arc	EL12	0.199 + flux	≈100	0.34	0.053
Gas metal-arc	E70S-3	0.243 + flux	92	0.31	0.048

*The volume of shielding or cover gas ranges from 25–45 cu ft/hr, and the maximum amount of solid flux is a weight about equal to the weight of deposited metal.

an operating factor of 30 percent, with labor plus overhead rate of $7.00 per hour, we have

$$CL = (CR)/S(OF)$$
$$CL = (7.00)/85(0.3)$$
$$CL = \$0.274/\text{ft} \qquad\qquad (13\text{--}20)$$

The use of equations gives a welding engineer an opportunity to study the effect of changes in the fixturing and material handling in the job. Perhaps the operating factor can be increased to 40 percent by such changes, in which case the cost as calculated by Eq. (13–20) becomes $0.206/ft. But there may be some doubt that an operating factor of 40 percent can be achieved by using stick electrodes, so semi-automatic processes may be proposed with the expectation of achieving an operating factor of 50 percent. The resulting costs from the same equation are given in Table 13–3.

As an example of the influence of different types of filler metal supply, let us compare the use of E8018-C3 (a low-hydrogen stick electrode coated with iron powder) and 9000 C-1 (a flux-cored wire) to weld low-alloy steels. The relevant data and conclusions are shown in Table 13–4, where it may be seen that the lower cost of stick electrodes is seriously offset by the cost of labor.

Figure 13–3 gives a compounded chart which demonstrates the relationship between various processes, electrode sizes, deposition rates, operating factors, hourly labor rates, and labor costs. The welding processes are divided into manual, semi-automatic, and automatic processes; the approximate ranges of operating currents for each electrode size are also incorporated. To make cost calculation more accurate, the deposition rates and welding current relations used for the various processes should be taken from accurate graphs.

Figure 13–3 consists basically of three parts, which can be used separately or all together:

1. *Welding processes.* The left-hand section demonstrates the variation of the deposition rates of the various welding processes with electrode size and current. The different vertical columns are the range of deposition rates for particular sizes of

TABLE 13–3　ESTIMATED COST TO WELD FOR VARIOUS PROCESSES

Process	Arc Speed (ft/hr)	Operating Factor	Labor and Overhead ($/hr)	Weld Cost ($/ft)
Shielded metal-arc	85	0.30	7.00	0.274
Shielded metal-arc	85	0.40	7.00	0.206
Gas metal-arc semi-automatic	80	0.50	7.00	0.175
Self-shielded flux-cored, semi-auto.	100	0.50	7.00	0.140
Submerged arc semi-automatic	110	0.50	7.00	0.127

TABLE 13–4   COMPARATIVE COSTS OF USING TWO TYPES OF FILLER STOCK

Factor		E8018 Electrode and Manual Welding	9000C-1 Wire and Semi-automatic Welding
Labor and overhead rate	(CR)	$20.00 per hour	$20.00 per hour
Deposition rate	(D)	3.11 lb/hr	5.00 lb/hr
Operating factor	(OF)	0.30	0.45
Cost of labor and overhead to deposit weld		$21.43/lb of weld	$8.88/lb of weld
Electrode cost	(CE)	$0.667 per lb	$1.019 per lb
Deposition efficiency	(DE)	0.68	0.78
Cost of deposited metal		$0.98/lb	$1.31/lb
Shielding gas flow rate	(VG)	none	40 cu ft per hour
Shielding gas cost	(CG)		$0.09 per cu ft
Cost of shielding gas			$0.72/lb dep. metal
Cost of welding		$22.41/lb dep. metal	$10.91/lb dep. metal

electrode, the different shading representing the different processes; the numbers alongside these columns are the maximum and minimum currents for the electrode size. The deposition rates are the points of mild steel weld metal deposited in one arc-hour, taking into account spatter and stub end losses and the like.

2. *Welding speeds,* shown in the right-hand section, relate the joint completion speeds (in./min) and times (arc-minutes for each foot of joint) to the deposition rates for varying weights of weld metal required for the joint. The lower horizontal scale (arc-minutes/ft) is the reciprocal of the upper scale, multiplied by 12.

3. *Labor costs,* shown in the bottom right-hand section, are related first to the operating factor and second to the hourly labor rate.

An example of utilization of this chart is given by the dotted line *a-b-c-d-e-f.* By using a ¼-in.-diameter conventional manual metal-arc electrode with solid flux attached at about 320 amperes, a deposition rate (point *b*) of 6 ¼ lb/arc-hour will be obtained. For a weld operation which requires 0.6 lb of weld metal per foot (point *c*), the welding speed (or joint completion speed in the case of multipass welds) will be just over 2 in./min, or alternatively, the joint completion time will be 5 ¾ arc-minutes per foot (point *d*). With an operating factor of 20 percent (point *c*) the time to complete 1 foot of weld is 28.75 minutes or 0.48 hours; the labor costs at $8 an hour will be $3.83 per foot of weld (point *f*).

As mentioned earlier in this chapter, there is often a conflict between the equipment *available* in the factory to do a job and *more desirable* equipment which would be better for the job. Thus elementary cost calculations of the sort given in this section may suggest that one process is far cheaper than another, but equipment for the cheapest process may not be available and would have to be bought. One means of making more reliable comparisons between welding processes is to add to the direct welding costs that portion of the capital cost attributable to depreciation of equipment.

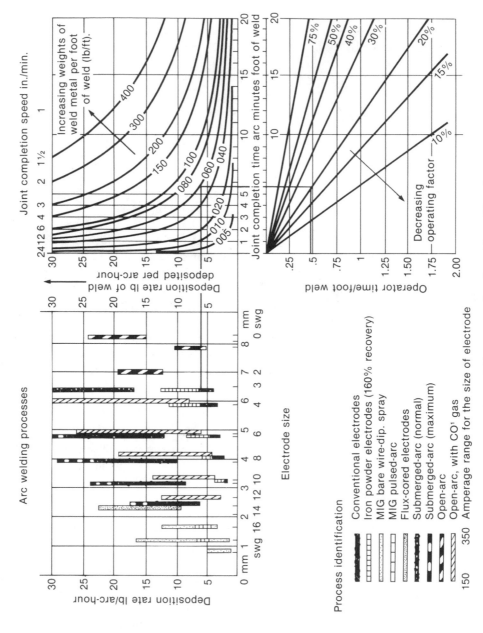

Figure 13-3 Weld cost comparison for welding mild steel.

13.4.2 Adhesive Bonding Costs

Half of the total costs in manufacturing relate to costs of assembling parts together, which fact encourages innovation in joining methods. A great deal of assembling is now done with the use of ''structural'' or ''engineering'' adhesive compounds (hereafter referred to simply as ''adhesives''). Apart from providing desirable product attributes, there are often cost savings to be gained by using adhesives, because parts to be assembled with adhesives need not be made to high accuracy.

For example, in the assembly of automobile gear shift levers, a metal sphere with a hole in it is slid onto a steel shaft and held in place by an adhesive, thereby replacing forging and machining operations. One can also choose between types of adhesives, as in one example of taillight assemblies. A hot-melt adhesive bonding process replaces an acrylic adhesive, which increases the production rate by about a factor of ten because the hot melt holds the parts together immediately upon cooling, whereas the acrylic adhesive sets by a more time-consuming evaporation of solvent. This reduces the labor costs and produces considerable overall savings, even though the hot melt adhesive costs about 30 percent more than the acrylic.

Another study involves the making of a box overarm for a small horizontal milling machine. Competing proposals were to make it from (1) cast iron, (2) welded steel, and (3) channels and I-beams adhesively bonded together. Section sketches of the three proposals are shown in Fig. 13–4. The traditional method of casting has several inherent disadvantages, one of them being that the section thicknesses must be governed by casting technology rather than by strength requirements. Casting is an old industry; it produces low-cost products, but it is slow and relatively inflexible. The manufacture of machine structures by building up of readily available bars and plates, and so forth, has some advantages, but it was not considered seriously in the past because of difficulties in fabricating the built-up overarm. For instance, welding produces residual stresses which causes structural distortions, a major drawback when accuracy is of prime importance. These residual stresses may be reduced by heat treatment, which is an unwelcome extra production operation, especially for large machines. Other fabrication techniques such as bolting and riveting are also impracticable and cause unequal stress distribution in the

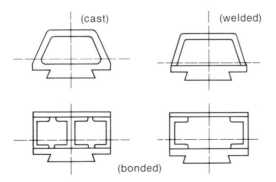

Figure 13–4   Cross sections of different overarms.

TABLE 13–5 ESTIMATED COST TO MAKE OVERARMS
BY THREE METHODS

Overarm	Cast Iron	Welded Sections	Bonded Sections
Cost	$85	$75	$44

joint. However, adhesive bonding can be used to good advantage in the manufacture of machine tool structures, and when it is combined with other manufacturing techniques, new possibilities in fabrication arise.

Realistic costs for innovations are difficult to find, since costs of all products depend on the scale of production. Estimates of the production costs of single overarms of the three types, as would be done by a shop that is not equipped for multiple production, are given in Table 13–5.

These 1975 figures cover material, labor cost, and overhead. For all cases the capital cost of equipment for efficient volume production—that is, jigs, fixtures, and the casting pattern—was excluded. When the cast-iron overam is produced in large quantity, its cost drops to $36. The cost of the welded and adhesively bonded types is not very much affected by numbers, since these processes are highly labor-intensive, in which relatively little can be saved by special jigs or tools. These cost figures cover production only. They do not take into consideration any interest to be "paid" on the cost of stock levels required for avoiding delays in delivery, which are particularly high for cast components and low for bonded ones. It would appear that bonding should be particularly attractive when the components are very large and/or when a large number of them is required at short notice.

13.5 COSTING OF NEW PROCESSES: ORBITAL FORGING

When new methods of manufacturing are proposed, accurate cost calculations may not be possible, particularly if related processes are not now used. An example is making pulley blanks from a cropped billet (short cylinder) by "orbital forging." In this process a cylindrical "slug" is progressively worked outwards to give the general shape shown in Fig. 13–5. A more conventional method would be to press blank a disk with a large hole in it, into which a hub is pressed. The attraction of orbital forging is the elimination of cleating the inner surface of the hole in the disk and then pressing the hub into that hole. The major disadvantage is the slowness of the orbital forge. The example also illustrates a not-uncommon situation in which many of the basic data for the new process were lacking at the time of a study, in particular the life of the tooling. In such circumstances,

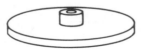

Figure 13–5 The pulley blank used in the example of orbital forging.

a "base case" is set up with some assumptions, and sensitivity analyses are carried out to investigate the significance of the various assumptions made.

Total material costs (including the hub cost and cost of the cleating operation in the traditional route) are less for the orbital forging method, as the following calculations show for one particular size pulley:

	Gross	Scrap	Net
1. *Blanked and cleated*			
(a) Weight (g) (disc only)	228	85	143
(b) Material cost ($/ton)	225	55	
(c) Unit material cost (¢) (a) × (b)	5.14	0.46	4.68
(d) Hub (purchases) (¢)			8.01
(e) Cleating (¢)	0.7		
(f) Total cost (¢)			13.24
2. *Orbitally forged*			
(a) Weight (g) (includes hub)	271	49	222
(b) Material cost ($/ton)	231	55	
(c) Unit cost (¢)	6.26	0.26	6.00

There is thus a saving per unit of $(13.24 - 6.00) = 7.24$¢ with regard to material for this particular size pulley.

Set against the reduced material cost, however, is the cost and life of the tooling on the orbital forge, and the cycle time of the operation on the forge. In general terms, if tool lives are short and orbital forge cycle times are long (that is, low production rates), the savings on the materials side will have to be large to justify the new method, whose costs must include depreciation of the new equipment and so on. On the other hand, if tool lives are long and cycle times are short, the materials savings can be less. Clearly it is required to establish the relationship between tool life and cycle time for various assumed materials savings costs. In this way, options for different size pulleys can be accommodated. Figure 13–6 plots the results of extended calculations for other assumed tool lives.

For the particular size pulley for which material savings of 7.24¢ were calculated above, it appears that orbital forging will be viable under any tool life/cycle time combinations to the upper left of a contour about midway between the 7¢ and 8¢ contours on the graph.

13.6 A METHOD OF ACCOMMODATING PART COMPLEXITY: SIMPLE FORGED PARTS

There are many manufacturing processes that are too complicated to analyze in useful detail. For such operations seemingly illogical cost figures in handbooks are given in terms such as ¢/lb. An example to be discussed is that of forgings. The process begins with a piece of metal, known as the billet, which is of crude shape and the proper weight. It is set onto a "lower" die, which is a piece of metal into which a shape has been cut and

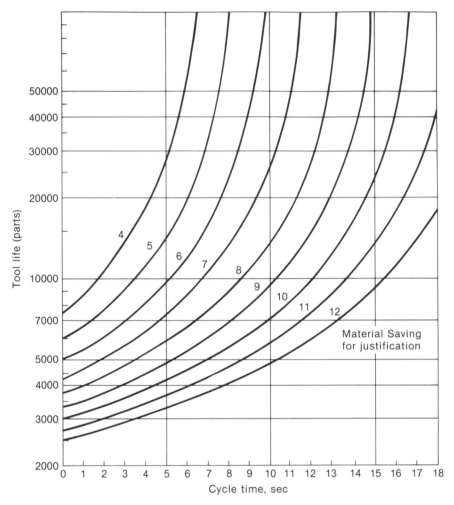

Figure 13–6   Economic factors in orbital forging.

ground. The upper die also contains a carefully made cavity. The closed die will have the shape of some desired part. The upper die may be forced against the billet slowly and with great force in a press, or it may be attached to a weight and dropped onto the billet. With continued squeezing or with repeated hammering, the billet is deformed into the shape of the cavities in the die segments. However, the billet is never perfectly located in the die, so there may be excess metal in some places and insufficient metal elsewhere. The solution to this problem is to use a billet with excess metal and to provide an area in the die set for excess metal to flow in all directions. This excess is called the *flash* or flashing, and it must be removed before final processing of the product.

It can be seen that forgings can be made in many different materials and in a wide range of complexity. Thus, a single cost to forge in terms of ¢/lb may not be realistic. An inexperienced product designer would have some difficulty in estimating a realistic cost for forgings. To aid in this exercise Poli and Knight* have developed a "classification of component features," and "materials to be used," to assess the relative costs implied in alternative design and processing routes for forged components.

The elements to cost in their forging model relate to those listed earlier:

1. material (including wastage)
2. capital charge on equipment
3. direct labor
4. heating and furnace costs
5. handling
6. set up costs
7. die costs
8. overheads

On average the material cost represents approximately 50 percent of the cost of forged parts, die costs around 15 percent and direct labor 15 percent. Material costs will increase, for a given workpiece material, with the complexity of the part, as more wastage (in the form of flash and so forth) will be necessary to produce the required shape. Die costs will increase as the number of dies is increased and the die life is reduced, which will be influenced by the complexity of the shape and the material being forged. Direct labor costs will increase with the size of the crew required to operate the press or hammer (that is, as the capacity of equipment increases). The precise proportion of each of these cost contributions to overall forging costs will vary from plant to plant, dependent to some extent on local practice and methods.

Since, in forging, setup time is a relatively small part of component cost, it is neglected, and the calculations condense down to costs (per unit produced piece) of

    (i) material (allowing for scrap), K_m

    (ii) "equipment operating costs" (that is, items 2, 3, 4, 5, and 8 above grouped together to represent the "most usual practice," with appropriate furnace handling equipment and crew size), K_p

    (iii) die costs, K_D

Using their symbols for convenience, we can express these components as

$$\text{(i)} \quad K_m = V\rho C_m K_i \tag{13–21}$$

where V is the volume of the part, ρ the material density, C the material cost per unit mass, and K_i the ratio of gross-to-net weight for the particular forging in question.

$$\text{(ii)} \quad K_p = C_p/P_R \tag{13–22}$$

where C_p is the operating cost of the equipment per unit time and P_R is the production rate, that is, ($/time) / (pieces/time) = $/piece.

*C. Poli and W. A. Knight, *Design for Forging Handbook*, (Amherst, Mass.: University of Massachusetts, 1981).

$$(iii) \quad K_D = \text{die cost/no. of parts in die life} \qquad (13\text{-}23)$$

where n/D is the number of sequential dies required to do the job at the forging cost center. The costs and lives of a series of dies employed sequentially in the complete forging operation may all be different, which complicates matters, so in order to simplify calculations it is customary to assume an average die cost C_D and an average die life N_L. Then

$$K_D = n_D C_D N_L \qquad (13\text{-}24)$$

The total unit cost to manufacture is the sum of $(K_m + K_p + K_D)$.

The detail of these calculations relies on information about blow rates, factors that influence initial die cost, and factors that influence die life, among other things, as discussed in the following:

For presses, which are comparatively slow-acting, the number of blows required is normally the same as the number of operations required, excluding flash removal. For hammers several blows per operation are used and an average of around three blows per operation is usual. The relative blow rate of the machines should be the effective values, that is, reflecting the handling of the forgings between operations, rather than the ratios of the maximum cycling rates. Thus for very large presses and hammers the effective rates will be governed largely by the speed with which large parts can be manipulated. Hammers tend to be used in preference to presses when large changes in workpiece cross section are required.

Die costs can be considered to be made up of

1. The material of the die blocks
2. The cost of machining and finishing the die cavities.

In general, for small dies, the latter makes up the major proportion of total cost, whereas for large dies, the material cost is more significant. Consequently, as the size of the part is increased, the proportion of the total cost attributable to the die material increases. It is reasonable to assume that the quantity of die material will be influenced predominantly by the size and material of the forging, and the effect of increasing the complexity of the part will be to alter mainly the machining cost of the dies.

Die life is influenced by a number of factors, including

1. The material to be forged.
2. The shape of the part.
3. The tolerances applied to the part.
4. The forging equipment to be used.

The cost of manufacturing a simple disc-shaped part from low-carbon steel has been chosen as the yardstick against which all other costs are referenced. Using the additional subscript 0 to refer to this basic part, we have the total cost made of component costs

$$K_{T0} = V_o \rho_o C_o K_{fo} + C_{po}/P_{Ro} + n_{Do} C_{Do} C_{Do}/N_{1o} \qquad (13\text{-}25)$$

Alternatively, Poli and Knight* refer to the relative values of K_m, K_p, and K_D taken separately. That is, for materials

$$K_{mR} = \frac{V\rho C_m K_f}{V_o \rho_R C_m K_{fO}} = \frac{V\rho_R C_R K_{fR}}{V_O} \qquad (13-26)$$

where K_{mR} is the cost of material in the forging of interest relative to the cost of material in the basic disc-shaped part, ρ_R is the density ratio of the (possibly different) materials, C_{mR} is the material cost relative to that of the reference low carbon steel, and K_{fR} is the relative ratio of gross-to-net weights.

Likewise

$$K_{pR} = \frac{C_p P_{RO}}{C_{pO} P_R} = \frac{C_{PR} N_p}{R_{pR}} \qquad (13-27)$$

where C_{pR} is the relative operating cost compared with the equipment used to produce the reference disc part, N_p is the number of blows on the forge required to produce the part, and R_{pR} is the relative blow rate of equipment used compared with that of the forge used to produce the reference disc part.

Again

$$K_{DR} = \frac{n_D C_D N_{LO}}{n_{DO} C_{DO} N_L} = \frac{n_{DR} C_{DR}}{N_{LR}} \qquad (13-28)$$

where n_{DR} is the relative number of dies compared with the standard operation, C_{DR} is their relative collective cost, and N_{LR} is their relative collective life.

If, on the basis of fully costing out the basic part in a given factory on given equipment, it appears that the breakdown of total costs K_{TO} is 50 percent for material costs, 30 percent operating costs, and 20 percent for die cost, then the total cost K_T of any other forging is given by

$$K_T = K_{mR} K_{mO} + K_{pR} K_{pO} + K_{DR} K_{DO}$$
$$K_T = K_{mR}(0.5 K_{TO}) + K_{pR}(0.3 K_{TO}) + K_{DR}(0.2 K_{TO})$$
$$K_T = (0.5 K_{mR} + 0.3 K_{pR} + 0.2 K_{DR}) K_{TO}$$

that is, the total relative cost per piece is

$$K_T / K_{TO} = 0.5 K_{mR} + 0.3 K_{pR} + 0.2 K_{DR}$$

(for the assumed initial 50 percent/30 percent/20 percent breakdown of K_{TO}).

An example of the application of these ideas relates to the part shown in Fig. 13–7. It may be demonstrated that the ribs are of such proportions as to be considered closely spaced and consequently of relatively high forging difficulty. Altering the design so that the ribs are less severe or so that a peripheral rib only is used for stiffening would lead to a reduction in forging difficulty. Table 13–6 shows the effect of these design alterations on the relative forging costs estimated by using the described procedure. In this way the

* Poli and Knight, *Design for Forging Handbook.*

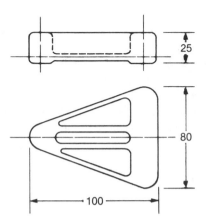

Figure 13–7 Flat nonround part with severe ribs—shape code 282.

product designer is given an indication of the relative forging costs of the different designs and hence the cost benefits of various design improvements.

TABLE 13–6 ESTIMATED FORGING COSTS FOR VARIATIONS IN DESIGN OF THE PART

	Relative Material Cost	Relative Production Cost		Relative Die Cost
		Hammer	Press	
Original design	0.41	9.0	8.8	58.0
Redesign with less severe ribs	0.38	6.0	6.3	25.0
Redesign with peripheral ribs	0.36	5.6	6.3	18.5

__ *14*
the economics of integrated design and manufacturing __

14.1 INTRODUCTION

In this chapter, we discuss how the original design of a product can influence the choice of what manufacturing process is most sensible. Ideally, design, materials, and processing should be considered as complementary and integrated activities.

Established rules, by which a component or assembly of components, may be designed to perform a desired function are well established. These relate, for example, to stressing, shape and proportioning, choice of material, environment in which the component will perform, and the like. The degree of complexity of this sort of design analysis depends on circumstances, ranging from elementary strength of materials calculations for simple components, to complicated finite element/finite difference calculations for crucial components in nuclear plant applications. Many books cover the functional performance aspects of design, and we shall not repeat the basics here. We shall assume that appropriate rules are being followed and that these rules may be different for components made from metals, polymers, or ceramics. But functional aspects of design are only part of the full picture. It is not much good if a brilliant design, incorporating some splendid material, cannot be *made, assembled,* or *fabricated,* especially in an economic sense. It seems less well-known that there are rules and procedures that should be followed so that the component can be *successfully* made and/or assembled.

How a part is going to be made, and how it may fit with another part should not be left until the last step of the overall design procedure. Exact reproduction of a part as-designed may prove to be a difficult and costly, if not impossible, task. Some slight alteration in the original design can often permit easier manufacture without detracting from the functional performance requirements. This pertains at both intermediate stages of manufacture, when semi-finished products are being produced, and at finishing stages, regardless of the process to be used.

The overall process of design should involve proper consideration of functional design, materials, and manufacturing method, as shown in Fig. 14–1. It is an interactive situation, but with few exceptions, these different areas have usually been treated separately both in industry and in academia. But there is a growing realization that this is not the best approach, and they are being brought together in more and more establishments.

There can be many combinations of the three elements of design, materials, and manufacture without there being one *best* route from the drawing board to the marketplace; it depends on circumstances. If one, or even two, of the three elements are fixed (for example, a fixed design with fixed materials), there may still be a number of acceptable alternative manufacturing routes to produce the desired design; this may depend on the number of parts to be made, the availability and quality of raw material, the number and capability of existing machines in the factory, and, of course, the resulting cost per part. Alternative designs for different manufacturing procedures are usually possible. Design changes may be required if CNC manufacturing methods rather than more traditional ones are used.

The in-service properties of the manufactured design may be influenced strongly by the production route employed. For example, a complicated shape, machined from the solid, will display the properties of the initial solid, whereas the same shape produced by forging will usually have properties that differ from the raw material, especially if cold working is involved. In many cases, desired in-service properties may be attainable only by certain thermomechanical processing sequences. Sometimes, it may be nearly impossible to manufacture an article using particular materials owing to production difficulties. Materials which are extremely difficult to machine or weld can introduce severe problems.

Clearly, there must be some interplay between what the designer specifies and what can actually be achieved economically by certain processes. Specifications of tolerances and surface finishes required, for example, must be considered when different manufacturing methods are to be compared as alternative methods of production. Linked to such considerations is the question of overall economics. Savings arising from one or two cheap operations early in a production sequence may be lost altogether if expensive finishing operations are required later.

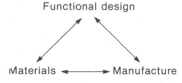

Figure 14–1 Interactive considerations in the design and manufacture of a product.

14.2 CUTTING OR FORMING: NEAR NET SHAPE

Expensive machining operations involve grinding and contour milling, and there will be clear advantages, in setting up production schedules, if the need for these two types of operations are avoided. On the other hand, drilling and turning typify less costly machining operations. Thus, in casting, for example, it is often more economical to drill holes subsequently instead of using cores for holes in the basic casting, since this can be a more expensive operation. Yet in the general field of design for production of metallic components, there is growing awareness that *all* metal removal processes are really wasteful in terms of time, money, and energy consumption. One example relates to the production of engineering components of complex shape from expensive raw material, where massive machining can involve the removal of as much as 98 percent of the metal in producing gas turbine components made from nickel alloys. Figure 14–2 shows typical types of advanced engineering components which have to be machined from solid stock having appropriate wrought metallurgical structure. Their shapes are so complicated (and they may "twist" in and out of the paper) that it would be impossible to forge them between closed dies owing to the difficulty of extracting them from the dies after the forging is completed. Open-die forging would not achieve the awkward shapes of these types of components with the precision needed for subsequent finish machining. Hence they are made with massive machining operations. In the case of some highly stressed parts of

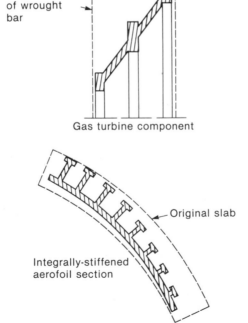

Gas turbine component

Original piece of wrought bar

Original slab

Integrally-stiffened aerofoil section

Figure 14–2 Types of components that require machining from solid stock.

TABLE 14–1*   TYPICAL ENERGY CONSUMPTION AND MATERIAL
UTILIZATION FOR VARIOUS PROCESSES

Process	Energy Required—MJ	Material Utilization—%
Cold or warm extrusion (forging)	41	85
Hot, closed-die forging	46–49	75–80
Process casting	30–38	90
P/M pressing and sintering	28.5	95
Metal removal processing	66–82	40–50

*Source: K. H. Kloos, VDI Berichte, Nr 277 193 (1977).

complicated shape which are not excessively large, such as turbine blades and connecting rods, precision casting processes may be used, if the desired metallurgical structure can be achieved. This is not always possible, so machining has to be used.

Forging and casting processes are examples of so-called *near net shape* processes, while true precision casting and powder metals methods are called *net shape* processes. Material wastage is greatly reduced, as illustrated in Table 14–1, which also includes information on the energy required per tonne* of finished product. For reference purposes, comparable data are included for traditional metal removal processes. It can be seen that conventional cutting methods consume about twice as much energy and at the same time display only half the material utilization of the forming methods.

These efficient net shape forming methods have been usefully classified into five categories.† The relevant chapters in this book where they are covered are also indicated:

1. Forming by solidification of molten or semi-molten metal — casting processing (Chapter 11)

2. Forming from solid billets by material displacement — hot, warm, and cold forging, extrusion, wire drawing, tube making and ring rolling (Chapter 8)

3. Forming from sheet material by displacement or perforation — blanking, deep drawing, and ironing (Chapter 8)

4. Forming from powder metal by pressing, sintering, and (sometimes) deformation — conventional powder metallurgy, coining, powder forging (Chapter 10)

5. Forming by building up components on a mandrel or former — electroforming, chemical and physical vapor deposition, spraying, spray peening, weld metal deposition with forging (Chapter 12)

Note that category 5 incorporates comparatively new ideas of *deposition forming*, in which a component is built up by the deposition of molten metal followed by the

* Note that tonne equals 1000 kg mass, or approximately 2200 lbm.

† Michael Neale and Dennis Waterman, Report TRS 289 for Science & Engineering Research Council/Department of Trade and Industry. (Farnham, Surrey, U.K.: Michael Neal and Associates, 1982).

immediate hot working of the solidified deposit. This idea is clearly the antithesis of metal cutting.

It will be evident from later coverage that there may be a cutoff or breakeven point, in terms of quantity of parts produced, beyond which it may be cost-effective to employ metal forming methods in place of metal removal methods. Such an alternative would be used even more widely were forming processes more accurate in meeting design requirements, thereby involving only limited finish machining or even satisfying the necessary precision directly. Questions of tolerances and surface finish, both desired and attainable, are discussed in Sec. 14.5

14.3 ELEMENTARY CONSIDERATIONS OF DESIGN FOR MANUFACTURE

Since metal-cutting methods have been with us for so long, one might think that designers would know by now how to design parts that are amenable to machining with no complications. But this is not the case, and parts to be machined are sometimes designed without regard to the limitations of the machine tools intended to be used. A number of considerations should be borne in mind, among which are (1) allowing ample space for the cutter, (2) adequate provision for chip removal, and (3) keeping the number of separate cutting operations as small as possible, all of which can reduce costs. Strasser* made the following points:

1. Tool holders, noncontacting peripheries of spinning cutting wheels, and other unavoidable obstructions in machine tools must clear the workpiece. Parts designed with ample tool clearance can be machined rapidly and without damage.

2. Machined chips, whether discontinuous or continuous, must be removed from the cutting tool/workpiece interface. Often, a coolant or a stream of air can be used to help the chips escape. Detail features of some components, such as long or blind holes, do not lend themselves to easy chip removal, and provision must be made in the design to accommodate a buildup of chips in order not to reduce the efficiency of cutting.

3. Material handling can become a significant part of the cost of a machined product. Scheduling and shifting a workpiece from one machining operation to another adds to the cost of the finished product. To keep machining costs down, the part should be designed to minimize the number of separate operations on the workpiece.

4. The volume of material removed and the distance the cutting tools travel should be kept as small as possible, in order that time and material are saved. In castings and forgings, these objectives may be achieved by including recesses.

* F. Strasser, *"Designing Parts that are Easy to Machine"*, Machine Design, (Aug. 1975), p. 65.

The Smallpiece Education Trust* published a list of do's and don'ts and a number of interesting examples of "before" and "after" designs which well illustrate "design for production" problems. All the component designs were the work of experienced engineers; however, they lacked manufacturing experience. While the examples were acknowledged to be rather elementary, they were intended to be just that because there is much more scope for redesigning for production, simple items which are usually produced in much larger quantities than more complicated parts. Furthermore, complicated parts and assemblies often incorporate numerous simple components.

In many cases, relatively minor changes from the original design, which do not in any way affect functioning of the component, facilitate machine setting, operation, and inspection to such an extent as to enable at least half the machining cost to be saved.

A number of related examples of changes in design to facilitate easier manufacture are given in the 1974 edition of the Fulmer Materials Optimizer.† The point is made that it is sensible to see whether a given massive part can be subdivided into components that are individually more easily made. For example, Fig. 14–3 shows a die-cast pulley wheel which, had it been made "solid," would have meant either casting a solid rim (bad practice owing to large changes in cross section) followed by machining out the V-groove, or the manufacture of a complicated die in which the pulley groove would have been formed by a withdrawable-mold section. Instead, the same result can be achieved more economically by forming two dished castings which are then riveted together with integrally cast pins. Advantage can also be taken of the possibility of welding two or more castings or forgings together to give final components. This is clearly of value when the size of the component exceeds the capacity of the forge or foundry, and smaller parts can be designed appropriately to make up the final article.

A specialist casting operation called "insert work" can simplify manufacturing procedures and eliminate alternative expensive routes. Figure 14–4 shows a universal ball joint made by *in situ* casting of a zinc alloy housing around an existing sintered steel ball; the ball is loosened in the housing by a blow as part of the trimming action after casting. Complicated precision-machining of the housing has been avoided.

An example which illustrates the need for the designer to be thoroughly familiar with different methods of manufacture, and their associated costs, is that of the pro-

* In *Machinery and Production Engineering,* 312, (Feb. 1968).
† Fulmer Research Institute, 1974, Slough, U.K.

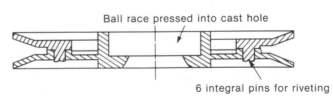

Figure 14-3   Schematic of a pulley wheel.

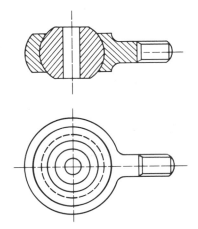

Figure 14–4 Universal ball joint made by casting a zinc alloy around a steel ball.

duction of a simple rocking arm by various methods.* Figure 14–5(a) to (d) shows different detail arrangements which permit manufacture by (a) casting, (b) drop stamping, (c) hand forging, and (d) welding. Although particular economic conditions are involved, if we *presume* that all of the capital equipment required for the separate methods of manufacture are *available,* welding is cheapest for batches up to 75, casting for batches of 75 to 700, and drop stamping for batches greater than 700. Figure 14–5 brings out an important point, namely that the cost curves for different processes *intersect,* so there are different ranges of numbers over which different methods are most economical. As we shall discover, this is a recurring theme in assessments of alternative methods of manufacture.

A number of particularly interesting examples of producing the same shape design by different methods concerned the employment of the fine-blanking process, in place of ordinary blanking combined with subsequent machining which is necessary to transform the sheared edges up to adequate standards. Fine blanking avoids the jigging and clamping inaccuracies which can arise between operations by the conventional route.

Table 14–2 gives details of the times taken to manufacture, by various blanking methods, the plunger shown in Fig. 14–6. Progressive fine-blanking tooling eliminated the time-consuming shaving operation required to achieve cleanly sheared surfaces. As the chamfers are coined before shearing, the costly milling operation could also be excluded. From Table 14–2 it can be seen that approximately 456 minutes of manufacturing time were saved in the production of 1000 parts.

The extra tooling costs (780 hours for building the progressive fine-blanking tool) were amortized after the production of 100,000 to a maximum of 200,000 parts, depending upon the hourly rates involved. As at least 500,000 plungers were required each year, the change of working method paid for itself in a short period. A second example concerned a pair of handed yokes, illustrated in Fig. 14–7. Conventional and

* See C. Ruiz and F. Koningsberger, *Design for Strength and Production* (London: Macmillan, 1970).

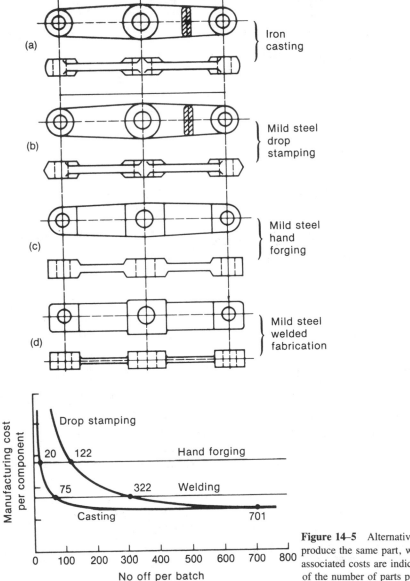

Figure 14-5 Alternative methods used to produce the same part, where relative associated costs are indicated as a function of the number of parts produced.

fine-blanking methods were again used. Expressed in time per 100 yoke components, 5234 minutes were saved, and the tooling costs for the old method of production were 82 hours higher. Servicing the tooling for the conventional method, and the preparation prior to production were also *more costly* than the newer method. The cost comparison shows that fine-blanking provided greater economy in producing these yoke components.

TABLE 14–2 COMPARISON OF PRODUCTION RATES AND TIMES FOR TWO BLANKING METHODS USED TO MAKE THE SAME PART

Operation	Conventional	Fine Blanking
Blank (single punch)	20	Fine blank (twin punch) 14.3
Shave	150	Deburr 30
Mill	300	
Deburr	30	
	500 min/1000 parts	44.3 min/1000 parts

Production Time for Tools		
Blanking tool	150	Fine blanking tool 1100
Shaving tool	120	
Milling jig	50	
	320 hours	1100 hours

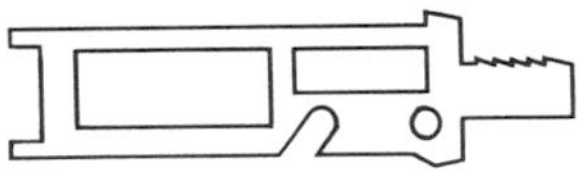

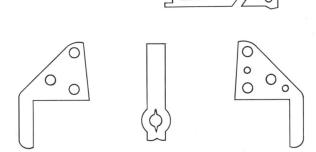

Figure 14–6 Plunger made by blanking methods detailed in Table 14–2.

Figure 14–7 Yokes made by blanking.

In an entirely different field, the detailed design of weldments has a marked effect on manufacturing costs as well as strength performance. Consider the various alternative designs of T-joints in 25 mm plate shown in Fig. 14–8. The full-penetration butt weld at (a) requires initial preparation of the vertical plate, and the root runs made at relatively low deposition rates to gradually build up the joint. Additional work would include (1) backcutting of the first side root and (2) NDE* to ensure that full penetration was achieved; this adds materially to the total fabrication cost. A twin fillet weld of equivalent static strength, Fig. 14–8(b), would have approximately the same volume, but considerably lower costs by the elimination of an angled preparation, backcutting, and inspection. Relatively high welding currents could be used for all runs. A simple twin fillet weld of equivalent fatigue strength, transferring the site of failure from the weld throat to the weld toes, would need to be much larger, as in Fig. 14–8(c), but might still be more economical than a full penetration butt. A considerable reduction in weld volume could be achieved,

* Short for nondestructive evaluation.

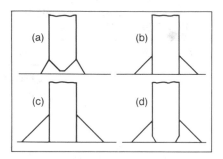

Figure 14–8 Different methods of welding used to produce a T-joint: (a) full penetration butt weld, and (b), (c), and (d), twin fillet welds of different sizes to produce joints of different strengths.

however, as indicated in Fig. 14–8(d) by simply chamfering the plate or ensuring that the root run was made by a deep penetration process.

For noncritically stressed joints, it is often convenient to specify *partial-penetration* butt welds, but the unfused edges can act as stress concentrators or cracks, thereby lowering both static and fatigue strengths. Of course, if a full-penetration weld is specified but not achieved, we have so-called lack of penetration (see Chapter 10).

At this stage, questions of *joint preparation* have to be addressed. It is possible to make full-penetration butt joints between adjacent sheets of material up to some 3 mm thick by using manual metal arc or Mig processes, and up to 12 mm with submerged arc welding, since the penetrating capacity of the arc is at least as deep as these sizes. For thicker materials, the edges must be *shaped* so that the abutting faces are no thicker than the limiting sizes just mentioned. A variety of joint preparations is shown in Fig. 14–9. Single V-bevels of about 60 degrees total included angle are the simplest type, followed by double-V chamfers used in thicker plate, say, > 18 mm, in order to limit the excessive amount of weld metal required to fill a single V cavity. In plates thicker than about 75 mm, the large amounts of weld metal required to fill even the double-V cavity can be reduced by using U-shaped joint preparations. Similar remarks apply to T-joints. Of course, use of double-V and double-U preparations necessitates that the joint be accessible from both sides, so that the component can be turned over.

It is clear that most preparation is concerned with access and that all the metal chamfered away has to be replaced subsequently. This affects overall welding productivity, so, in addition to increasing metal deposition rates as one means of optimizing

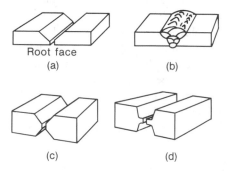

Figure 14–9 Various ways to produce a butt joint by welding.

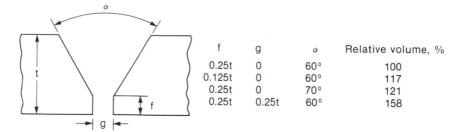

f	g	α	Relative volume, %
0.25t	0	60°	100
0.125t	0	60°	117
0.25t	0	70°	121
0.25t	0.25t	60°	158

Figure 14–10   Effect of joint preparation on the volume of metal deposited to produce a satisfactory joint.

fabrication methods (Section 13.4.1), reduction in joint volume is also a worthy target. The majority of commonly used preparations have been developed for use with low penetration manual metal arc welding, for which it is essential that the arc impinge on the side wall at a minimum angle of 30 degrees if lack of fusion is to be avoided. Thus, a joint included angle of 60 degrees has become traditional. When tolerances on preparation angle, root face, and gap are taken into account, much greater angles are often encountered. The effect of these changes on joint volume, and therefore on welding time, can be considerable, as shown in Fig. 14–10.

A related area concerns the physical *size* of weld specified by a designer. In order to make the structure safer, welds larger than really needed may be specified, and costs in labor, materials, and inspection can get out of proportion. For example, a fillet weld of 4 mm leg thickness has a weight of weld metal of 73 grams per meter and requires an arc time of 2 min per meter; a 6-mm leg has 174 grams per meter and requires an arc time of 3.92 min per meter; an 8-mm leg has 304 grams per meter run and requires an arc time of 7.64 min per meter.

It was pointed out in Chapter 10 that, unlike many other manufacturing processes, which do not drastically affect the basic mechanical properties of components and assemblies of components, welding affects the metallurgical structure of the materials being joined and can also produce *variations* in properties around the joint region. Fabrication by welding is thus an area in which design, materials, and manufacture not only come together but are inseparable. Owing to large temperature gradients and the presence of pools of molten metal in welding, problems of distortion and residual stress have often to be anticipated. Welding can, for example, induce far higher internal stresses than does the cooling of whole components as in casting or hot working, and the designer should consider such difficulties.

14.4 BASIC DESIGNS FOR FORMING IN PLACE OF CUTTING

Much traditional design anticipates manufacturing routes which involve substantial metal cutting. Indeed, the familiar shape of many common engineering components is a direct result of such types of processing. It does *not* follow that, if a component were to be designed for manufacture by forming in place of cutting, the shape of the component would be the same as before. Alternative shapes, well suited to manufacture by metal forming, may be as good or better than those made by cutting. These considerations are relevant to near *net shape forming*.

It is valuable in this context to divide engineering components into *sections* which will experience bending moments, torques, tensile loads, rubbing contacts, and so on, and to consider what shape or particular feature is appropriate to meet each functional requirement. By considering a range of simple components, each of which carries one of these functional requirements, it is possible to see what would be an ideal shape for such a component if it were to be produced by a metal-forming process. This can then serve as a guide towards the application of this approach to more complicated components. It can also reveal where manufacture to finished size by metal forming, instead of cutting, can enable improved design of components as well as lower costs.

14.4.1 Tension Links

The shape and size of the main body of a tension link must provide sufficient cross-sectional area, in conjunction with the tensile yield strength of the material, so that the required load can be carried without the member's distorting. However, the shape and size of the *end fittings,* which usually connect to a pin or some form of bearing support tube, are more affected by the method of manufacture, as illustrated in Fig. 14–11. The various tension links in that figure are also generally suitable for compression, provided the body is short enough to avoid buckling. They are suitable for applications where the pivot pins at the ends, or the link itself, can be withdrawn.

If they cannot be withdrawn, the end connection must be split. Since an inserted bearing in the end of the link needs to be fitted with an interference fit, the clamping arrangement across the split needs to provide large controlled clamping forces. This is conveniently obtained from screw-threaded connections, as in a conventional engine connecting rod. The big end housing also needs to be rigid and of substantial radial thickness, because oil film thicknesses are small and large loads are generated. Due to these factors, the design of items like connecting rods would be unlikely to change even if they were formed to size. For lighter duty, they could be formed to size at low cost with a suitable integral bearing material, and some design changes might be possible.

14.4.2 Torsion Shafts

Shafts are usually of circular section, since this is an efficient shape in terms of torque capacity per unit mass, and is particularly efficient if the section is a tube. Also, shafts of circular section have a surface which remains in a fixed position relative to the shaft axis when rotating; this enables bearings and seals to operate directly on this surface.

The choice of an appropriate shaft cross section in terms of torque capacity is a straightforward application of strength-of-materials theory, but the design of *end connections* to transfer the torque to other components is critical, particularly when these con-

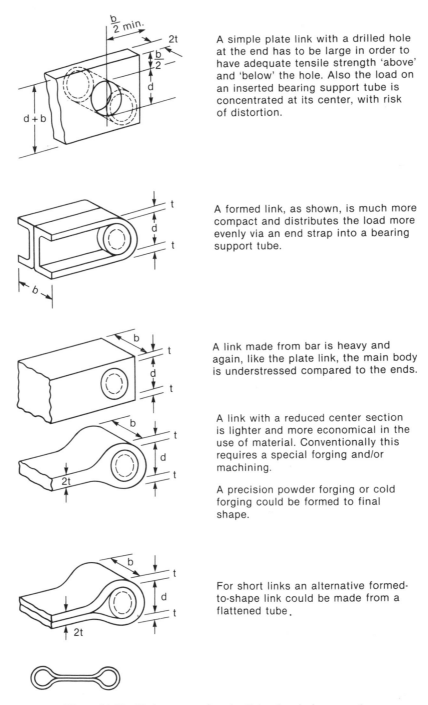

A simple plate link with a drilled hole at the end has to be large in order to have adequate tensile strength 'above' and 'below' the hole. Also the load on an inserted bearing support tube is concentrated at its center, with risk of distortion.

A formed link, as shown, is much more compact and distributes the load more evenly via an end strap into a bearing support tube.

A link made from bar is heavy and again, like the plate link, the main body is understressed compared to the ends.

A link with a reduced center section is lighter and more economical in the use of material. Conventionally this requires a special forging and/or machining.

A precision powder forging or cold forging could be formed to final shape.

For short links an alternative formed-to-shape link could be made from a flattened tube.

Figure 14–11   Various types of tension links of equivalent strength.

nections need to be made demountable. Some common end connections with their advantages and disadvantages are shown in Fig. 14–12.

14.4.3 Torsion Discs

Many engineering components include a number of disc-shaped members which carry torsional loads. Driving flanges, gears, sprockets, and turbine or compressor discs are typical. Nominally, for constant torsional shear stress, such discs would be expected to be made with a thickness which decreases toward the outside. In practice, however, due to attachment loads on the driving flanges, to the required width of gear or sprocket teeth, or blade roots, the periphery of such discs has to be made thicker than torsional shear stresses alone might require. Constant thickness discs are therefore commonly used in many cases. Only where centrifugal stresses are important in high-speed machines are profiled discs normally used.

 With gears, the loads are transmitted across a small number of teeth at one time. On sprockets, although the number of teeth in action is larger, loading is still carried by a limited number. Consequently there are substantial bending loads on teeth of this type and the meshing profile contact regions experience high contact stresses. Teeth which are cold formed rather than cut will tend to have superior strength. However, many gears still have to be cut and ground, owing to severe requirements on tooth profile accuracy, associated surface finish, and the relative peripheral location of the teeth.

14.4.4 Beams and Plates in Bending

For a given bending load, flat plates and beams of solid rectangular section are not as light or economical in the use of material as a section such as an I-beam or rectangular hollow section. Such sections have material remote from the central or neutral axis to carry tension and compression forces together with a connecting web or webs to carry shear forces. Sections like these are usually produced by forming processes such as hot rolling or extrusion. This is a direct example of net shape forming, but it is noted that the tolerances required are fairly lax, since they strike a balance between adequate strength and excess weight. While structural sections are formed to final shape, the shape is not particularly precise.

14.4.5 Springs

Springs are usually in the form of coils, which deflect by torsion, or beams, cantilevers, or diaphragms, which deflect by bending. They usually need to be made to some fairly precise stiffness, and since this varies as the cube of the material thickness, this dimension has to be accurate owing to the multiplying effect of this third-power relationship. Precision forming methods such as wire drawing or cold rolling usually give tolerances of sufficient accuracy and, additionally, elevate the yield strength, thereby giving a somewhat larger elastic strain range.

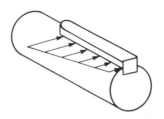

Keys and keyways are the most commonly used method since they are relatively easy to manufacture by metal cutting operations, although the accuracy of fit is critical.

The torque capacity of keys and keyways is limited by the fact that the torque is carried across a single region of the shaft periphery. Also due to shaft torsional elasticity the transmitted forces tend to be further concentrated towards the inner end of the key.

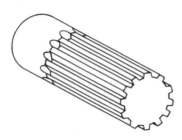

Splines involve more complex metal cutting but have a higher torque capacity because the transmitted forces are distributed around the periphery of the shaft.

They can also be made by metal forming processes.

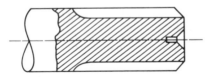

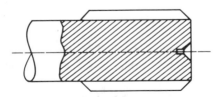

To enable spline manufactured from simple shafts they are usually cut in below the existing shaft diameter, although ideally they should be cut into a raised boss at the shaft end. This, however, requires more extensive machining.

Figure 14–12   Various types of end connections on shafts as used to transmit torque.

14.4.6 Tribological Contacts

Tibological contacts are of two types called *conforming* and *nonconforming*. With the conforming type, the contact takes place over a considerable area such as between a plain bearing and shaft, or between a piston and cylinder. If the contact is lubricated, the film thickness tends to be of the order of 25 μm, but because of the size of the contact area, there are risks of edge loading due to misalignment, and also of the trapping of dirt between the surfaces. To deal with this situation, one surface is made hard and the other relatively soft. In service the soft one tends to bed in to the hard one, and therefore, it is the finish and accuracy of the hard surface that is particularly critical.

In nonconforming contacts such as between the balls and races of a ball bearing, between a cam and tappet, or between two involute gear teeth, the contact occurs over a very small area, and high contact pressures result. These high pressures cause the formation of elastic flats on the surfaces, together with a considerable increase in the viscosity of any lubricant which may be present. The resultant film thicknesses are of the order to 2.5 μm. To carry the loads without plastic yield, both surfaces need to be hard and, owing to the low film thicknesses, a high standard of finish is required. Where there is a line of contact, accuracy *along* that line is important, but in other directions it may not be as critical.

14.5 TOLERANCES AND SURFACE FINISH REQUIREMENTS

In Chapter 3, the typical surface finishes which can be achieved by using different manufacturing processes were discussed. Just what are *required* for certain applications and what will the *cost* be?

In broad terms, there are three levels of tolerances which are important in designed products.* These are

1. *Cross-sectional* tolerances to meet the need for adequate strength and stiffness without excessive weight or volume.
2. *Positional* tolerances on the location of those parts of the component surface which mate with adjacent components.
3. *Fit* tolerances on the actual mating surfaces where there are tribological interactions or precise fits for static connections.

These three requirements are in an order of increasing dimensional accuracy. Examples of their application to a selection of varied engineering components are shown in Fig. 14–13.

With regard to surface finish, the greatest demand for the finest finish arises with tribological surfaces owing to the small oil film thicknesses at which these surfaces are

* Taken from Neale and Waterman, Report TRS 289 for SERC/DTI, 1982.

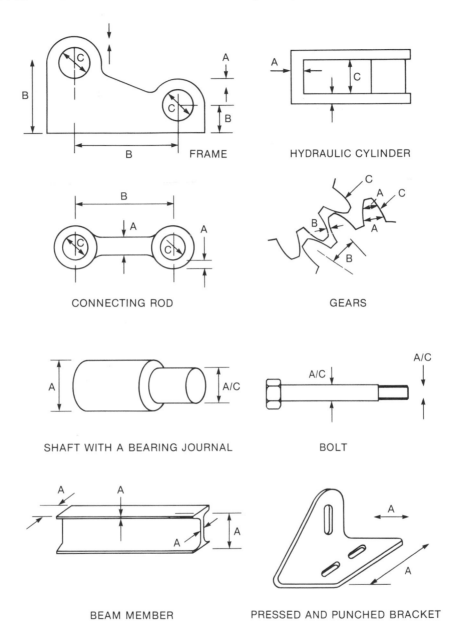

Figure 14–13   Comparative tolerances on dimensions of typical engineering components.

intended to operate. Sometimes, highly stressed components are required to have a high standard of surface finish to avoid the risk of rough surfaces giving rise to stress raising discontinuities. Some components used in applications in food processing require smooth finishes to make them easy to clean so as to maintain their sterility or appearance.

Figures 14–14 and 14–15 show how tolerances relate to component size and surface finish for different processing methods. Thus, in Fig. 14–15, open die forging is typically used on components with a characteristic size bigger than about 20 mm to over a meter; tolerances obtained are from 2 to 20 mm. Ordinary powder forming is used for objects between 5 and 100 mm in size, with tolerances between 10 to 100 μm. Other processes can be analyzed in a similar manner. It is seen that a number of processes can directly

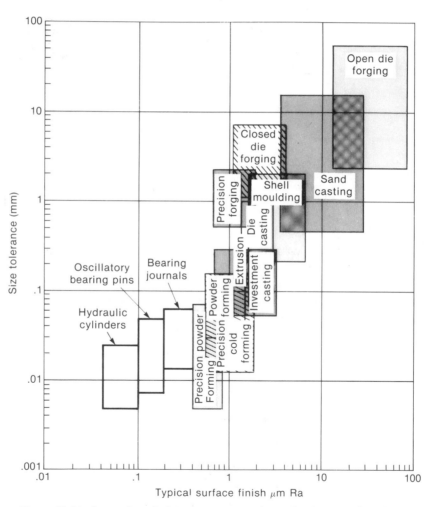

Figure 14–14   Interraction of tolerance, component size, and various manufacturing processes.

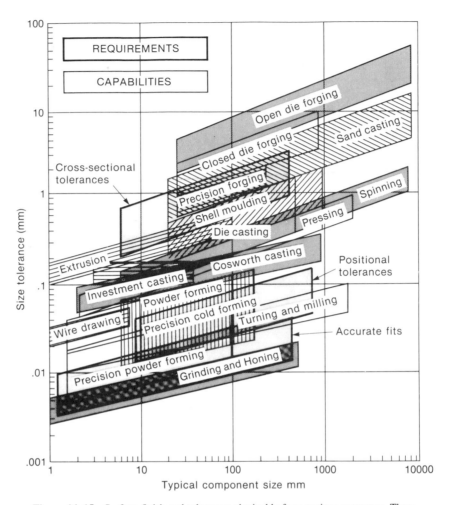

Figure 14–15   Surface finish and tolerances obtainable from various processes. These are compared to requirements for some sample components.

satisfy some of the tolerances demanded, whereas second operations are required to attain closer tolerances. Thus, cross-sectional tolerances can be achieved directly by casting or closed die forging, positional tolerances by precision cold forming, and accurate fits by precision powder forming or grinding. Table 14–3 gives approximate relative costs of achieving different grades of surface finish by different cutting operations. The standard of comparison in the table is a very rough machining operation, which is taken as having unit cost.

The increase in cost of attaining a greater degree of accuracy and finer surface finish is illustrated by the curve rising to the right in Fig. 14–16. However, when a manufactured component consists of an assembly of parts, the costs of assembly and fabrication will usually be *reduced* if more accurate parts are used. This is reflected in the curve falling

TABLE 14–3 SURFACE FINISH AND COST COMPARISON FOR VARIOUS CUTTING OPERATIONS

Surface Finish	RMS, um	Relative Cost
Very rough	50	1
Rough	26	3
Semi-rough	13	6
Medium—Machined	6.5	9
Semi-fine	3.2	13
Fine	1.6	18
Coarse	0.8	20
Medium—ground	0.4	30
Fine	0.2	35
Super fine—lapped	0.1	40

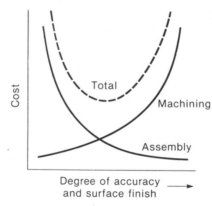

Figure 14–16 Typical type of cost curve showing the individual costs of machining and assembly as the degree of accuracy of finer surface finish is increased.

to the right in Fig. 14–16. The combination of the two effects leads to an optimum, where the overall cost of manufacture and assembly is least. Note the similarity of this curve with optimum values in machining economics, Fig. 13–1. A similar type of optimum occurs regarding reliability, in that the cost of design and manufacture must increase with increased reliability of the product, but the costs after delivery (to do with warranties, and the like) fall with improved reliability. It does *not* follow, of course, that manufacturers can always work at the optimum, as it may be necessary, for component operational design requirements for example, to have degrees of accuracy to the right of the minimum in overall costs.

14.6 DESIGN OF PRODUCTION ROUTES

14.6.1 Introduction

Within a given general manufacturing method, there will usually be a choice for the sequence of operations performed and what takes place in each operation. For example, in machining from a rough casting, how many cuts are to be taken and what speeds, feeds, and depths of cut are proper for each? Appropriate conditions applying to repetitive mass-production machining have been given in Sec. 13.3 relating to optimum cutting combinations to achieve either maximum production rates or minimum cost per piece. In solid forming processes, what sequences of forward or backward extrusion, upsetting, heading, piercing, die angles, reductions, and so on are possible, and which is best? Or in sheet forming, what sequences of cupping, redrawing, ironing, blanking, and with what tooling and so on are possible and best? Similar thoughts apply to other manufacturing and fabrication processes.

The *best* sequence is that which is most economical but which does not prejudice the in-service performance of the component. Yet even for relatively simple components the number of feasible alternative sequences can be great, and laborious calculations are often involved. Computer-aided calculations can help greatly in these circumstances. In this section we explore options available in forging and how sequences of operation are designed to achieve certain aims.

Forging operations can produce a wide range of shapes and sizes of components. Even with simple articles, a variety of sequences of operations to obtain the required shape and size are nearly always possible. As shown in a subsequent example, the ease of production and final part cost can be markedly influenced by early decisions on component and process design.

The determination of forging schedules is still based to a great extent on the experience, imagination, and intuition of skilled personnel. In the past, it has been difficult to propose rules for other than simple shapes. But instead of relying totally on the skills of experienced individuals, component and process design can be approached systematically. Such an approach is sometimes called group technology. The core of the method is to identify groups of related parts by means of an appropriate classification system.

14.6.2 Some Concerns in Forging

One book* builds on and systematizes earlier techniques of classification used by a variety of groups in the international forging industry, and is intended to give guidance for schedules for easy manufacture. The attributes of a part which may cause difficulties in forging are indicated by a coding system with the aim of avoiding undesirable features in the design finally decided upon. Most forgings are (1) *compact,* with similar dimensions in the three orthogonal directions, (2) *flat,* such as discs, or (3) *long.* Of forged parts manufactured, less than 10 percent are compact, about 20–30 percent are flat, and the rest are long. Each type is characterized by a basic sequence of operations. Compact parts are made from billets or blanks and, in general, have a simple forging sequence, but a proportion of parts within this category may require multiple-action machines. Flat parts are also produced from billets or blanks with upsetting (causing material flow) being predominant, and with a significant proportion of parts requiring pierced-through holes.

* C. Poli and W. A. Knight, *Design for Forging Handbook,* (Amherst, Mass.: University of Massachusetts, 1981).

Long parts are produced directly from bar stock and generally require initial elongation and drawing-down stages, prior to the impression die forging sequence. One or more bending stages may be necessary if the longitudinal axis of the part is not straight. In general, more deformation is required in going from compact to flat to long parts.

The major factors which influence forging difficulty in a particular case are the complexity of the part and the material used. These factors are strongly interrelated, since a shape that is relatively easy to forge from one material may be difficult or impossible to obtain in another.

As an illustration of the use of the *Design for Forging Handbook* coding system, Fig. 14–17 shows two flat round forgings, together with the code numbers allocated. One is a simple part which has a low-valued code number and is consequently relatively straightforward to produce as a forging. The second part is increased in complexity by the addition of closely spaced radial ribs, which are difficult to produce by forging. This requires additional forming stages, with reduced die life and greater material losses, and

(a)

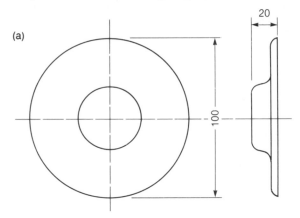

(b)

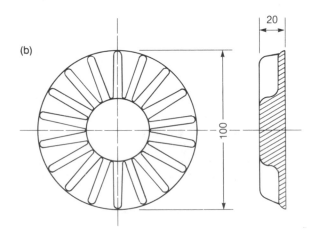

Figure 14–17 Two forgings described by code numbers from the *Design for Forging Handbook*, (a) Code 102, (b) Code 181. All dimensions in mm.

is given a higher-value code number within the classification. Allocation of a code number enables relevant data sheets to be consulted, which contain recommendations for forging design features appropriate to various classes of parts.

For the various classes of parts defined by coding systems, standardized alternative processing sequences can be developed, with the number of stages determined by the basic design limitations. Consideration of load capacity, limiting strains, need for interstage heat treatment, and the like are involved. Once likely alternative operations are established, the subsequent laborious calculations required to optimize the processing sequence are ideally suited for the computer. This gives the designer and the manufacturing engineer more scope to ensure that the sequence selected is best for the given conditions. Computer-aided-design (CAD) of this sort permits rapid comparisons to be made between alternatives under conditions prevailing at the time, and also rapid ''sensitivity'' assessments of the effects of future changes in material costs, energy costs, labor rates, and other cost factors.

14.6.3 Alternative Routes in Forging

Calculations of this sort have already been presented in Chapter 13 for *prescribed* manufacturing forming sequences. But what about the assessment of a large number of alternative sequences under greatly varying conditions? One such example follows and relates to Fig. 14–18. First, a number of alternative sequence of operations was devised which appeared feasible for this particular geometry. To retain a reasonable clarity only thirteen alternatives are shown in Fig. 14–18; this is only a small proportion of the total studied.*

Depending on the level of work hardening, the second annealing and lubrication might not be required for alternatives 9 and 13. Similarly, the annealing of the slugs would not be necessary if the bar from which they are made is supplied in a soft condition. Of the sequences illustrated in Fig. 14–18, 11 and 12 utilize hollow machined slugs, while solid slugs produced by a number of alternative methods are used for the rest. There are also alternative methods for slug production from the solid, such as cutting off by machining, sawing, cropping or, if the length/diameter ratio of the slug is less than a specified value, cropping followed by axial compression or blanking from sheet metal. Furthermore, depending on the actual value of angle α in the particular case, it might be better to extrude partially at the optimum die angle and obtain α by an additional operation. The thirteen alternatives indicated in Fig. 14–18 represent well over a hundred different sequence of operations, although the differences in some cases might be the addition of just one further operation to an otherwise unchanged sequence.

Each sequence is characterized by a combination of operations carried out in a specific order and, for it to be feasible, each operation must be possible in the order corresponding to that sequence. Every operation is subjected to a set of constraints which must be satisfied before proceeding to the next operation, as shown in Fig. 14–19. Typical

*B. Lengyel and T. Venkatasubramanian, *Proceedings of the 18th Machine Tool Design and Research Conference* [MTDR] **153** (1977).

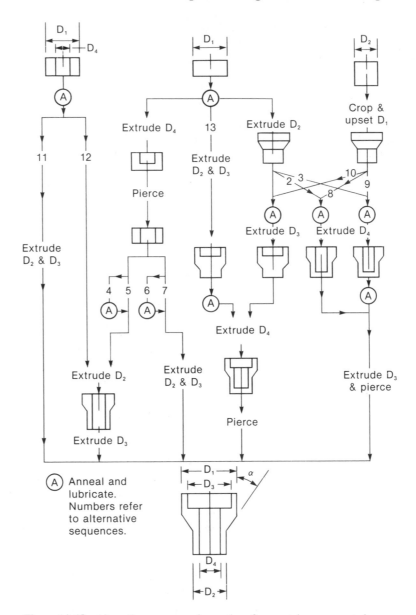

Figure 14–18   Alternative sequence of operations for a certain component shape.

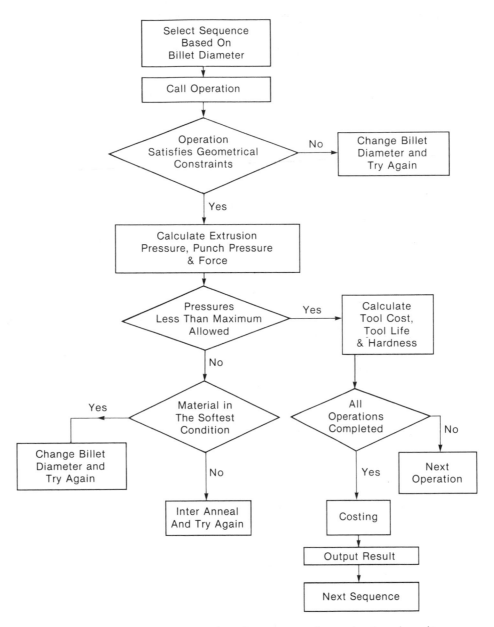

Figure 14–19 Steps to determine the optimum sequence for a part such as shown in Fig. 14–18.

constraints are the permissible level of work hardening, the maximum practicable extrusion pressure, or the largest feasible length/diameter ratio of the extrusion punch. As a result of this approach two questions are answered initially.

1. Is it feasible to cold forge a particular component by applying a certain sequence?
2. Is it possible to manufacture that component at all by cold forging?

It is essential to estimate the average level of work hardening after each operation, because that determines whether the next forming operation may be carried out and what will be the properties of the forging at that stage. Whether the next forming operation could be carried out or not is influenced by prior work hardening because, if the hardness exceeds a predetermined limit, the deforming material could fracture, or the permissible level of tool stress could be exceeded; in either case that particular sequence must be rejected.

Figure 14–20 shows two versions of the part presented in Fig. 14–18, while Fig. 14–21 illustrates a few comparative cost results. In order to have a reasonably small number of alternatives in this illustration, computed data are shown only for a vertical mechanical press with a single station tool, for either hand or hopper feeding. Costs are given as a function of the number of components required per calendar month. For clarity, only a few of a large number of curves are shown, but even then a large variation is obvious, both from one process to another and with quantity requirements for any one sequence. In general, those sequences requiring the least number of operations tend to give lower unit cost for small batches, because both labor and capital cost contents are then lower. Also, for small batches, tool utilization could be increased and tool costs lowered by allowing higher tool stresses, for example, by using large deformations in a single stage or by planning for combined operations. For large quantities, increasing the number of operations was found not to incur penalties, because automatic feeds and

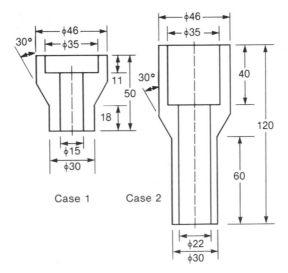

Figure 14–20 Alternative dimensions for the part in Fig. 14–18.

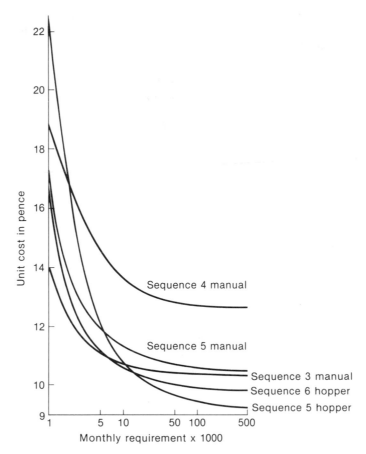

Figure 14–21 Comparative costs for alternative methods of producing the part in Fig. 14–18 as influenced by batch size.

transfer tools could be used to ensure better equipment utilization and lower labor costs. At large batch sizes, tool lives must be maximized by reducing tool stresses; otherwise, frequent tool replacement increases the unit cost by decreasing the utilization of the more expensive equipment owing to extensive machine downtime.

14.7 ECONOMICS OF ALTERNATIVE METHODS OF MANUFACTURING

In Chapter 13, the procedures for costing articles were outlined and examples were given for a variety of different components manufactured by different prescribed routes. What, though, if there are alternative manufacturing methods to achieve the identical or essentially the same result? Are some processes always more expensive or cheaper than others, or are there ranges, depending on the numbers being produced, over which one process is more economical than another? A hint that the latter can be true was shown in Fig. 14–5. In this section we shall explore the conditions under which different methods are possibly more economical than others.

The National Engineering Laboratory (NEL) in Great Britain commissioned in 1965 a pioneering survey into the relative economics of machining methods versus combinations of cold extrusion and forging.*

Six components were selected for investigation. Their choice centered on the fact that *originally* they had been designed for production by normal machining methods in reasonable quantities, but *subsequently* they had been produced by cold forming. Appropriate production information was thus available for both modes of manufacture. The selected components included

1. A double can component
2. A spur gear
3. A synchronizing gear
4. An annular component
5. A flanged component
6. A stepped can component

Two of these components are shown in Fig. 14–22.

They were all typical light engineering products, capable of manufacture by standard production machine tools; none was of an unusual asymmetrical shape, which might have been expected particularly to favor cold extrusion. The sequences of extrusion operations used in producing the final forms of the components were not necessarily the only ones which were feasible, but were selected by qualified production engineers in accordance with ruling conditions of plant, tooling, and operational experience.

Figure 14–23, which shows the relative cost of manufacture against the number of the synchronizing gears produced, typifies the type of findings. A breakeven, or crossover, point occurs, and we see that cold extrusion is more expensive at low batch sizes but improves relative to machining with increase in batch size. Similar curves are given in the NEL Report for the other components, with the crossover batch quantities in Table 14–4.

The crossover batch quantity at which extrusion becomes cheaper is generally low, with the exception of the annular component. That particular component used tube as the starting stock with consequent little material waste, but in all other cases extrusion makes large savings in raw material and, at large batch numbers, this outweighs all other factors, such as power costs for annealing furnaces, chemicals for lubrication, and the like. At low production numbers, of course, these savings are offset by the increased prorated cost of tooling; indeed, tooling cost is the only significant reason for the economic disadvantage of forming processes compared with traditional machining at these low quantities. If parts were to be made on a lathe, an entire small production run could be finished in the time it takes to design, manufacture, and deliver dies suitable for cold forging. Even so, it is

---

* B. P. Clapp *Economics of Cold Extrusion,* NEL Report No. 195, East Kilbride, Scotland, 1965.

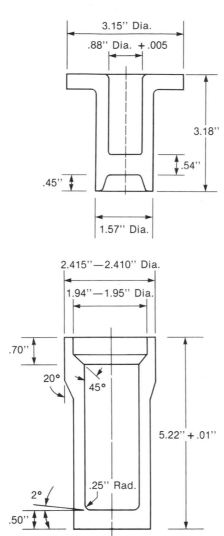

Figure 14-22 Two of six components selected for comparison of the relative economics involved in machining or cold forming these parts.

clear that under appropriate conditions, manufacturing methods involving a large proportion of cold extrusion/forging are highly competitive. For all six components, no further improvement in cost advantage occurred for batch sizes greater than about 30,000, and savings over machining settled to the figures shown in Table 14-5.

Another study conducted at NEL in 1969 involved a steel bearing race sleeve as shown in Fig. 14-24;* this is a hollow axisymmetric part well suited to cold forging. Manufacturing costs were calculated on the basis of

---

* B. P. Clapp, A. M. Evans, and T. H. Pasteur, *Cold Forging Study,* NEL Report No. 468, East Kilbride, Scotland, 1969.

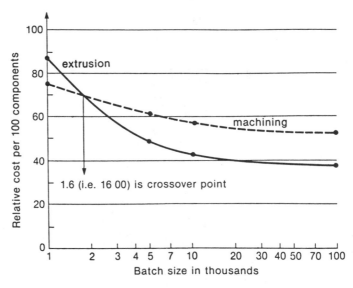

Figure 14–23   Comparison of costs versus batch size for machining or cold forming and machining the gear in Fig. 14–22.

1. Machining from an as-purchased hot-forged blank
2. Cold forging a purchased billet to a semi-finished condition followed by finish machining.

Using sequence (2) cut costs almost 50 percent compared with sequence (1).

TABLE 14–4

Component	Crossover Batch Quantity
Double can component	1400
Spur gear	1100
Synchronizing gear	1600
Annular component	5000
Flanged component	1100
Stepped can component	1300

TABLE 14–5

Component	Saving on Machining
Double can component	47%
Spur gear	63%
Synchronizing gear	29%
Annular component	8%
Flanged component	56%
Stepped can component	43%

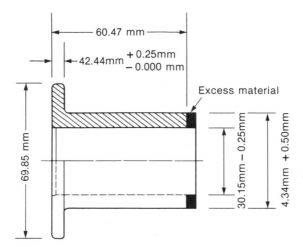

Figure 14-24 A bearing race sleeve.

Other interesting cases of choice of alternative production methods on the basis of quantity and cost are described by the ASM Committee on Cold Heading.* One example involves the case of the production of lawnmower wheel bolts shown in Fig. 14–25. These bolts were originally produced by (1) heading the slug, and simultaneously extruding the opposite end to 0.525-in. diameter; (2) coining and trimming the round head to hexagon shape; and (3) turning the bolt blank to 0.331-in. diameter in a secondary operation prior to rolling the thread. With the improved method shown in Fig. 14–25, the slug was first extruded to form two diameters on the shank end, then headed, coined, and trimmed. By this procedure the minor extruded diameter was ready for thread rolling, and no turning was required. The improved method not only reduced cost by 40 percent, by eliminating the secondary turning operation, but also produced a stronger part, because metallurgical flow lines were not interrupted at the shoulder. Owing to the turning operation, production by the original method was only 300 pieces per hour. By the improved method, 3000 pieces could be produced per hour. The ASM Committee quoted another example of a study at one plant which compared the costs of producing the pin shown in Fig. 14–26 in the same quantity (25,000), using the same material (AISI 8740 steel) by three different processes, machining, hot heading, and cold heading. Table 14–6 gives the results. Although tool and setup costs were greater for cold heading than for the two competitive methods, these higher initial costs were outweighed by the lower costs for material and production when cold heading was used. The lower production cost for cold heading resulted from the high production rate. For larger quantities, the cost differential would have increased in favor of cold heading; for smaller quantities, the cost advantage of cold heading would have decreased.

All these sorts of cost estimates from different sources lead to the following main conclusions:

* Source Book on Cold Forming (Metals Park, Ohio: American Society for Metals, 1975).

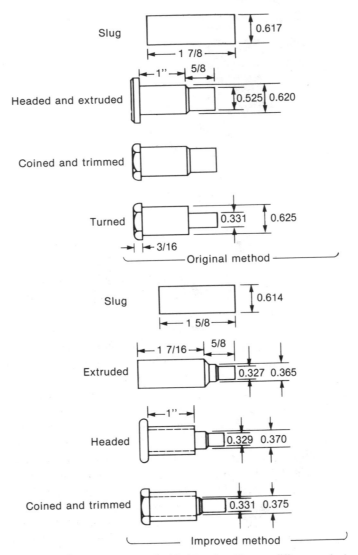

Figure 14-25 A lawnmower wheel bolt produced by two different methods.

1. Machining costs are usually less for cold than for hot forging, because cold forgings are nearer net shape.
2. Cold forging costs decrease with increasing batch size.
3. Cold forging requires less raw material.

A remarkable example of economic savings which can be achieved when bulk machining is replaced by cold forging concerns the manufacture of helical planetary

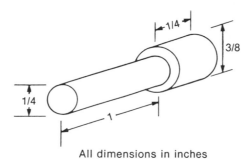

Figure 14-26 A pin produced by three different methods; see Table 14–7 for cost comparisons.

All dimensions in inches

pinions of automobile transmissions. In 1976, about 60 million of these were being made per year in the United States, all by traditional gear cutting methods. An alternative method was developed at the Ford Motor Co. by which the gears can be cold extruded with fully formed teeth, have virtually no defects, and require no subsequent machining. Merely on the basis of using lower-cost feedstock with minimal scrap generation, the new process showed cost savings of between 30 and 50 percent, and savings are further increased when production levels and rates are increased. Much greater production rates are achievable by the forming route. Hobbing *one* gear takes between 1 and 2.7 minutes, whereas gears can be extruded at the rate of 10 per minute. A Ford report* indicates that the saving per car is around $7.

In place of machining versus forming of solid billets (which has been the basis of most examples so far), other manufacturing routes may be possible. Consider Fig. 14–27, which shows a spur gear that is traditionally made by cutting from a blank but could be made alternatively by using the techniques of powder metallurgy. Specifically, these involved powder compaction and sintering of a suitable preform, followed by forging and grinding to the precise finished dimensions. A detailed comparison of the various operations required for both the machining and the powder metallurgy/forging routes is given

* S. K. Samanta, *Proc. NAMRC-IV,* **199** (Battelle Labs., Columbus, Ohio, 1975).

TABLE 14–6   COST OF PRODUCING 25,000 STEEL PINS BY THREE DIFFERENT METHODS

Item	Machining	Hot Heading	Cold Heading
Material cost	$295.00	$212.50	$192.50
Tooling cost*	60.00	90.00	150.00
Setup cost	30.00	35.00	50.00
Production cost	382.50	432.50	65.00
Total cost, 25,000 pcs.	$767.50	$770.00	$457.50
Cost per piece	$0.0307	$0.0308	$0.0183
Production in pcs/hour	285	430	5000

*Amortization of machines not included.

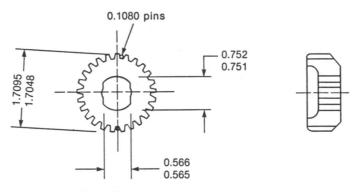

Figure 14–27 A typical spur gear. Comparative costs to produce by machining and by powder metallurgy are given in Table 14–18.

in Table 14–7. A cost advantage is shown for the latter method, mainly due to reduction in the quantity of raw material required.

When it is possible to completely *redesign* a component, or an assembly of components, there can be considerable savings in achieving the same end result, that is, a different design of component or assembly that still performs as appropriately as the original design. This is, of course, the essence of *design for production*. By way of illustration of the principles involved, Fig. 14–28 shows how a shaft can be redesigned on the basis of equivalent functional requirements and suitability for cold forming. The redesign on the right-hand side of the illustration satisfied both these criteria, and additionally saved the material shown crosshatched.

Another pertinent example is shown in Fig. 14–29. The part (a) is a simple shaft with a central hub that might later have gear teeth machined in. To turn it on a lathe (b) requires 18 cu in. of solid barstock, whereas the finished component is only 8 cu in. An alternative manufacturing route (c) would be to redesign the component so that a ma-

TABLE 14–7 COSTS TO PRODUCE A SPUR GEAR BY MACHINING
AND POWDER METALLURGY FORMING

Machining		Powder Metallurgy Forming	
Material	26.00	Material	14.00
OD turning, boring, drilling, parting	0.34	Compaction	0.76
Gear cutting, broaching	4.83	Sintering	2.61
Case hardening	2.00	Forging	5.68
Grinding	3.50	Machining, grinding	4.54
Inspection	1.89	Inspection	1.89
Scrap	1.12	Scrap	3.78
Total cost*	39.68		33.26

*These are in cents/part, with labor and overhead excluded.

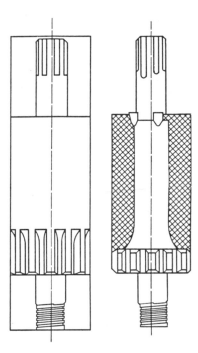

Figure 14–28 Two designs of a shaft made by machining (left) and cold forming (right). Crosshatched area on right indicates material savings.

chined sleeve is pressed over the shaft. But following the theme of earlier examples in this section, much less material would be wasted were the hub hot-upset-forged integrally with the shaft (d) or, better still, produced by cold extrusion/cold forging.

We conclude this section by repeating an observation made earlier in Sec. 13.2: When saying a certain process is the best to produce a component, we must recognize that the equipment appropriate for the task may not be available in a given organization. Even if suitable equipment is available, it may be fully committed to the manufacture of other products. If the projected costings relate to large production volumes, there may be financial justification in purchasing new equipment. However, for smaller quantities, it may be cheaper overall to use less optimum, but available, equipment.

14.8 ENERGY COSTS AND MANUFACTURING

The production of metals, polymers, and ceramics used in manufacturing processes requires energy, so this cost must influence the direct starting price of materials before any manufacturing operations are carried out. Table 14–8 shows the energy/ton required for overall metallurgical and materials extraction and processing, where some materials require proportionately more energy in the original making stage from the raw materials and less on subsequent processing, as seen in Table 14–9. Also shown in Table 14–8 is the energy content involved in providing equivalent stiffnesses and strengths in a body de-

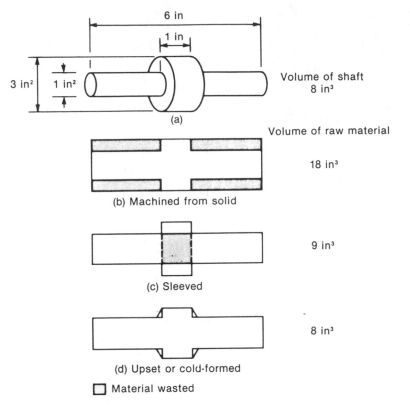

Figure 14–29 Various methods used to produce a stepped shaft to illustrate the interaction between design and manufacturing on material utilization.

TABLE 14–8 ENERGY CONSUMPTION IN THE MANUFACTURE OF VARIOUS MATERIALS

Material	Energy/Tonne-GJ	Energy Per	
		Unit Stiffness	Unit Strength
Aluminum	200	2.8	870
Cast iron	45	0.3	300
Low-carbon steel	50	0.24	240
Glass	20	0.30	333
Brick	6	0.10	30
Concrete	2	0.05	80
Wood	1	0.07	10
Polyethylene	45	90	1500
Carbon fiber- reinforced composite	4000	27	6000

TABLE 14–9   ENERGY REQUIRED OR AVAILABLE AT DIFFERENT
STAGES IN THE MANUFACTURE OF STEEL AND ALUMINUM
COMPONENTS

Stage	Carbon Steel	Aluminum
Manufacture of material	36	300
Primary working	3	40
Machining	1	1
Swarf	40	340
Energy saved by using scrap	10	295
Energy required to melt	1	1

signed to be manufactured from a given material. On this basis, the pursuit of materials based on exotic fibers is quite inefficient in terms of energy. Clearly, those materials requiring high energy for production are most vulnerable to changes in energy prices.

The oil crisis of 1973–1974 resulted in significant increases in the price of oil, and hence increases in the cost of all processes utilizing oil. This applied to processes involving oil both directly and indirectly. Oil-fired furnaces, electricity generated by steam turbines fed by oil-fired boilers, and the products of the petrochemicals industry all demanded higher costs.

It is clear that any increasing shortages, with the subsequent rising prices of energy, will necessarily lead to a shift to production methods that save energy. The saving will be on a global basis, although actual changes in methods of processing and manufacturing seen in different industries will vary with the size of the enterprise. Some who buy all of their starting raw materials will be concerned primarily with changes in traditional methods of manufacture to more energy-saving routes. However, their decisions will also be based on increased prices of the starting material, increases which may depend on the type of product. As a rough guide, 25 percent of the costs of common metal feedstock are energy-related costs.

We have already seen that all machining processes are wasteful in energy terms. It costs money to produce the feedstock in the first place; energy is consumed in the manufacturing operation, and much of the raw material becomes scrap, the value of which is far less (<20 percent) than that of the cost of starting material. It also takes energy to recycle the scrap back to good feedstock, and that is not always possible. Forming processes, on the other hand, are more energy-efficient in that there is far less scrap. Energy is required to conduct the forming operation, of course, and the capital cost of forming equipment is often greater than that of metal cutting equipment. The examples in Sec. 14.7 show that metal cutting has always been cost-effective for small batch sizes, but forming wins out for long production runs. There will be a breakeven point beyond which it will be cost-effective to alter the manufacturing method, as indicated in Fig. 14–23.

Given all these favorable indicators to change away from metal cutting methods, we may wonder why machining processes "hang on." One reason relates to the fact that many factories are already equipped with machine tools, which are still being paid for, and also that the labor force is more skilled and experienced at cutting than at forming.

Even so, economics must rule at the end of the day and cheaper manufacturing methods will predominate.

When anticipated increases in energy costs are considered, the breakeven point moves more in favor of forming processes. Of course, there are some operations which, owing to the geometries involved, are unlikely to be displaced by forming; drilling, tapping, slotting, and the generation of thin-walled sections are examples. Also finish-turning and grinding operations, which produce fine tolerances and fine surface finishes, will inevitably remain important.

Acknowledgements, continued

Figure 4–5, 4–6, 4–8, 4–9
From K. H. Moltrecht and R. M. Caddell, "How to Determine Production Tolerances–Part 2," *The Tool Engineer* (Nov. 1957), pp. 85–89. Reprinted courtesy of The Society of Manufacturing Engineers.

Figure 6–1, 6–2, 6–3, 6-4, 6–5, 6–6, 6–8, 6–9, 6–11, 8–1, 8–10, 8–11, 8–12, 8–13, 8–14, 8–15, 8–16, 8–17, 8–18, 8–21, 8–22, 8–23, 8–24, 8–25, 8–26, 8–27, 8–28, 8–29, 8–30, 8–31, 8–32, 8–33, 8–34, 8–35, 8–36, 8–37, 8–38, 8–39, 8–40, 8–41, 8–42, 8–43, 8–44, 8–45, 8–47, 8–48, 8–49, 8–50, 8–51, 8–52, 8–53, 8–54, 8–55, 8–56, 8–58, 8–59, 8–60
From William F. Hosford and Robert M. Caddell, *Metal Forming: Mechanics and Metallurgy* (1983). Reprinted by permission of Prentice-Hall, Inc., Englewood Cliffs, N.J. pp. 2–86.

Figure 6–10
From R. M. Caddell and R. Sowerby, *Bulletin of Mechanical Engineers Education,* **8** (1969), pp. 31–43. Used with permission of the *International Journal of Mechanical Engineering Education* and Pergamon Press.

Figure 6–7, 6–10, 7–33, 7–34, 7–35, 7–36
From Robert M. Caddell, *Deformation and Fracture of Solids,* (1980). Reprinted with permission from Prentice-Hall, Inc. Englewood Cliffs, N.J.

Figure 6–13, 6–16
From *Materials Science and Engineering,* American Society for Metals (1975), pp. 213–20. Used with permission ASM.

Figure 6–13, 6–14, 6–15, 6–16, 7–4, 7–5, 7–10, 7–11, 7–24, 7–25
From David K. Felbeck and Anthony G. Atkins, *Strength and Fracture of Engineering Solids,* (1984), p. 45. Reprinted by permission of Prentice-Hall, Inc., Englewood Cliffs, N.J.

Figure 6–14, 6–15
Reprinted from R. Raghava, R. M. Caddell, L. Buege, and A. G. Atkins, *Journal of Macromolecular Science and Physics,* **B6,** 4 (1972), pp. 655–65, courtesy of Marcel Dekker, Inc., N.Y.

Figure 7–1, 8–57, 8–58
From *Metals Handbook,* Vol. **7,** 8th ed., American Society for Metals, (1972), p. 4–14. Used with permission ASM.

Figure 7–4, 7–5
From *Metals Handbook,* Vol. **8,** 8th ed., American Society for Metals, (1973), p. 275–76. Used with permission ASM.

Figure 8–11, 8–50
From D. J. Meuleman, Ph.D. thesis, The University of Michigan (1980).

Figure 8–12
From G. Dieter, *Mechanical Metallurgy*, 2nd ed., (New York: McGraw-Hill Book Co., 1976). Used with permission.

Figure 8–17
This was adapted from J. Wistreich, *Metals Review,* **3** (1958), pp. 97–142.

Figure 8–28, 8–29
From R. M. Caddell and A. G. Atkins, *Journal of Engineering for Industry Transactions of ASME,* Series B, No. 90 (1968). Used with permission of ASME.

Figure 8–30
From B. B. Hundy and A. R. E. Singer, *Journal of the Institute of Metals,* **83** (1954–55), p. 402. Used with permission of The Institute of Metals.

Figure 8–31
From H. C. Rogers, General Electric Report No. 69-C-260 (July, 1969).

Figure 8–32
Reprinted with permission from *Transactions of the Metallurgical Society,* Vol. **175** (1948), pp. 337–54. A publication of The Metallurgical Society, Warrendale, Pa.

Figure 8–33
From D. J. Blickwede, *Metals Progress,* **97** (May 1970), pp. 76–80. American Society for Metals. Used with permission ASM.

Figure 8–36
From M. G. Cockroft and D. J. Latham, *Journal of the Institute of Metals,* **96** (1968), pp. 33–39. Used with permission of The Institute of Metals.

Figure 8–37
From A. L. Hoffmanner, Interim Report, Air Force Contract F 33615-67-C-1466, TRW (1967). See also G. E. Dieter in *Ductility,* ASM (1968), pp. 1–30. Used with permission.

Figure 8–38
From H. C. Rogers and L. F. Coffin, Final Report, Contract NOW-66-0546-d, Naval Air Systems Command (June 1967). See also H. C. Rogers in *Ductility,* ASM (1968), p. 48. Used with permission.

Figure 8–39, 8–40
From H. A. Kuhn, ''Formability Topics–Metallic Materials,'' ASTM STP 647 (1978), pp. 206–219. Used with permission of ASTM.

Figure 8–45
From C. T. Yang, *Metals Progress,* **98** (Nov. 1970), pp. 107–110. Used with permission of American Society for Metals. See also J. Datsko, *Materials Properties and Manufacturing Processes* (New York: John Wiley and Sons, 1966), p. 318

Figure 8–47
From W. F. Hosford and W. A. Backofen, ''Strength and Plasticity of Textured Metals,'' *Fundamentals of Deformation Processing*, Proc. 9th Sagamore Army Materials Research Conference, eds. W. A. Backofen, et al. (Syracuse, N.Y.; Syracuse University Press, 1964). Used by permission of the publisher.

Figure 8–49
Reprinted with permission from *Formability: Analysis, Modeling and Experimentation,* ed. by W. F. Hosford, ed. Used with permission of The Metallurgical Society, Warrendale, Pa.

Figure 8–51, 8–53
From D. V. Wilson and R. D. Butler, *Journal of the Institute of Metals,* **90** (1961–62), pp. 473–83. Used with permission of The Institute of Metals.

Figure 8–54, 8–55
From S. S. Hecker, *Sheet Metal Industries,* 52 (1975), pp. 671–75. Used with permission of Sheet Metal Industries.

Figure 8–56
From A. K. Ghosh, *Journal of Engineering Materials and Technology,* ASME, H, 99 (1977), p. 269. Used with permission of ASME.

Figure 8–60
From C. E. Pearson, *Journal of the Institute of Metals,* **54** (1934), p. 111. Used with permission of The Institute of Metals.

Figure 8–63, 8–64, 8–65, 9–1, 9–14, 9–16, 9–17, 9–18, 9–19(a), 9–21, 9–24 (a), 9–36(a), 11–5, 11–6, 11–9, 11–10, 11–12, 11–13
From Lawrence E. Doyle, Carl A. Keyser, James L. Leach, George F. Schrader, and Morse B. Singer, *Manufacturing Processes and Materials for Engineers,* 3rd ed. (1985). Reprinted with permission from Prentice-Hall, Inc., Englewood Cliffs, N.J.

Figure 9–2, 9–10(a), 9–19(b), 9–25
From *Fundamentals of Tool Design,* ASTME (1962). Reprinted by courtesy of the Society of Manufacturing Engineers.

Figure 9–4(b), 9–5, 9–24(b)
From Amstead, Ostwald and Begeman, *Manufacturing Processes,* 7th ed. (1977). Used with permission of John Wiley and Sons, Inc.

Figure 9–20
Photo orginally supplied by the Carborundum Co.

Figure 9–22
Used with permission of Warner and Swasey.

Figure 9–40, 9–41
Adapted from *Manual on Cutting of Metals,* 2nd ed., ASME (1952), pp. 270–71.

Figure 11–1
Courtesy of Central Foundry Division, General Motors Corporation. Used with permission.

Figure 11–2, 11–16, 11–17
From *The Making, Shaping, and Treating of Steel,* 9th ed. U.S. Steel Corp. (1971). Used with permission of the U.S. Steel Corp.

Figure 11–18(a), 11–18(c)
From D. S. Clark and W. R. Varney, *Physical Metallurgy for Engineers,* 2nd ed. (D. Van Nostrand and Co. Inc., 1962). Reprinted by permission of PWS Publishers, Boston.

Figure 11–18(b), 11–19
From M. L. Begeman and B. H. Amstead, *Manufacturing Processes,* 6th ed. (New York: John Wiley and Sons, Inc., 1969). © John Wiley and Sons, Inc. Used with permission.

Figure 11–22
From F. R. Samp and M. E. Shank, *Aircraft Turofans: New Economical and Environmental Benefits,* (Sept., 1985), Mechanical Engineering, p. 47. Used with permission of Pratt and Whitney Engineering Division, United Technologies, CT. Photograph courtesy of D. M. E. Shank.

Figure 13–2, 13–3
From B.P. McMahon, ''Welding and Metal Fabrication,'' (1970). Used with permission of Fuel and Metallurgical Journals Ltd., Industrial Press, Sutton, Surrey.

Figure 14–3, 14–4, 14–6, 14–7
Courtesy of the Fulmer Research Institute Ltd., England.

Figure 14–8, 14–10
R. G. Salter, FWP Journal (August 1983). Used with permission.

Figure 14–9
J. G. Hicks, ''Machine Design and Control,'' 1970. Used with permission.

Figure 14–25
From ''Source Book on Cold Forming'', used with permission of The American Society for Metals, Metals Park, Ohio.

Figure 14–29
From George E. Dieter, *Engineering Design* (New York: McGraw-Hill Book Co., 1983), p. 210. © 1983 McGraw-Hill Book Co. Used with permission.

index _____

i

j

k